POWER CIRCUITS AND ELECTROMECHANICS

K.R. Davis
M. A. Pai
Dept. of Electrical and Computer Engineering
Texas A&M University
University of Illinois

ISBN 978-1-64617-334-1

Published by

Stipes Publishing L.L.C.
www.stipes.com

Contents

To

George W. Swenson, Jr.

Preface to the New Edition

The original textbook by Professor M.A. Pai has been used extensively in undergraduate power engineering curricula. It is currently being taught as ECEN 340 Electromechanical Energy Conversion at Texas A&M, a required junior-level course that is a gateway to the energy and power area. When I taught this course at University of Illinois at Urbana-Champaign, Professor Pete Sauer was the course director and my mentor. I am fortunate to know Prof. Pai and Professor Pete Sauer, and it is a tremendous privilege and honor to be able to work with them to complete the update of this textbook. I will always be grateful for their support. They are amazing people and significantly inspired and encouraged me at critical parts of my journey. These were parts of my experience at the University of Illinois that were critical in leading into my ideal role as a professor at Texas A&M. A favorite story about him and this course– one time I was leading a project review meeting, and Pete was there, too, and I was going to need to leave early to go teach this class. Exactly as I was about to leave, Pete says "You stay. I'll go teach your class. Where did you leave off?" and walked right over and taught it (with no prior warning). Pete's teaching notes accompanying Pai's book have served as my guidance in teaching the course over the years. I would also like to thank Prof. Sauer for writing the new chapter on induction generators. In addition, I am grateful for the guidance and inspiration from the power faculty group at UIUC: Professor Alejandro Dominguez-Garcia, Professor George Gross, and Professor Phil Krein. This new edition of the book aims to capture some of the course's evolution over time, from the iterative process of teaching it to undergraduates. In particular, the course also reflects also the inclusion of power electronics and renewable energy resources.

It is essential for me to thank Dr. Mack Grady and Howard Neil who have deeply influenced me since I was an undergraduate at the University of Texas at Austin. I hope to one day to have the same influence on my students. Dr. Grady's mentorship was pivotal; it sparked my love for power systems research, which led me to graduate school and to meet Professor Tom Overbye. I would like to thank my students and colleages at Texas A&M, especially Won Jang, Prasad Enjeti, and Tom Overbye for their support in revising this book. I thank all my colleagues who I have taught alongside. The timely support of Miroslav Begovic and the Electrical and Computer Engineering Department is gratefully acknowledged. I would also like to thank Samantha Papesh, an undergraduate student at Texas A&M, who helped me with identifying and correcting errata in the textbook, as well as creating some of the figures, and Joyce Mast who helped us with continuity between the two editions. Their dedication and attention to details are greatly appreciated.

I would like to thank the support of my family especially Matt, Mary Evelyn, Maud, and Margaret. You guys are my world.

K.R. Davis
College Station, Texas
June 2022

Original Preface

This book covers the topics of the ECE 430 Power Circuits and Electromechanics course in the Electrical and Computer Engineering curriculum at the University of Illinois at Urbana-Champaign. It is one of five electives at the senior level from which the student must choose three. The traditional circuit theory course at the sophomore level has been replaced by an introductory course in electrical engineering at the first-year level, followed by a course on signals and systems at the sophomore level. Hence, the material in this book is a combination of sinusoidal steady-state analysis, power circuits, (previously taught in the traditional circuit theory course), and basic electromechanics.

Power Circuits and Electromechanics begins with a review of phasors and then examines in detail the concepts of power in single- and three-phase balanced circuits. Magnetic circuits and single-phase transformers are treated in a circuit-theory framework. The derivation of forces of electric origin using the concept of energy–co-energy in electromechanical systems (magnetic) is based on the quasi-static approximation of Maxwell's equations. The magnetic equivalent circuit concept is also used as an alternative in solving problems. Next, the complete dynamics of the electromechanical system are discussed in a state space framework. Computation of static equilibria, their stability, and the time-domain solution of the state space equations are explained. Stability via linearization and the energy function method are explored. Three-phase synchronous machines, induction machines, and DC machines are considered in the framework of energy and co-energy. The final chapter on single-phase machines discusses the principles behind their operation.

The development of the dynamics portions in Chapters 5–9 is inspired by the work of Woodson and Melcher, *Electro Mechanical Dynamics Part I: Discrete Systems* (John Wiley, 1968, and Krieger Publishing, 1990 [reprint]). The author is grateful to Professor H.H. Woodson and Ms. Janet D. Melcher, wife of the late Professor James R. Melcher, for their permission to use some of the problems, illustrations, and examples from the book. It provides a fundamental framework for the analysis of electromechanical systems including machines based on the concepts of energy and co-energy. The discussion of synchronous and induction machines has been extended to the three-phase case. I would also like to thank Professor R.J. Turnbull for writing Chapters 9 and 10 on DC and single-phase machines.

I also thank many of my colleagues for their useful feedback while teaching the course when it was in notes form and Dr. Tony Nguyen, who, as a teaching assistant to this course, solved several numerical problems. A special set of thanks I owe to Karen Chitwood, who typed several versions of this book in a notes format each semester for the past several years, and Nina Parsons and Terry Peterson for doing many of the drawings. I would like to thank Frances Bridges for typing the final version of this manuscript.

Student feedback for the notes was quite favorable. Their preference for a science-based approach to electromechanics encouraged me to convert the notes into the present book form. I would like to thank Joyce Mast for her excellent job of co-ordinating the various aspects in bringing the notes into the present form. Her attention to detail and useful suggestions have been very helpful. The timely support and encouragement of Professor R.E. Blahut, Head of the Electrical and Computer Engineering Department is gratefully acknowledged.

I would like to thank Professor Phil Krein, Director, Grainger Center for Electric Machinery and Electromechanics, for his constant support in this project. Assistance of the Center in bringing the work into the present form is greatly appreciated.

Finally, I would like to thank my family for their support of this work.

M.A. Pai
Urbana, Illinois
August 2002

Chapter 1

STRUCTURE OF POWER SYSTEMS

1.1 Overview of an Integrated Power System

An electric power system consists of four principal components: 1) the generating system, 2) the transmission system, 3) the distribution system, and 4) the loads as shown in Figure 1.1.

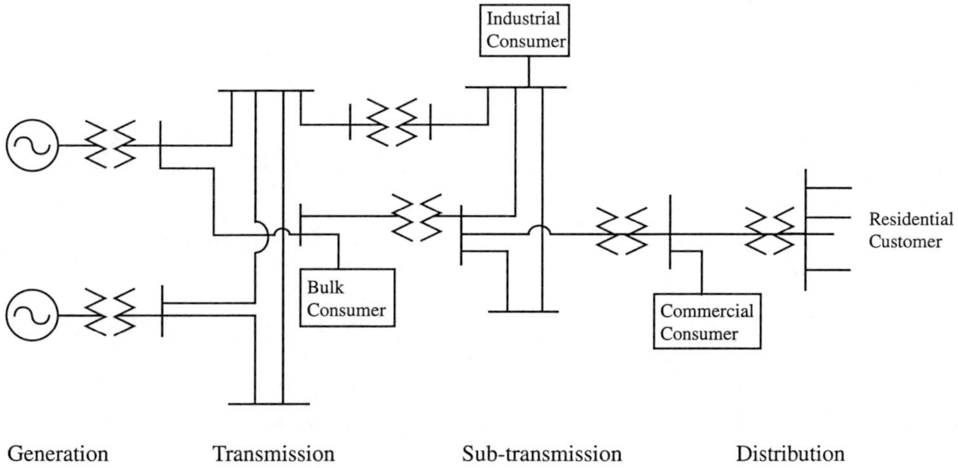

Figure 1.1: Power system structure

The generating system consists of a fuel source such as coal, water, and natural gas or nuclear power. The wind and the sun are also sources of power that can be converted to electric energy and made available to the grid. Water power through hydroelectric power plants and nuclear power are also non-polluting sources of energy. Hydropower is considered renewable since periodic rainfall will replenish the source. Nuclear power requires elaborate safety systems and extensive training of plant personnel. About 61% of U.S. electric energy production comes from fossil fuels, i.e., coal, natural gas, petroleum, and other gases. About 20% is from nuclear energy, and about 20% is from renewable energy sources, i.e., wind, hydropower, solar, biomass, and geothermal [1]. Wind and hydropower comprise most of the renewable generation, at about 8.4% and 7.3%, respectively.

In a generating system, the turbine acting as a prime-mover converts the mechanical energy to electric energy through a three-phase synchronous machine. The three-phase voltages of these generators on a line-line basis range from 13.8 kV to 24 kV. There is a three-phase transformer at each generator that steps up the voltage to a high value ranging from 138 kV to 765 kV. The transmission lines carry power over these lines to load centers and, where appropriate, transformers step it down to lower voltages up to 34 kV at the bulk-power substations. Some industrial customers are supplied from these substations. This is known as the transmission—sub-transmission system.

Figure 1.6 on page 13 shows the electrical power grid, or just the major transmission lines 230 kV and above, plus the HVDC lines criss-crossing the country. The U.S. power grid has been called the largest man-made system. It consists of three loosely interconnected system grids: one in Texas, and two more splitting the bulk of the country roughly along the continental divide. Each interconnect has its own distinct characteristics. These systems are far from orderly; each grid appears like a tangle of transmission lines operated by a number of owners, ranging from federal to independent small-scale power producers.

The power systems we have today can be appreciated by exploring their growth over time, and their history shows how more localized grids came to be interconnected [2]. Electrification began with initial appearance of a diverse variety of non-interoperable systems. Frequency control has played a major role in interconnection abilities, allowing real power outputs of generators to be coordinated to maintain control of frequency according to an area control error (ACE) equation.

Fig 1.5 on page 11 shows the U.S.'s electricity balancing authorities (BAs), based on data from the U.S. Energy Information Agency (EIA), a centralized resource of operating data on the high-voltage bulk electric power grid. Circle sizes are roughly indicative of each balancing authority's system size. Balancing authorities play a functional role defined by the North American Electric Reliability Corporation (NERC), primarily responsible for balancing electricity supply, demand, and interchange on their electric system in real-time [1].

The EIA classifies utilities into three main ownership types: investor-owned utilities, publicly owned utilities, and cooperatives. Investor-owned utilities, or IOUs, are large electric distributors that issue stock owned by shareholders. Publicly owned utilities, or POUs, include federal-, state-, and municipal-run utilities. Cooperatives, or co-ops, are not-for-profit member-owned utilities.

Large investor-owned utilities (IOUs) serve a large customer base per company and tend to be dominant in densely populated areas. A few large IOUs have most of the customers, whereas most places in the US are served by publicly owned power providers. The EIA data indicates the numbers of each type of electricity provider in the United States, with 2003 publicly owned utilities, 856 cooperatives, and 178 IOUs in 2019. There are 9 federal power agencies, 15 behind-the-meter electricity providers, 18 community choice aggregators, and 315 power marketers.

About 4,007 billion kilowatthours (kWh) of electricity was generated from utility-scale electricity generation, and an additional 42 billion kWh came from small-scale solar photovoltaic systems in the U.S. in 2020. Utility-scale generation is defined as power plants with at least one megawatt of capacity.

Power marketers are business entities engaged in buying and selling electricity. They do not usually own generating or transmission facilities. Behind-the-meter entities install, own, and/or operate systems (usually solar PV) and sell or lease production under a long-term power purchase agreement (PPA) to a homeowner or business with a net metering agreement. Community choice aggregators are local governmental entities allowed to negotiate with power providers and procure electricity on behalf of retail customers [3].

There were about 155 million electricity customers in 2019, where IOUs served 66.9% of electricity customers. Publicly owned utilities served 14.5%, and cooperatives served 13.2%. The small remaining portion of the customers are served by power marketers, behind-the-meter entities, and federal power agencies (4.8%, 0.5%, and 0.0%, respectively). The American

Public Power Association (APPA) publishes annual reports containing various graphs and summaries of pertinent utility industry data; the statistics above are from the 2021 report (with 2019 EIA data) [4].

The reliability of the electric power grids in the U.S. is amazingly high. With the onset of deregulation, the transmission system or grid belongs, in some sense, to anybody who can generate power and sell it to someone else. This selling of power over long distances has caused problems of its own. The transmission system or grid gets overloaded. Hence, even though the generation picture is improving, the grid can prove to be a bottleneck unless it is able to be strengthened in a way that can keep up with the changing technologies and changing resource mix.

While U.S. grids have historically been highly reliable, a recent emphasis is on resilience. For example, power system operation considers the environment and economics. The optimal power flow (OPF) [5] minimizes operational costs while meeting reliability requirements. The security-constrained optimal power flow (SCOPF) has been used to ensure the safety of power systems with postulate contingencies while minimizing operational costs [6]. The SCOPF focuses on specific contingencies, so unexpected contingencies can still result in system stress and loss of load. A recent incident, the 2021 Texas Winter Energy Crisis [7], caused large-scale blackouts due to unexpected extreme weather leading to generator unavailability. This poses a critical question: can existing economics-driven power system operation survive such unexpected extreme disturbances?

While reliability is considered a part of resilience, resilience includes more. Resilience is about ensuring the system can survive while maintaining its critical functions over a long period of time, and over a wide range of disruptions, with engineered system improvements that span the entire life cycle of events. Improving grid resilience requires a multifaceted approach that also includes lessening the likelihood that outages will occur, restoring power rapidly afterwards, and learning from events to plan the system better. Threats to electric grid operation and resilience include catastrophic events such as extreme weather, which are the most prevalent, as well as other types of high-impact low-frequency events such as electromagnetic disturbances and cyberattacks [8]. Such disruptive events can cost billions of dollars and disrupt the basic necessities of modern life. It is thus a critical need to plan and operate critical infrastructure systems particularly the electric grid to be resilient to hazards. How to best quantify and achieve resilience in large-scale cyber-physical systems is an ongoing challenge.

The need to strengthen the grid has been addressed in recent years by a wide range of "Smart Grid" and "Grid Edge" initiatives, which encompass different philosophies and approaches. One major theme of these approaches is making the customer more involved – by giving consumers a more proactive role in their own energy monitoring and control; this is supported by smart metering and by enabling customers to sell back power to the utility. Renewable energy sources such as wind and solar systems support these approaches to grid strengthening, and they are connected via power electronics converters. Together, such emerging distributed systems are often called distributed energy resources (DERs), and they are beginning to transform the grid landscape and future smart grid architectures.

Distribution of power to residential and commercial users takes place through a distribution system consisting of substations where transformers step down the voltage to a range of 2.4 kV to 69 kV. Power is carried by main feeders to specific areas where there are lateral feeders to step it down further to customer levels, such as 208, 240, or 600 volts. The distribution transformers serve anywhere from 1 to 10 customers. This complex transmission and distribution system encompasses large geographical areas. Currently, customers in an area are served by a utility company. In many places in the U.S., customers are free to choose their electricity service provider just as with the communication industry. The resulting problems are quite complex.

Utilities are interconnected with each other through high-voltage transmission lines (known as tie-lines) for the sake of reliability. In times of emergency, one utility comes to the rescue of others because of these tie-lines. Energy transactions between utilities through tie-lines are routinely done on a daily basis to minimize the cost to the customer. Bulk power transfer over very long transmission lines is common on the west and east coasts. On the west coast, power from B.C. Hydro in Canada and the Bonneville Power Authority (BPA) in Portland is transmitted over 765 kV lines as well as HVDC lines (500 kV) to Southern California. On the east coast, power from Hydro-Quebec and Ontario-Hydro in Canada is transmitted to New York.

At the consumer level, power is consumed in various ways, such as industrial loads, commercial loads (offices and malls), and ordinary households. The average cost of electric energy per kWH at the consumer end ranges from 8.9 cents to 30.6 cents, shown in Figure 1.2.

Both efficient transmission, distribution, as well as energy conservation

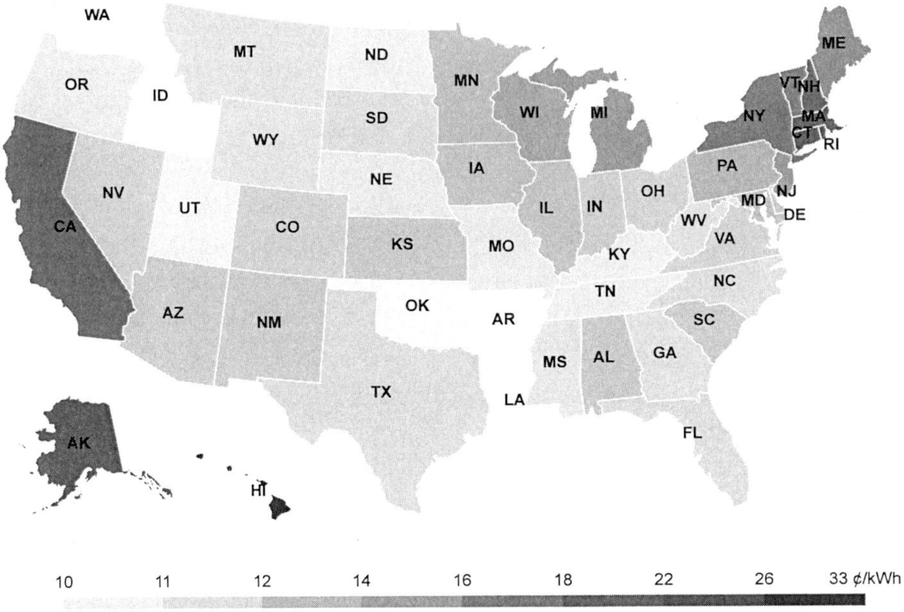

Figure 1.2: Average retail electricity prices by state, 2021 (cents per kilowatt-hour). Data source: Energy Information Administration (EIA), www.eia.gov; figure source: www.electricrate.com.

coupled with lower oil costs are contributing factors to the low electricity costs. Another general aspect to be appreciated in the electric power industry is the remarkable way in which the frequency is maintained at 60 Hz (50 Hz in Europe and in most of Asia and Africa) in spite of varying load demand on a minute-to-minute basis. Keeping the frequency constant, and maintaining a constant voltage free of harmonics is an essential requirement for the information technology industry.

In this complex power system structure, electromechanics plays a key role at several levels. First, at the generation level, the synchronous machines convert mechanical energy to electric energy. At the level of loads, the bulk of the energy is consumed through motors, mostly induction (three-phase and single-phase). At various stages at the transmission-distribution level, circuit breakers trip via protection systems if they detect a fault in the area they are supposed to protect. These circuit breakers are operated by electromechanical devices. There has been a trend toward replacing electromechanical relays with digital relays for better reliability and coordination among relays. A single modern microprocessor-based digital relay box can also perform numerous protection functions that would have each required its own electromechanical relay previously. The settings and configurations of digitial relays are also easy to download, import, study, and change with the help of specialized computer software for protection coordination studies. In the industrial sector, electromechanics also plays an important role, with devices that make use of translational and rotational forces. All the transformers in these systems operate on the same principle—namely, mutual coupling of coils through an iron core. The stability of the power system despite disturbances large or small is extremely important in order to avoid blackouts or outages.

1.2 Restructuring the Power Industry

The Public Utilities Regulatory Act of 1978 effectively initiated the restructuring process in the United States (often called "deregulation") by allowing non-utility generators (NUG) to enter the wholesale power market. The act required utilities to buy power from the NUG. From an initial capacity of 42,000 MW in 1989, the total NUG capacity grew to about 98,000 MW by 1998. In April 1996, the Federal Energy Regulatory Commission (FERC) issued final rules on Open Transmission Access, making the transmission network accessible to all in a non-discriminatory manner. In Sept. 2020, FERC order 2222 was issued, which opens wholesale markets to distributed resources. This action breaks down previous barriers for emerging grid edge technologies and promises to boost competition.

Electricity markets are a complex topic. Emerging work continues to study how to best structure, monitor, and operate them. The challenge is further increased by the evolving architectures of today's grids characterized by increased DERs, power electronics, and cyber-physical features that cause previous fundamental assumptions about power system operations to need to be re-examined. At a basic level, the classes of entities are the following:

1. Generator companies
   The old vertically integrated utilities (VIUs) were forced to divest their generation portions and make them into separate entities. These were sometimes bought by other companies. Thus, we have the generating companies (GENCOs), also called independent power producers (IPPs).

2. Transmission companies
   The transmission companies, often called TRANSCOs, own and operate the transmission wires. Their primary responsibility is to transport power from generators to customers and make the transmission system available to all.

3. Distribution companies
   There are local companies, often called DISCOs, owning and operating the local distribution network. They buy wholesale power on the spot market or enter into long-term contracts with the GENCOs.

4. Independent system operator (ISO)
   The ISO is charged with ensuring the reliability and security of the entire system. It is a critical entity that does not own generation but acts as a nodal agency for clearing contracts between parties.

A pictorial view of electricity system operation before and after electricity market restructuring is given in Figures 1.3 and 1.4. The idea of the transformation can be seen by considering the two extremes: moving from one end with a traditional vertically owned utility with 'obligation to serve,' to the other end, with comprehensive wide-spread customer choice and free electricity markets.

This restructuring has produced varied degrees of experience across the globe. While it has succeeded, for example, in Nordic countries, in other countries, it has either not benefited the consumers dramatically or, as in the case of California, it has caused concern. The lesson learned in the 2001

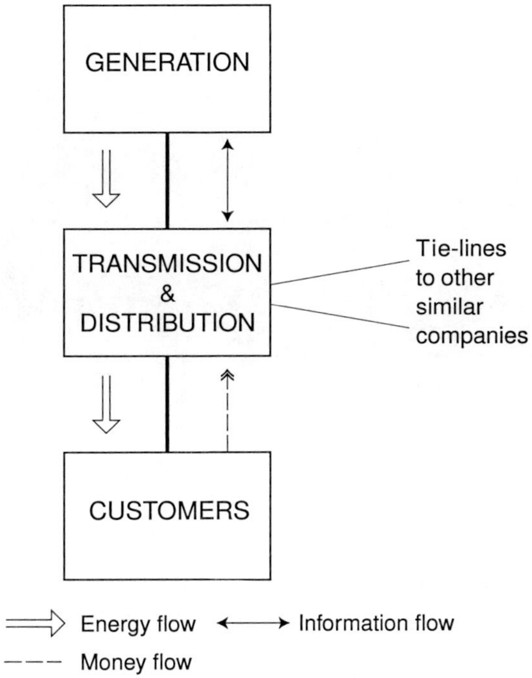

Figure 1.3: Vertically integrated utility

Energy Crisis in California is that operating the grid based on a more free-market approach does not work where there is a deficit of available power generation. The lessons from California are clear: unless you have excess generating capacity, competition simply will not work.

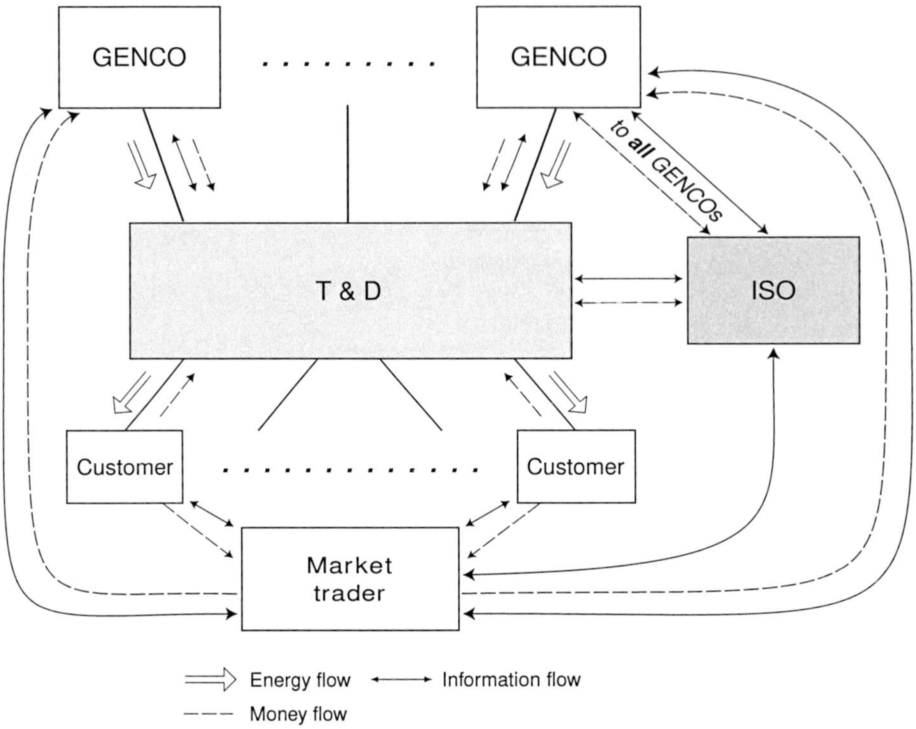

Figure 1.4: Restructured power system [9]

1.3 Dynamics of Power Systems and Components

The entire power system is a dynamic one, i.e., it is described by a set of vector differential equations of the form

$$\dot{x} = f(x, u) \tag{1.1}$$

where the state vector x and the input vector u are n and r dimensional vectors, respectively. The dimension of x is very large, and the time scales in the response of (1.1) span from milliseconds, in the case of electromagnetic transients, to a few seconds in the control of frequency, and a few hours in the case of boiler dynamics. Even at the device level of electromechanical components, a similar description applies, except that the dimensions of x and u are small. Therefore, we analyze such equations for time-domain response, steady-state sinusoidal behavior, equilibrium points, stability, etc.

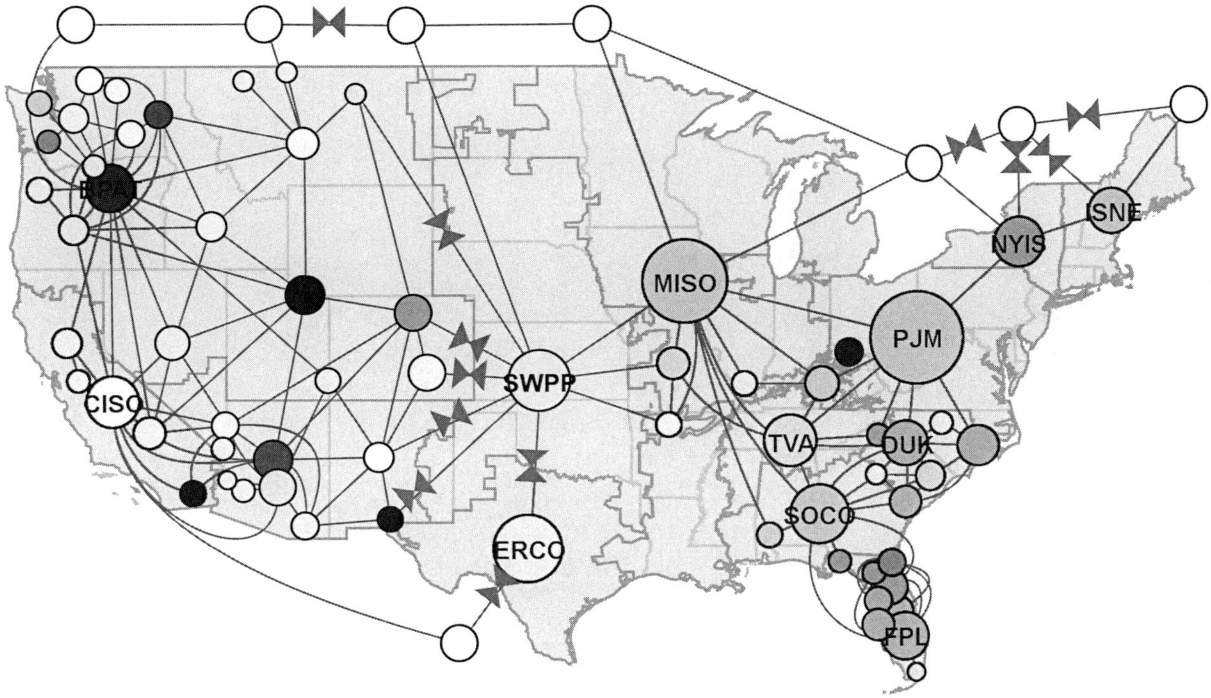

Figure 1.5: Balancing authorities, 2021 (Source: Energy Information Administration (EIA), www.eia.gov [1])

A critical step in the analysis process is the modeling from physical principles and quasi-static form of Maxwell's equations. After the model is derived, we put the dynamic model in the state-space form (1.1). Although the emphasis is on sinusoidal steady-state analysis, a brief introduction to time-domain response will be given. For sinusoidal analysis, a firm understanding of phasors and complex power is essential. We will also discuss the computation of equilibria of dynamic systems and their stability.

1.4 Electromechanical Systems

We consider two types of electromechanical systems, translational and rotational. Translational systems are used in electromechanical relays, and plungers and are generally easy to analyze. Rotational systems are more complicated. Although both follow the same principles analytically, the analysis of rotational systems is limited to sinusoidal steady-state analysis using phasors and equivalent circuits. Once the equivalent circuit is derived, the mechanical aspects are also reflected in it. This applies to synchronous, induction, and DC machines. For single-phase machines, only a qualitative analysis is done.

1.5 Conclusion

In this chapter, we discussed the power system structure, its different components, and their interaction. Being perhaps the largest physical system ever created on the planet, it has several dimensions to its study. Market restructuring aspects and how the current structures differ from previously existing structures have been discussed. In the new scenario that is evolving, a complex interchange of information and money needs to be carefully designed and managed, along with the judicious and secure monitoring and control of the electric system, thus affecting the power flows from sources to loads.

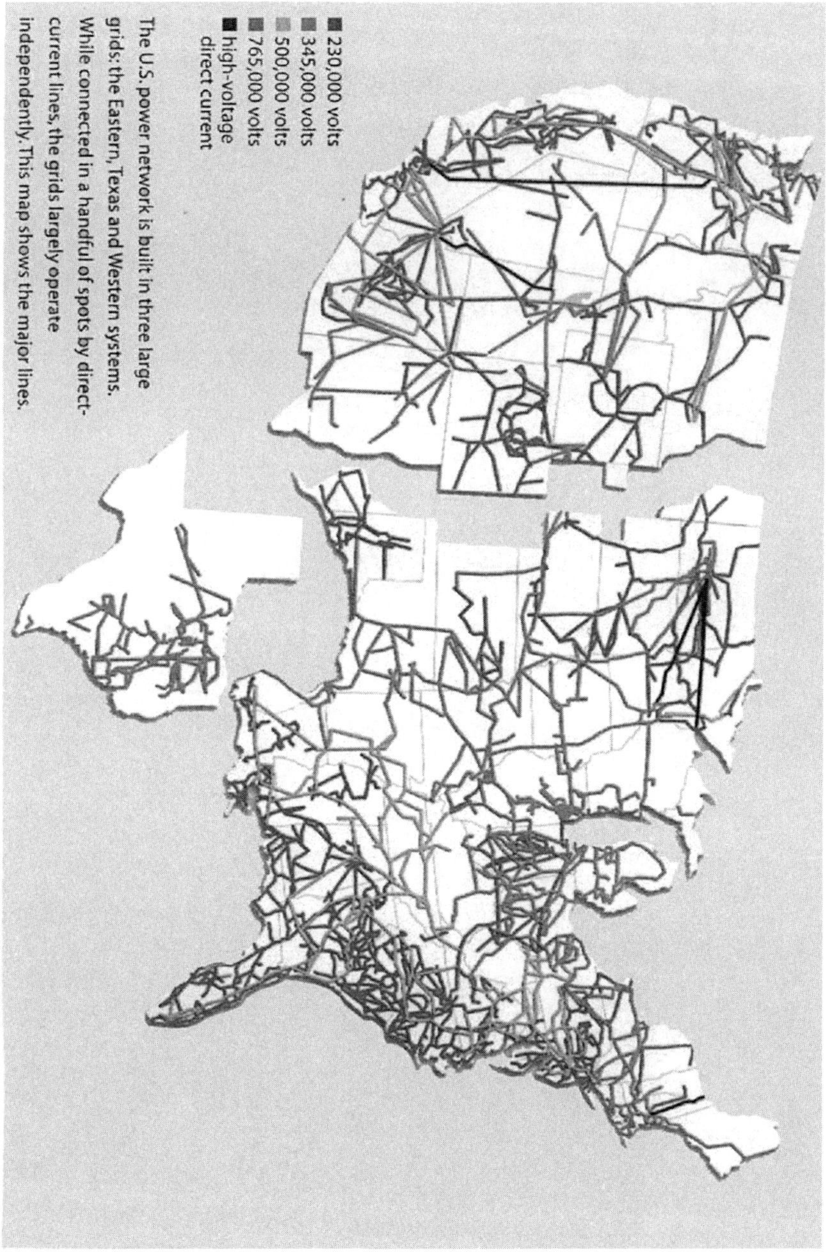

Figure 1.6: U.S. power grid: The electricity lifeline (illustration by Betsy Hayes from PowerWorld of Urbana, IL).

Chapter 2

PHASORS AND THREE-PHASE POWER CIRCUITS

2.1 Introduction

This chapter discusses sinusoidal voltage-current relationships, phasor representation, and complex power in single-phase and balanced three-phase circuits. Sinusoidal steady-state analysis forms the basis of most power system studies.

2.2 Two-Terminal Network Relationships

Consider the two-terminal network of Figure 2.2. The passive sign conven-

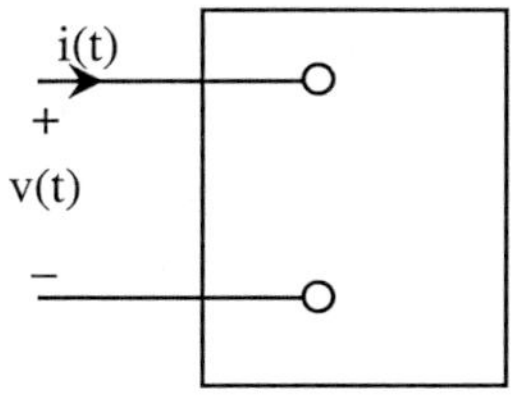

Figure 2.1: Two-terminal network

tion assumes a passive circuit element, which thus defines currents and powers as positive going into the load. Using the passive sign convention, the instantaneous power $p(t)$ into the network is

$$p(t) = v(t)i(t) \tag{2.1}$$

Let

$$\begin{aligned} v(t) &= V_m \cos(\omega t + \theta_v) \\ i(t) &= I_m \cos(\omega t + \theta_i) \end{aligned} \tag{2.2}$$

where θ_v and θ_i are the phase angles of v and i, and $\omega = 2\pi f$ with frequency f in Hz.

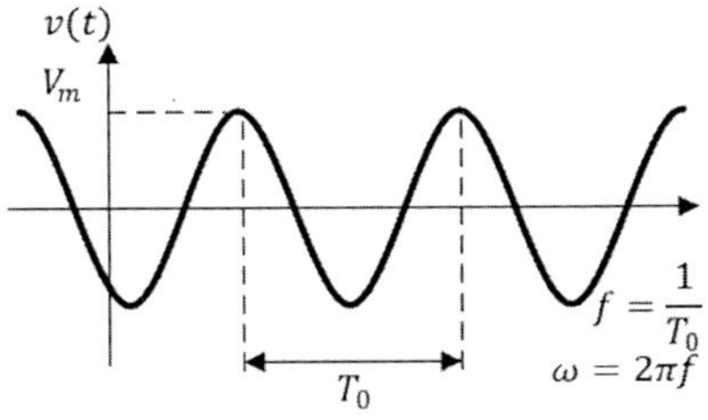

Figure 2.2: Sine wave parameters for AC circuit analysis with phasors

In an AC circuit, both current and voltage vary with time. In the US, the grid frequency is 60 Hz, so one cycle (period T_0 of sine wave) is 1/60Hz = 0.01666 sec. The angular frequency is omega $\omega = 2\pi f$ (rad/s). If you were to measure the voltage at a wall outlet, it would appear similar to Fig. 2.3. Looking at this AC voltage waveform, we can see that there are several possible measurements that could be taken of the voltage. This is important to note, and to recall that votlages are always with respect to a reference. A voltage is always between two points. Here, the peak-to-ground, peak-to-peak, and root mean square (RMS) voltages are noted. The RMS is the *square root* of the *mean* of the *square* of the waveform (see 2.8). We often use the RMS voltage in power systems.

Phase angles are important in power systems, and to interpret them, one must recognize that any time reference can be chosen. To see this, consider the following time domain waveforms. Without a reference provided, or a $t = 0$ axis, we could potentially define $t = 0$ to be anywhere. Hence, let us do a thought experiment, and draw an axis for $t = 0$ at the place where the sine wave crosses the x axis, such that this corresponds to a sine wave with zero phase shift (shown in bottom waveform). Then, making sure to use the

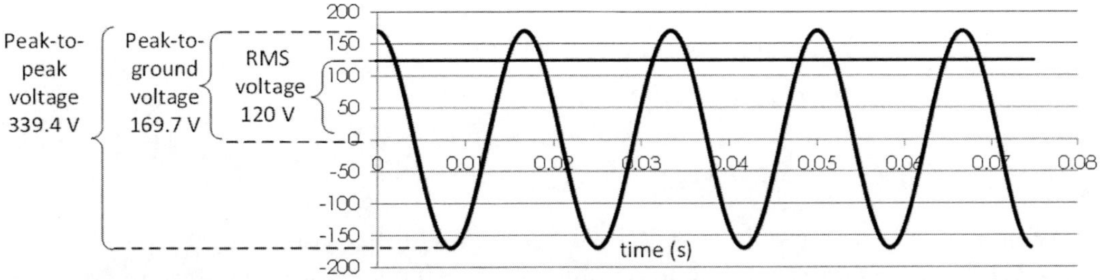

Figure 2.3: AC wall outlet - examples of types of voltage measurements

same vertical axis as the reference, look to see the angle of the sine wave on the other waveform. In this case, when the bottom waveform is at 0 the top waveform is at $\pi/4$.

The most important note here is that this phase difference will always be the same for these waveforms, regardless of where we draw the vertical axis; hence, it is easy to see that all phase angles are relative and must be measured with respect to a reference to have meaning. This is shown in Figure 2.4.

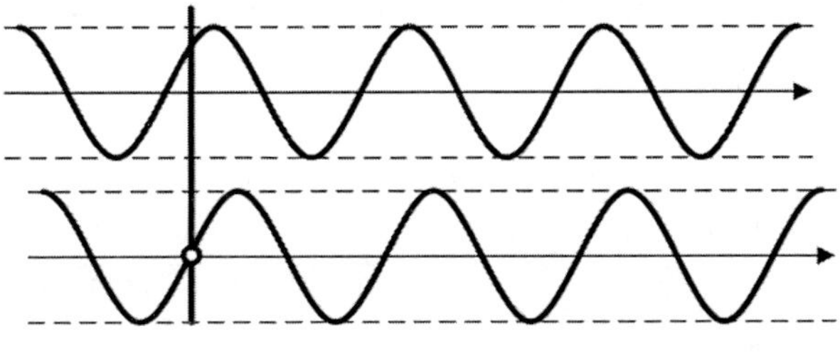

Figure 2.4: Phase angle differences

In the example in Figure 2.4, we first chose the bottom waveform v_2 as the angle reference, as shown. This gives the top waveform v_1 an angle of $\pi/4$. Then, the equations for these waveforms are $v_1 = Vsin(\omega t + \pi/4)$ and $v_2 = Vsin(\omega t + 0)$. The key insight is that if the top waveform is chosen as the angle reference instead, the same angle difference still exists, $\theta_{v_1} - \theta_{v_2} = \pi/4$. Hence, the choice of angle reference is arbitrary.

From the phase angle discussion, one angle is always chosen as the reference when comparing sinusoids of the same frequency. Hence, instead of writing both θ_v and θ_i as the phase angles in $v(t)$ and $i(t)$, respectively, we can shift the angles and let $\theta_i = 0$. Then, we apply the this shifted time variable $t' = t + \theta_i/\omega$ to (2.2), and the voltage angle of $v(t')$ becomes $\theta_v - \theta_i$

$$\begin{aligned} v(t') &= V_m \cos(\omega t' + \theta_v - \theta_i) \\ i(t') &= I_m \cos \omega t' \end{aligned} \tag{2.3}$$

While performing this phase shift is helpful to simplify calculations, it is not necessary, and the results will be the same. Hence, one can also proceed directly with (2.2) as follows. The instantaneous power is

$$p(t) = V_m I_m \cos(\omega t + \theta_v)(\cos \omega t + \theta_i) \tag{2.4}$$

The average power P is

$$P = \frac{1}{T} \int_o^T p(t)dt \tag{2.5}$$

where $T = \frac{2\pi}{\omega}$ is the period in seconds. Evaluating (2.4) and simplifying, we get

$$p(t) = \frac{V_m I_m}{2} \cos(\theta_v - \theta_i) + \frac{V_m I_m}{2} \cos(2\omega t + \theta_v + \theta_i) \tag{2.6}$$

Note that in (2.6), the first term on the right side is constant, whereas the second term is time-varying, with a period of twice the period T. Using this expression for $p(t)$ in (2.5), the average power P is obtained as

$$P = \frac{V_m I_m}{2} \cos(\theta_v - \theta_i) \tag{2.7}$$

In power system calculations, it is easier and more conventional to deal with root mean square (rms) values instead of peak values. We therefore define the root mean square values of voltage and current as

$$\begin{aligned} V_{rms} = \left(\frac{1}{T} \int_o^T v^2(t)dt \right)^{1/2} &= \left(\frac{1}{T} \int_o^T V_m^2 \cos^2(\omega t + \theta_v)\, dt \right)^{1/2} \\ &= \frac{V_m}{\sqrt{2}} \end{aligned} \tag{2.8}$$

Similarly

$$I_{rms} = \frac{I_m}{\sqrt{2}} \tag{2.9}$$

Thus from (2.7)

$$P = V_{rms} I_{rms} \cos(\theta_v - \theta_i) \tag{2.10}$$

$\theta_v - \theta_i$ is called the *power factor angle* θ and $\cos\theta$ is called the *power factor* (PF).

2.3 Phasor Notation

Phasors are used in electrical engineering, and particularly power systems, to represent sinusoids of the same frequency by an amplitude and phase angle. Phasors can be quickly derived from Euler's Identity.

Euler's identify can be written in different ways, all of which are useful.

$$e^{jx} = cos(x) + jsin(x) \tag{2.11}$$

$$e^{-jx} = cos(x) - jsin(x) \tag{2.12}$$

or

$$cos(x) = \frac{1}{2}(e^{jx} + e^{-jx}) = Re\{e^{jx}\} \tag{2.13}$$

$$sin(x) = \frac{1}{2j}(e^{jx} - e^{-jx}) = Im\{e^{jx}\} \tag{2.14}$$

$$A(t) = A_p cos(\omega t + \theta) \tag{2.15}$$

where *Re* stands for "real part of," and *Im* stands for "imaginary part of." Then, apply this identity to A(t) above to obtain

$$A(t) = \frac{A_p}{2}(e^{j(\omega t+\theta)} + e^{-j(\omega t+\theta)}) \tag{2.16}$$

Then, with (2.13), (2.16) becomes

$$A(t) = A_p Re\{e^{j(\omega t)}e^{j\theta}\} \tag{2.17}$$

Sinusoids with the same frequency are thus completely described by amplitude and phase angle, and a phasor is used to represent them.

Using the above definitions, the time-domain expressions in (2.2) can be expressed in phasor form as $\overline{V} = V_{rms}\angle\theta_v$ and $\overline{I} = I_{rms}\angle\theta_i$. Note that we have taken the rms values instead of peak values as magnitudes of the phasors. There are several ways to represent phasors in shorthand. Here, we denote a phasor by using capital letter with a bar over the top. Regardless of the representation you use, it is important to be consistent.

Depending on the sign of $(\theta_v - \theta_i)$, the current is either lagging the voltage or leading the voltage. We use the convention that "current leads" or "current lags" the voltage. If $\theta_i < \theta_v$, current lags the voltage (Figure 2.5(a)),

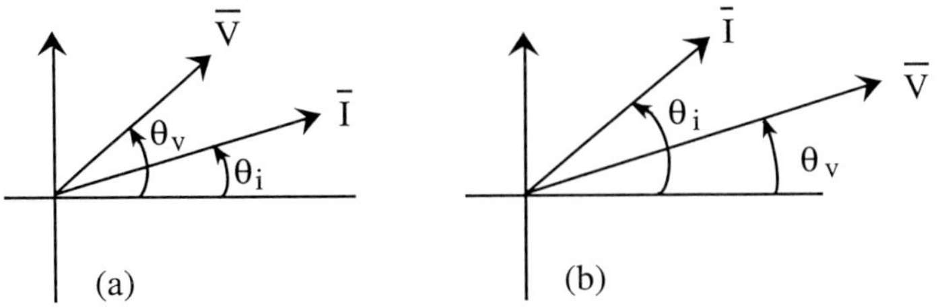

Figure 2.5: Lagging (a) and leading (b) power factor representations

and it is termed a *lagging* power factor. If $\theta_i > \theta_v$, current leads the voltage and it is termed a *leading* power factor (Figure 2.5(b)). If the two-terminal box in Figure 2.2 represents a load, we say the load is *inductive* for a *lagging* power factor (PF) and is *capacitive* for a *leading* PF.

If $\overline{V}$ is taken as the angle reference, which is common, then $\overline{V} = V_{rms}\angle 0°$ and $\overline{I} = I_{rms}\angle(\theta_i - \theta_v) = I_{rms}\angle - \theta$. Visually, this rotates the phasors in Figure 2.5 clockwise by θ_v. The term $\cos\theta$ is the PF, and $\theta = \cos^{-1}$ (PF) is the phase difference between $\overline{V}$ and $\overline{I}$. Since $\theta = \theta_v - \theta_i$, $\theta > 0$ represents a lagging PF, and $\theta < 0$ represents a leading PF.

Example 2.1 If in Figure 2.2, $v(t) = \sqrt{2}(10)\cos(\omega t + 30°)$ and $i(t) = \sqrt{2}(5)\cos(\omega t - 20°)$, write the phasor forms of $v(t)$ and $i(t)$. Find the average power P into the network.

Solution:

$$\begin{aligned} \overline{V} &= 10\angle 30° \text{ V}, \overline{I} = 5\angle - 20° \text{ A} \\ \theta_v &= 30°, \theta_i = -20° \\ \theta_v - \theta_i &= \theta = 30 - (-20) = 50° \end{aligned}$$

Since $\theta > 0$, we have a lagging PF.

$$P = (10)(5)\cos\ 50° = 32.14 \text{ W}$$

Example 2.2 If in Example 2.1 $i(t) = \sqrt{2}(5)\cos(\omega t - 90°)$ A, what is the value of P?

$$\begin{aligned} \overline{V} &= 10\angle 30° \text{ V}, \overline{I} = 5\angle - 90° \text{ A} \\ \theta_v - \theta_i &= 30 - (-90°) = 120° \\ P &= (10)(5)\cos 120° = -25 \text{ W} \end{aligned}$$

This means real power is delivered by the network. Note that the power factor is defined only for $|\theta| < 90°$. Hence, in this case we cannot define a power factor between $\overline{V}$ and $\overline{I}$.

When a sinusoidal function is expressed as a sine function, it must be converted to a cosine function first, for easier computation. Our phasors are referenced from a cosine wave. Thus, for a sine wave, we must convert it, with $sin(x) = cos(x - 90)$. An easy rule to remember is this: The positive and negative x axis represent cos and $-$cos, respectively. The vertical downward y axis represents sin and the vertical upward y axis represents $-$sin, respectively, as shown in Figure 2.6.

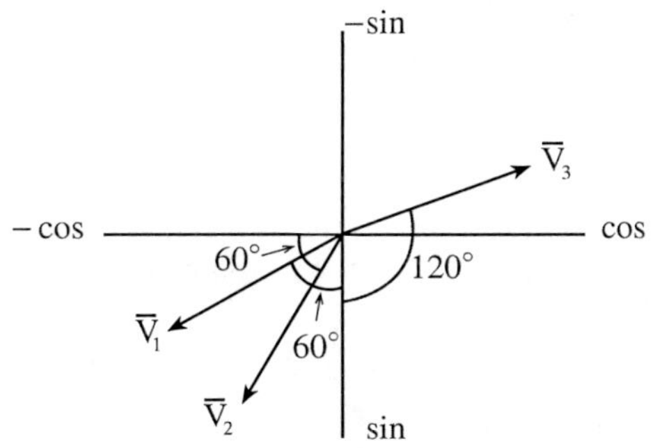

Figure 2.6: Phasors

Thus, any sinusoidal function can be quickly represented as a phasor in the complex plane. Note, angles are measured as positive *anti-clockwise* and as negative *clockwise.*

Example 2.3 Find the phasors for $v_1(t) = \sqrt{2}(10)\sin(\omega t - 60°)$, $v_2(t) = -\sqrt{2}(10)\cos(\omega t + 60°)$, $v_3(t) = \sqrt{2}(10)\sin(\omega t + 120°)$. Plot them in the complex plane.

Solution: Following the rule above, the phasors are shown in the complex plane in Figure 2.6. $\overline{V}_1 = 10\angle - 150°$, $\overline{V}_2 = 10\angle - 120°$, $\overline{V}_3 = 10\angle 30°$.

Example 2.4 Find the time domain waveform for phasors $\overline{V_1} = 120\angle 10$, $\overline{V_2} = 1\angle - 90$.

Solution: These phasors represent the following cosine waveforms: $v_1(t) = \sqrt{2}(120)\cos(\omega t + 10^\circ)10^\circ$, $v_2(t) = \sqrt{2}\cos(\omega t - 90^\circ)$ which is equivalent to $v_2(t) = \sqrt{2}\sin(\omega t)$.

2.4 Complex Power

Let us consider the evaluation of instantaneous power in (2.6), where a time-varying component is present as

$$\begin{aligned}\frac{V_m I_m}{2}\cos(2\omega t + \theta_v + \theta_i) &= \frac{V_m I_m}{2}\cos(2\omega t + 2\theta_i)\cos(\theta_v - \theta_i) \\ &\quad -\frac{V_m I_m}{2}(\sin 2\omega t + 2\theta_i)\sin(\theta_v - \theta_i) \quad (2.18)\end{aligned}$$

Define

$$Q \triangleq \frac{V_m I_m}{2}\sin(\theta_v - \theta_i) = V_{rms} I_{rms}\sin(\theta_v - \theta_i) \quad (2.19)$$

Q is called the *reactive power*. Thus (2.6) becomes

$$\begin{aligned}p(t) &= P + P\cos(2\omega t + 2\theta_i) - Q\sin(2\omega t + 2\theta_i) \\ &= P(1 + \cos(2\omega t + 2\theta_i)) - Q\sin(2\omega t + 2\theta_i) \quad (2.20)\end{aligned}$$

Then, from (2.7), P and Q and (2.19) can be expressed respectively, as

$$\begin{aligned}P &= \frac{V_m I_m}{2}\cos(\theta_v - \theta_i) \\ &= V_{rms} I_{rms}\cos(\theta_v - \theta_i) \\ &= Re(V_{rms}e^{j\theta_v} I_{rms}e^{-j\theta_i}) \quad (2.21)\end{aligned}$$

and

$$Q = Im(V_{rms}e^{j\theta_v} I_{rms}e^{-j\theta_i}) \quad (2.22)$$

Since

$$\begin{aligned}V_{rms}e^{j\theta_v} &= \overline{V} \text{ (phasor)} \\ I_{rms}e^{j\theta_i} &= \overline{I} \text{ (phasor)} \quad (2.23)\end{aligned}$$

we have from (2.21) and (2.22)

$$\begin{aligned}P = Re(\overline{V}\,\overline{I}^*) &= V_{rms} I_{rms}\cos(\theta_v - \theta_i) \quad (2.24) \\ Q = Im(\overline{V}\,\overline{I}^*) &= V_{rms} I_{rms}\sin(\theta_v - \theta_i) \quad (2.25)\end{aligned}$$

$$\text{Define complex power } \overline{S} = \overline{V}\,\overline{I}^* = P + jQ$$

In power calculations, one always uses rms values. Hence the subscript *rms* will be dropped hereafter.

Thus, $P = VI\cos(\theta_v - \theta_i)$ and $Q = VI\sin(\theta_v - \theta_i)$ are the real and reactive power, respectively, and $S = VI$ is the magnitude of the complex power called the apparent power. Since $\theta_v - \theta_i \stackrel{\Delta}{=} \theta$ is the PF angle and the PF is $\cos\theta$, we have, as discussed earlier, $\theta = +$ for lag and $-$ for lead. Hence $Q > 0$ for lag and < 0 for lead.

The units of complex power are voltamperes. However, to distinguish the role of S, P, and Q, they are called *voltamperes* (VA), *watts* (W), and *voltampere reactive* (VAR), respectively.

Alternate forms of complex power

Consider a passive component with a complex impedance $\overline{Z} = R + jX$ connected across the source $\overline{V}$ (Figure 2.7). R is the resistance and X is the reactance. Complex admittance is defined as $Y = 1/Z$. By Ohm's law,

$$\begin{aligned} \overline{V} &= \overline{Z}\,\overline{I} \\ \overline{S} &= \overline{Z}\,\overline{I}\,\overline{I}^* \\ &= I^2\overline{Z} \\ &= I^2(R + jX) = P + jQ \end{aligned} \tag{2.26}$$

Hence,

$$P = I^2R, \quad Q = I^2X \tag{2.27}$$

If $\overline{Z} = R$, then $\overline{S} = P = I^2R = \frac{V^2}{R}$. If $\overline{Z} = jX$, then $\overline{S} = jQ = jI^2X = j\frac{V^2}{X}$. Hence, (i) Q will be positive for an inductor ($X = \omega L$), and (ii) Q will be negative for a capacitor $\left(X = -\frac{1}{\omega C}\right)$. From the passive sign convention, the + Q for an inductor indicates that it consumes reactive power, while -Q for a capacitor indicates that it supplies reactive power.

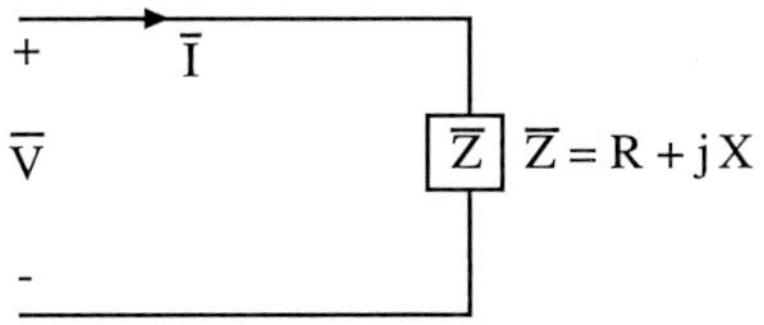

Figure 2.7: Impedance

Example 2.4 Suppose in Figure 2.2 $v(t) = \sqrt{2}(10)\cos(\omega t + 10°)$ and $i(t) = \sqrt{2}(20)\sin(\omega t + 70°)$. Find the complex power.

Solution: We first compute the phasors $\overline{V}$ and $\overline{I}$ as $\overline{V} = 10\angle 10° V$ and $\overline{I} = 20\angle{-20°} A$, $\overline{S} = \overline{V}\,\overline{I}^* = (10\angle 10°)(20\angle 20°) = 200\angle 30° = 173.2{+}j100$ VA

$$\begin{aligned} P &= 173.2 \text{ W} \\ Q &= 100 \text{ VAR} \end{aligned}$$

Example 2.5 In Figure 2.8, find (a) the complex power in each branch,

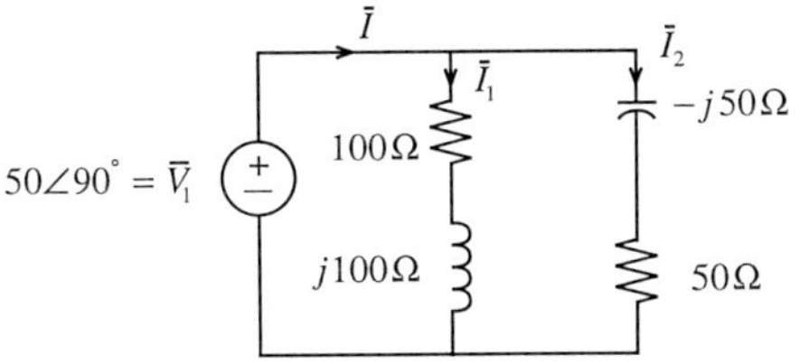

Figure 2.8: Complex power

(b) total complex power, and (c) individually the powers (real or reactive) in each of the four elements and verify if the sum agrees with part (b).

Solution:

$$\begin{aligned} \text{(a)}\quad \overline{I}_1 &= \frac{\overline{V}_1}{\overline{Z}_1} = \frac{50\angle 90°}{100 + j100} = \frac{50\angle 90°}{(\sqrt{2}\,)100\angle 45°} = 0.354\angle 45° \text{ A} \\ \overline{I}_2 &= \frac{\overline{V}_1}{\overline{Z}_2} = \frac{50\angle 90°}{50 - j50} = \frac{50\angle 90°}{(\sqrt{2}\,)50\angle - 45°} = 0.707\angle 135° \text{ A} \\ \overline{S}_1 &= \overline{V}_1\overline{I}_1^* = (50\angle 90°)(0.354)\angle - 45° \\ &= 17.68\angle 45° = 12.5 + j12.5 \text{ VA} \\ \overline{S}_2 &= \overline{V}_1\overline{I}_2^* = (50\angle 90°)(0.707)\angle - 135° \\ &= 35.35\angle - 45° = 25 - j25 \text{ VA} \end{aligned}$$

(b) Total complex power $\overline{S}$ is given by

$$\begin{aligned} \overline{S} = \overline{S}_1 + \overline{S}_2 &= (12.5 + j12.5) + (25 - j25) \\ &= 37.5 - j12.5 = 39.53\angle - 18.43° \text{ VA} \end{aligned}$$

(c)

$$\begin{aligned}
\text{Real power in the } 100\Omega \text{ resistor} &= I_1^2(100) \\
&= (0.354)^2 100 \\
&= 12.5 \text{ W} \\
\text{Real power in } 50\Omega \text{ resistor} &= I_2^2(50) \\
&= (0.707)^2(50) \\
&= 25 \text{ W} \\
\text{Total real power} &= 12.5 + 25 = 37.5 \text{ W}
\end{aligned}$$

$$\begin{aligned}
\text{Reactive power in the inductance} = I_1^2 X_L &= (0.354)^2 100 \\
&= 12.5 \text{ VAR}
\end{aligned}$$

In complex notation it will be $j12.5$ VAR since it is inductive.

$$\begin{aligned}
\text{Reactive power in the capacitor} &= I_2^2 X_c \\
&= (0.707)^2(-50) \\
&= -25 \text{ VAR} \\
\text{In complex notation it will be} &= -j25 \text{ VAR}
\end{aligned}$$

Total complex power will be the sum of the complex powers in all the elements. Hence

$$\begin{aligned}
\overline{S} &= (12.5 + 25) + j(12.5 - 25) \\
&= 37.5 - j12.5 \text{ VA}
\end{aligned}$$

This agrees with part (b).

Example 2.6 Consider two loads in parallel with $\overline{I}_1 = 50 - j40$ in one load and $\overline{I}_2 = 50 + j60$ in the second load. The voltage across them is $\sqrt{2}(100)\angle 45°$. Find the total complex power and the input power factor.

Solution: The input voltage is $\overline{V} = \sqrt{2}(100)\angle 45° = (100 + j100)$ V. The complex power in the two branches are

$$\begin{aligned}
\overline{S}_1 = \overline{V}\,\overline{I}_1^* &= (100 + j100)(50 + j40) \\
&= 1000 + j9000 \text{ VA} \\
\overline{S}_2 = \overline{V}\,\overline{I}_2^* &= (100 + j100)(50 - j60) \\
&= 11,000 - j1000 \text{ VA}
\end{aligned}$$

Total complex power is

$$\overline{S} = \overline{S}_1 + \overline{S}_2 = 12000 + j8000 \text{ VA}$$

Input current is given by $\overline{I}$ where $\overline{S} = \overline{V}\,\overline{I}^*$. Hence

$$\begin{aligned}\overline{I} &= \frac{\overline{S}^*}{\overline{V}^*} = \frac{12000 - j8000}{100 - j100} = \frac{120 - j80}{1 - j1} \\ &= 100 + j20 \\ &= 102\angle 11.3 \text{ A}\end{aligned}$$

Current lags the voltage by $(45 - 11.3) = 33.7°$. PF is $\cos 33.7° = 0.83$ lag.

2.5 Conservation of Complex Power

2.5.1 Series circuit

Consider the series circuit shown in Figure 2.9

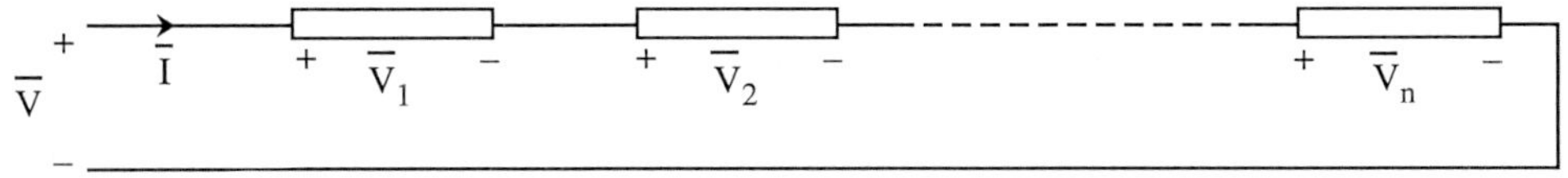

Figure 2.9: Series circuit

$$\begin{aligned}\overline{V}\,\overline{I}^* &= (\overline{V}_1 + \overline{V}_2 + \ldots + \overline{V}_n)\overline{I}^* \\ &= \overline{V}_1\overline{I}^* + \overline{V}_2\overline{I}^* + \ldots + \overline{V}_n\overline{I}^* \\ \overline{S} &= \overline{S}_1 + \overline{S}_2 + \ldots + \overline{S}_n \end{aligned} \quad (2.28)$$

2.5.2 Parallel circuit

Consider the parallel circuit shown in Figure 2.10 The total input complex power is

$$\begin{aligned}\overline{S} = \overline{V}\,\overline{I}^* &= \overline{V}(\overline{I}_1 + \overline{I}_2 + \ldots + \overline{I}_n)^* \\ &= \overline{V}\,\overline{I}_1^* + \overline{V}\,\overline{I}_2^* + \ldots + \overline{V}\,\overline{I}_n^* \\ &= \overline{S}_1 + \overline{S}_2 + \ldots + \overline{S}_n \end{aligned} \quad (2.29)$$

From (2.28) and (2.29), we see that the total complex power is the sum of the complex powers in individual elements. Most loads are always connected

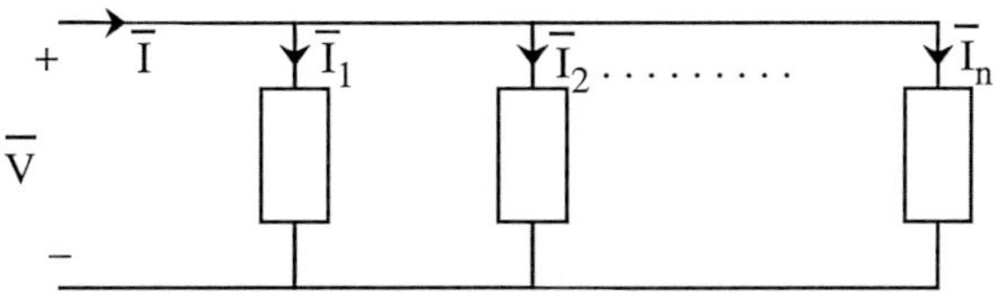

Figure 2.10: Parallel circuit

in parallel. Equation (2.29) is also a statement about the conservation of complex power. Thus if P and Q represent the total power, then

$$P = P_1 + P_2 + ... + P_n$$

and

$$Q = Q_1 + Q_2 + ... + Q_n$$

2.5.3 Power triangle

In the formula for complex power $\overline{S} = \overline{V}\,\overline{I}^* = P + jQ$, P is the real power, Q the reactive power, and S is the magnitude of $\overline{S}$. S is called the apparent power. It can be verified that $S = VI$, $P = VI\cos\theta$, and $Q = VI\sin\theta$ where $\theta \triangleq \theta_v - \theta_i$. For an inductive load (lagging PF), $\theta = \theta_v - \theta_i$ is > 0. Hence Q is > 0. For a capacitive load (leading PF), $\theta = \theta_v - \theta_i$ is < 0 and $Q < 0$. In some applications, it is easier to do computations via the "power triangle" method, as shown in the following example.

Example 2.7 If $\overline{V} = 100\angle 10° \, V$ and $\overline{I} = 10\angle -26.8° \, A$. Find the complex power and represent S, P, Q in the form of a power triangle.

Solution:

$$\begin{aligned}
\overline{S} &= \overline{V}\,\overline{I}^* \\
&= (100\angle 10°)(10\angle 26.8°) \\
&= 1000\angle 36.8° \\
&= 800 + j600 \text{ VA}
\end{aligned}$$

Hence, $P = 800$ W, $Q = 600$ VAR, $VI = 1000$ VA, $\theta = \theta_v - \theta_i = 10° - (-26.8°) = 36.8°$. $\theta > 0$ and, hence, this is a lagging PF case. The power triangle can be constructed as shown in Figure 2.11.

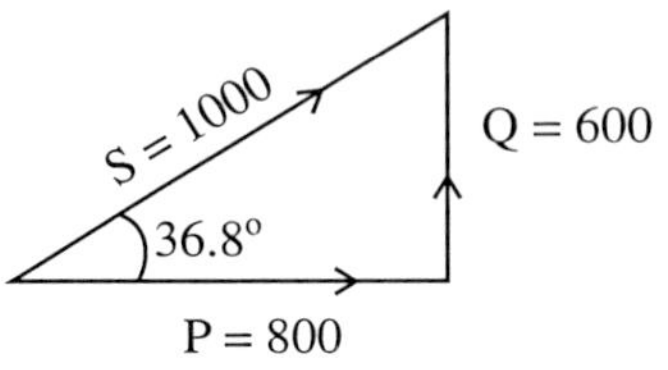

Figure 2.11: Power triangle

Example 2.8 If $\overline{V} = 100\angle 10°$ V, $\overline{I} = 5\angle 40°$ A, find $\overline{S}$, P, Q and draw the power triangle.

Solution:

$$\begin{aligned} \overline{S} &= \overline{V}\,\overline{I}^* = (100\angle 10°)(5\angle - 40°) \\ &= 500\angle - 30° \\ &= 433 - j250 \text{ VA} \end{aligned}$$

$P = 433$ W, $Q = -250$ VAR, $S = 500$ VA, $\theta = \theta_v - \theta_i = 10° - 40° = -30°$. This is a leading PF case. The power triangle is shown in Figure 2.12.

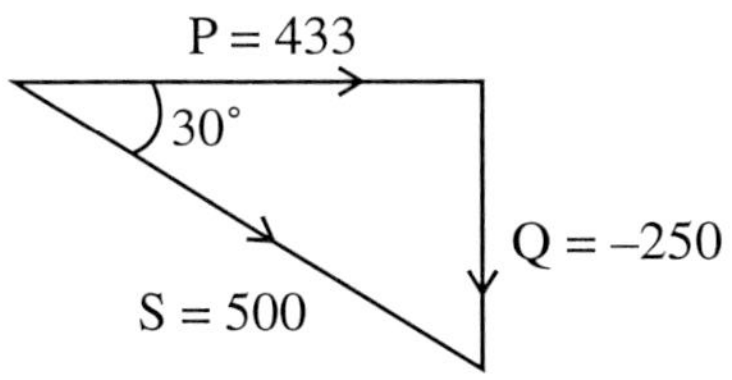

Figure 2.12: Power triangle

Example 2.9 Suppose the loads in Examples 2.7 and 2.8 are connected in parallel, as shown in Figure 2.13. Find the total power and current by (a) the current method and (b) the power triangle method.

Solution (current method):

By KCL, the input current $\overline{I}$ will be the phasor sum of the two currents, $\overline{I}_1$ and $\overline{I}_2$.

$$\begin{aligned} \overline{I} &= 10\angle - 26.8° + 5\angle 40° = 12.82\angle - 5.796° \text{ A} \\ \overline{S} &= (100\angle 10°)\overline{I}^* \\ &= 100\angle 10°(12.82\angle 5.796°) = 1282\angle 15.797° = 1234 + j349 \text{ VA} \end{aligned}$$

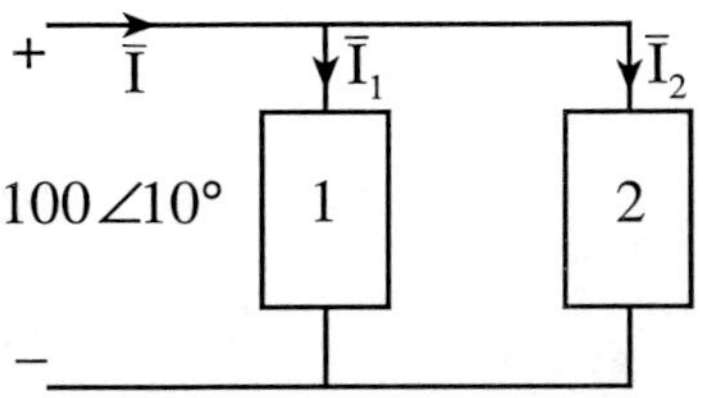

Figure 2.13: Two parallel loads

Solution (power triangle method):

From Example 2.7, $\overline{S}_1 = 800 + \jmath 600$ VA as shown in Figure 2.14(a). From Example 2.8, $\overline{S}_2 = 433 - \jmath 250$ VA as shown in Figure 2.14(b).

The total power $\overline{S}$ is given by

$$\begin{aligned} \overline{S} &= \overline{S}_1 + \overline{S}_2 \\ &= (800 + 433) + \jmath(600 - 250) \\ &= 1233 + \jmath 350 \text{ VA} \end{aligned}$$

This is shown in Figure 2.15. The input current is obtained from $\overline{S} = \overline{V}\,\overline{I}^*$ as $\overline{I} = \frac{\overline{S}^*}{\overline{V}^*}$. Hence

$$\overline{I} = \frac{\overline{S}^*}{\overline{V}^*} = \frac{1233 - \jmath 350}{100\angle - 10^\circ} = \frac{1282\angle - 15.847^\circ}{100\angle - 10^\circ} = 12.82\angle - 5.847^\circ \text{ A}$$

Thus, both the power triangle and current methods agree.

2.5.4 Reactive power support and power factor correction

Utilities do not like power factors that deviate substantially from unity. "Unity" power factor means that the load appears resistive, the current and the voltage are in phase, and no net reactive power is consumed or supplied. To supply VARs increases the current flowing down the line, which requires larger lines and transformers, additional generation, and results in greater line losses. Hence, it is desirable to supply reactive power locally to "correct" the power factor. Power factor correction results in less voltage drop across a line and lower power losses. Hence, capacitor banks are used throughout the system to supply reactive power and correct the power factor. Since capacitors have costs, and utilities must balance the costs of larger capacitors with their resulting benefits and savings, the power factor is often corrected

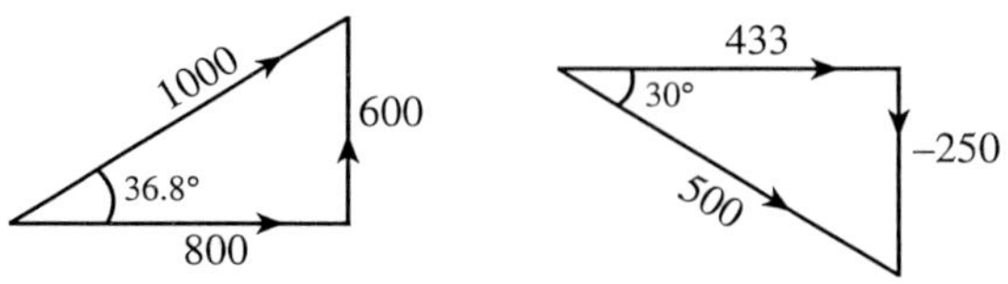

(a) $\overline{S}_1 = 800 + j600$ VA

(b) $\overline{S}_2 = 433 - j250$ VA

Figure 2.14: Individual power triangle for Example 2.9

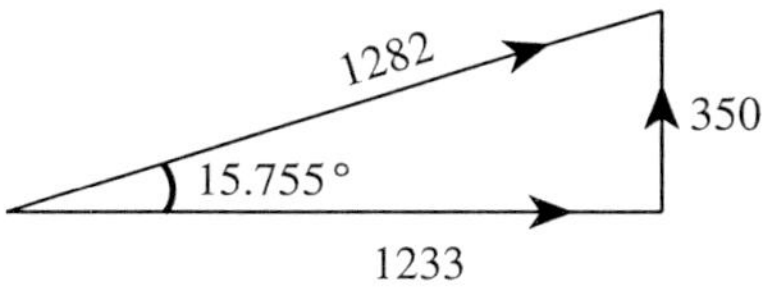

$$\overline{S} = 1233 + j350 \text{ VA}$$

Figure 2.15: Composite power triangle for Example 2.9

to something less than unity. Using the methods and techniques that have been described above, we can compute the reactive power, and hence also the capacitance, that would be needed to correct the power factor to any arbitary value. To visualize how this works with power triangles, Fig. 2.16 shows the main idea.

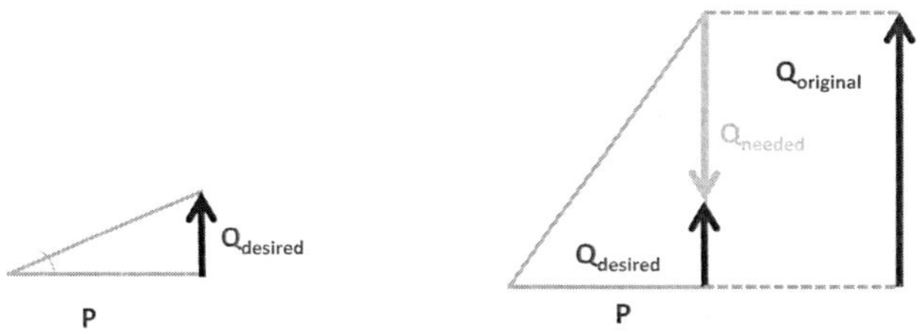

Figure 2.16: Power triangles showing (a) the system at desired power factor with $Q_{desired}$ and (b) the original system at $Q_{original}$ and power factor correction of Q_{needed} that achieves $Q_{desired}$

Example 2.10 In Example 2.9, (a) What is the power factor of the total load? (b) If the PF is to be improved to 0.98 lag by adding a capacitor, what is the kVAR needed of the capacitor? (c) If the PF is to be improved to unity, what is the kVAR needed? (d) What capacitance C would provide the kVAR found in part (c)?

(a) From Example 2.9

$$\overline{V} = 100\angle 10^\circ \text{ V}, \quad \overline{I} = 12.82\angle -5.847^\circ \text{ A}$$

The current lags the voltage, and the PF angle is $10 - (-5.847^\circ) = 15.847^\circ$. The PF is $\cos 15.847^\circ = 0.962$. The power triangle is shown in Figure 2.17.

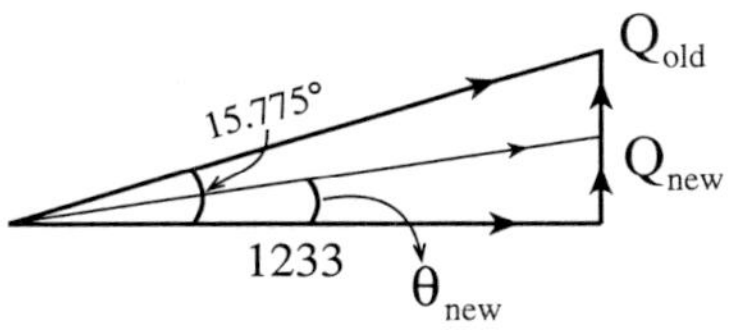

Figure 2.17: Old and new power triangles

(b) New desired PF = 0.98 lag

$$\theta_{\text{new}} = \cos^{-1}(0.98) = 11.48^\circ$$

The capacitor does not absorb any real power, and, hence, real power remains the same. From the power triangle

$$\frac{Q_{\text{new}}}{1233} = \tan 11.47^\circ$$
$$Q_{\text{new}} = 250 \text{ VAR}$$

Therefore $Q_{\text{cap}} + Q_{\text{old}} = Q_{\text{new}}$, i.e., $Q_{\text{cap}} = 250 - 350 = -100$ VAR. The negative sign of Q_{cap} confirms that it is capacitive.

(c) $Q_{\text{new}} = 0$. Therefore, the VAR needed is $Q_{\text{cap}} = -350$ VAR, and it is capacitive.

(d) Since Q_{cap} is known from part (c), capacitance can be easily found from

$$Q_{\text{cap}} = \frac{V^2}{X_{\text{cap}}} \qquad \text{where} \qquad X_{\text{cap}} = \frac{-1}{\omega C_{\text{cap}}}$$

Solving for capacitance needed, we thus require voltage and reactive power ratings and frequency; here $f = 60$ Hz

$$\begin{aligned} C_{\text{cap}} &= \frac{350 \times 10^3}{100^2(2\pi 60)} \\ C_{\text{cap}} &= 92.8 \text{ mH} \end{aligned}$$

2.5.5 Specifying power in a load

In solving power problems, the power absorbed by the load may be specified in a number of ways. It can be specified by a combination of three of the following six quantities:

$$V, I, \text{PF (lag or lead)}, S, P, Q$$

Since there is need for a reference in phasor analysis, we take $\overline{V} = V\angle 0°$. The power factor $PF = \cos\theta$, and $\theta = \pm\cos^{-1}$ (PF)(+ for lag and − for lead). The current is given by $\overline{I} = I\angle -\theta$.

1. If $\overline{V}$ and $\overline{I}$ are specified it is equivalent to specifying V, I, and the PF.
2. Another way is to specify V, PF, and P. Then

$$\begin{aligned} \theta &= \pm\cos^{-1}(\text{PF})(+ \text{ for lag and } - \text{ for lead}) \\ P &= VI\cos\theta \\ I &= \frac{P}{V\cos\theta} \\ Q &= VI\sin\theta \\ \overline{S} &= P + jQ \end{aligned}$$

Alternatively, from the values of V, PF, and P, we can compute

$$\begin{aligned} S &= \frac{P}{PF}, \theta = \pm\cos^{-1}(\text{PF})(+ \text{ for lag and } - \text{ for lead}) \\ \text{and } \overline{S} &= S\angle\theta \end{aligned}$$

3. A third way is to specify, V, PF, and S. Then $I = \frac{S}{V}$, $\theta = \pm\cos^{-1}$ (PF) (+ for lag and − for lead), $P = VI\cos\theta$, $Q = VI\sin\theta$.

$$\begin{aligned}\overline{S} &= P + jQ \\ \overline{S} &= S\angle\theta (\theta \text{ is } + \text{ for lag and } - \text{ for lead})\end{aligned}$$

4. Finally, we can also specify V, P, and Q. Then $\overline{S} = P + jQ$ and $PF = \frac{P}{S}$. $\theta = \pm\cos^{-1}(PF)$ (+ for lag and − for lead)

Example 2.11 In Figure 2.18, the two loads are

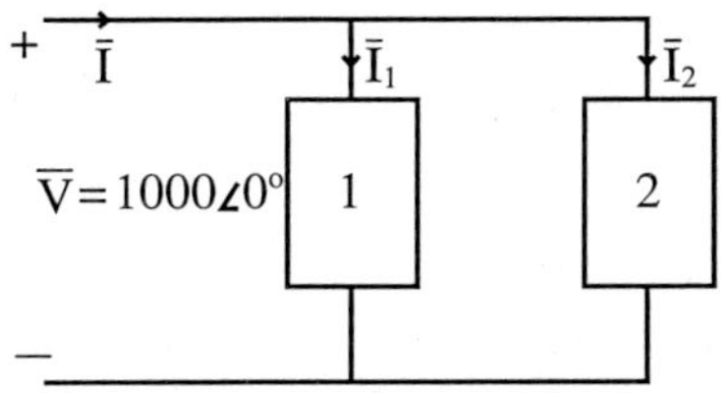

Figure 2.18: Two parallel loads

Load 1: 20 kVA at 0.6 leading PF
Load 2: 4 kW at 0.8 lagging PF

Solve for $\overline{I}$, input PF, and input complex power using (a) the current method and (b) the power triangle method.

(a) Current method
Load 1: Since it is a leading PF, $\theta = -\cos^{-1}(0.6) = -53.13°$. Hence

$$\begin{aligned}\overline{S_1} &= 20 \times 10^3 \angle -53.13° \text{ VA} \\ &= 20 \times 10^3 (0.6 - j0.8) \text{ VA} \\ &= \overline{V}\,\overline{I_1}^* \end{aligned} \tag{2.30}$$

Hence,

$$\begin{aligned}\overline{I_1} &= \frac{20 \times 10^3 (0.6 + j0.8)}{1000\angle 0°} \text{ A} \\ &= 12 + j16 \text{ A}\end{aligned} \tag{2.31}$$

Alternatively,

$$\begin{aligned} I_1 &= \frac{S}{V} \\ &= \frac{20 \times 10^3}{1000} \\ &= 20 \text{ A} \end{aligned} \tag{2.32}$$

Since PF is leading,

$$\begin{aligned} \overline{I_1} &= 20\angle 53.13^\circ \text{ A} \\ &= 20(0.6 + j0.8) \text{ A} \\ &= 12 + j16 \text{ A} \end{aligned} \tag{2.33}$$

Load 2: Since PF is lagging, $\theta = \cos^{-1}(0.8) = 36.87^\circ$.

$$\begin{aligned} P_2 &= VI_2 \cos\theta \\ 4 \times 10^3 &= (1000)I_2(0.8) \\ \text{Therefore, } I_2 &= 5 \text{ A} \end{aligned} \tag{2.34}$$

$$\begin{aligned} \overline{I_2} &= 5\angle -36.87^\circ \\ &= 5(0.8 - j0.6) \\ &= 4 - j3 \text{ A} \end{aligned} \tag{2.35}$$

The total current is

$$\begin{aligned} \overline{I} &= \overline{I_1} + \overline{I_2} \\ &= (12 + j16) + (4 - j3) \\ &= 16 + j13 \text{ A} \\ &= 20.6\angle 39.1^\circ \text{ A} \end{aligned} \tag{2.36}$$

The total complex power is

$$\begin{aligned} \overline{S} &= \overline{V}\,\overline{I}^* \\ &= (1000\angle 0^\circ)(20.6\angle -39.1^\circ) \\ &= 1000(16 - j13) \\ &= 16000 - j13,000 \text{ VA} \\ &= 16 - j13 \text{ kVA} && (2.37) \\ &= 21\angle -39.1^\circ \text{ kVA} && (2.38) \end{aligned}$$

Input PF $= \cos 39.1^\circ = 0.776$ leading.

(b) Power triangle method

Load 1: From (2.30)

$$\begin{aligned} \overline{S_1} &= 20 \times 10^3 \angle - \cos^{-1}(0.6) \text{ VA} \\ &= 20 \times 10^3 (0.6 - j0.8) \text{ VA} \\ &= 12000 - j16000 \text{ VA} \end{aligned} \tag{2.39}$$

Load 2: Since PF is lagging, θ is $+$.

$$\begin{aligned} Q_2 &= V I_2 \sin\theta \\ &= (1000)(5)\sin(36.87^\circ) \\ &= 5000 \times (0.6) \\ &= 3000 \text{ VAR} \end{aligned} \tag{2.40}$$

$$\overline{S_2} = 4000 + j3000 \text{ VA} \tag{2.41}$$

The individual and total power triangles are shown in Figure 2.19.

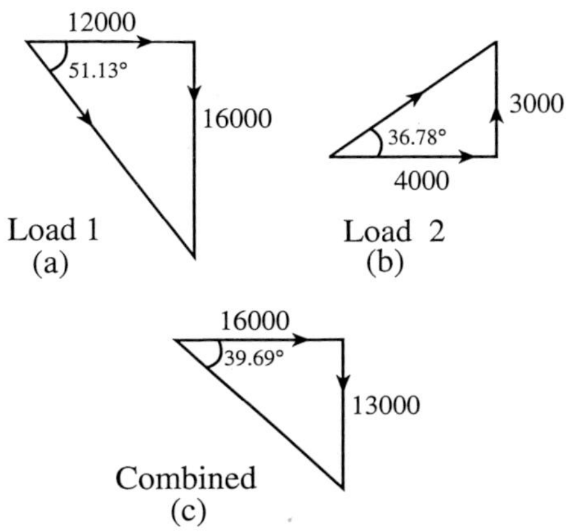

Figure 2.19: Power triangles for Example 2.11

2.6 Three-Phase Power

For reasons of economy as well as to obtain a smooth rotating field, power is generated as a three-phase system with the voltage in each phase differing from the other phase by an angle of 120°. The three voltages $v_{aa'}$, $v_{bb'}$, and $v_{cc'}$ are given by

$$\begin{aligned} v_{aa'} &= V_m \cos \omega t \\ v_{bb'} &= V_m \cos(\omega t - 120^\circ) \\ v_{cc'} &= V_m \cos(\omega t + 120^\circ) \end{aligned} \tag{2.42}$$

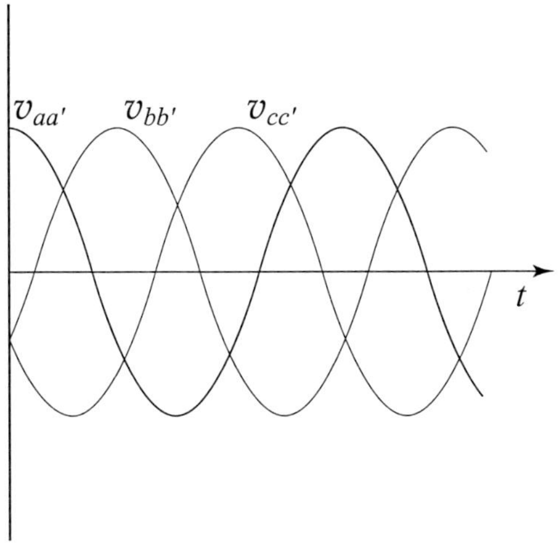

Figure 2.20: Three-phase voltage

This is called the phase sequence $a-b-c$, (*positive phase sequence* in power systems terminology), since along the positive time axis, voltage in phase a becomes maximum first followed by phases b and c.

2.6.1 Three-phase connections

The three single-phase voltages can be connected to form a practical three-phase system in two ways: 1) wye connection, or 2) delta connection. In Figure 2.21(a), the three windings are represented by coils aa', bb', and cc'. In the wye connection, terminals a', b', and c' are joined and labeled as terminal n, also called the *neutral terminal*. If we represent the coils by voltage sources, the resulting diagram is as shown in Figure 2.21(b). It is conventional to show the wye connection as in Figure 2.21(c) because of the geometrical similarity of the diagram to the fact that the phase difference between any two of the voltages is 120°. Currents i_a, i_b, and i_c are termed the line currents. *Line current* refers to current coming down the line, and

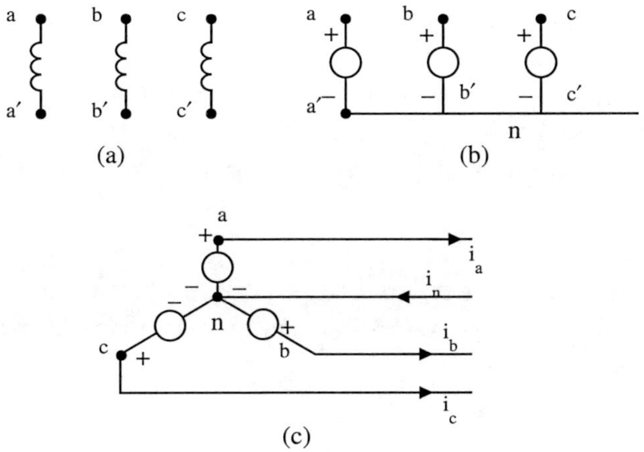

Figure 2.21: Three-phase wye connection

phase current refers to the current within an element of a phase. In the case of three-phase wye connections, the line currents are also equal to the *phase* currents. The neutral current i_n is the current in the neutral wire.

In the delta connection as shown in Figure 2.22(b), terminal a' is con-

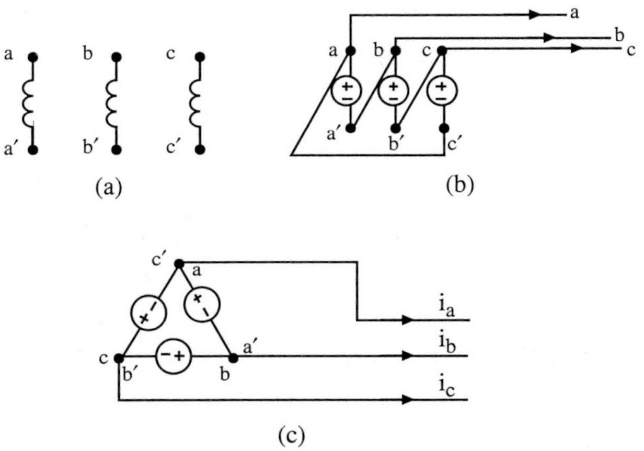

Figure 2.22: Three-phase delta connection

nected to b, and b' to c. Finally, c' can be connected to terminal a. The voltage across $v_{ac'} = v_{aa'}(t) + v_{bb'}(t) + v_{cc'}(t)$, and is identically equal to zero, as can be verified by summing the three voltages in (2.42). Again, it is conventional to represent the delta connection as in Figure 2.22(c).

We have represented the voltages by means of idealized sources and have not included effects such as resistance or reactance of the windings. We are justified in this kind of an approximation because in actual practice the voltages at the terminals of a generator are kept constant and the Thevenin impedance at the terminals of the generators is kept small compared to the load impedance.

2.6.2 Relation between line and phase voltages, and line and phase currents, under balanced load conditions

In a three-phase system, there are altogether four possible ways of supply and load connections. With the three-phase wye-connected supply, the loads may be connected either in wye or delta. Similarly, with the three-phase delta-connected supply, the load may be connected in wye or delta. The three-phase wye-connected source is the most common way of supplying power. The four possibilities are shown in Figure 2.23. Of these only Figure 2.23 (a) and (b) are commonly used configurations.

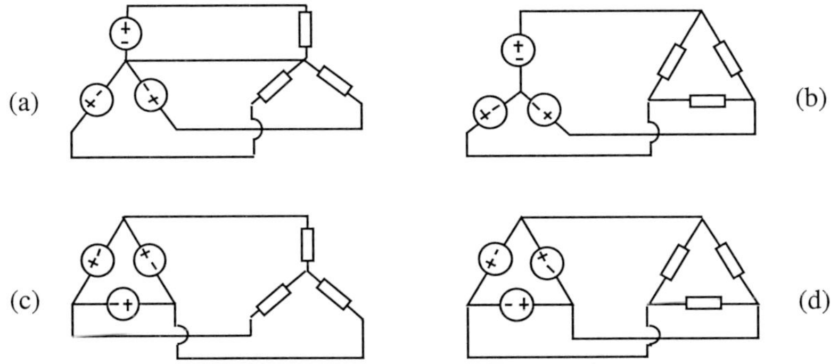

Figure 2.23: Different schemes of supply and load connections

Wye-connected source and wye-connected load

A balanced load operation is defined as one where all the load impedances are equal. Some interesting and useful relations exist among the voltages and currents under such an operation. Consider a wye-connected three-phase power supply connected to a wye-connected load as shown in Figure 2.24. Assuming a phase sequence $a - b - c$, we have $v_{an} = V_m \cos \omega t$, so that

$v_{bn} = V_m \cos(\omega t - 120°)$, and $v_{cn} = V_m \cos(\omega t - 240°)$. The three-phase line-to-neutral voltages in phasor form are

$$\begin{aligned}\overline{V}_{an} &= V_\phi \angle 0° \\ \overline{V}_{bn} &= V_\phi \angle -120° \\ \overline{V}_{cn} &= V_\phi \angle -240° \\ &= V_\phi \angle +120°\end{aligned}$$

where V_ϕ denotes the rms value of the phase-to-neutral voltage $= V_m/\sqrt{2}$.

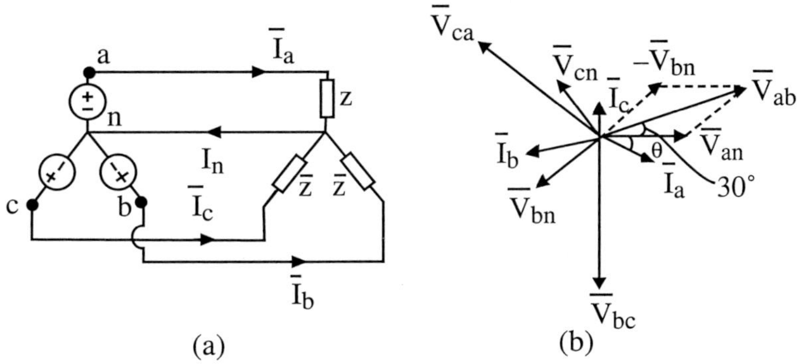

Figure 2.24: Phasor diagram for a balanced three-phase wye connection

The line-line voltages $\overline{V}_{ab}$, $\overline{V}_{bc}$, and $\overline{V}_{ca}$ are given by a simple application of KVL as

$$\begin{aligned}\overline{V}_{ab} &= \overline{V}_{an} - \overline{V}_{bn} \\ \overline{V}_{bc} &= \overline{V}_{bn} - \overline{V}_{cn} \\ \overline{V}_{ca} &= \overline{V}_{cn} - \overline{V}_{an}\end{aligned} \tag{2.43}$$

$\overline{V}_{ab}$ is shown in Figure 2.24(b) as the sum of $\overline{V}_{an}$ and negative of $\overline{V}_{bn}$. From simple geometrical considerations, the angle between $\overline{V}_{an}$ and $-\overline{V}_{bn}$ is 60°. Since both $\overline{V}_{an}$ and $\overline{V}_{bn}$ have the same magnitude equal to V_ϕ, it is easily verified that

$$V_{ab} = 2V_\phi \cos 30° = \sqrt{3}V_\phi = V_L$$

Hence, as a phasor

$$\overline{V}_{ab} = \sqrt{3}V_\phi \angle 30°$$

The reader may verify this trigonometrically as

$$\overline{V}_{ab} = V_\phi \angle 0° - V_\phi \angle -120° = V_\phi \angle 0° + V_\phi \angle 60° = \sqrt{3}V_\phi \angle 30°$$

Similarly,

$$\overline{V}_{bc} = \sqrt{3}V_{\phi}\angle - 90^{\circ}$$

and

$$\overline{V}_{ca} = \sqrt{3}V_{\phi}\angle 150^{\circ}$$

The *line-to-line voltages* (or simply the line voltages) $\overline{V}_{ab}$, $\overline{V}_{bc}$, and $\overline{V}_{ca}$ are again a balanced set of voltages with equal magnitudes given by $\sqrt{3}$ *times the phase voltage*. The line-to-line voltages differ from one another in phase by 120°, as shown in Figure 2.24.

If the load impedances are denoted by $Z = |Z|\angle\theta$, then the currents $\overline{I}_a$, $\overline{I}_b$, $\overline{I}_c$ are equal in magnitude but differ in phase by 120°. Thus $i_a = I_m \cos(\omega t - \theta)$, $i_b = I_m \cos(\omega t - 120^{\circ} - \theta)$, and $i_c = I_m \cos(\omega t + 120^{\circ} - \theta)$. The rms value of the line currents is $\frac{I_m}{\sqrt{2}} = I_L$. The corresponding phasors are $\overline{I}_a = I_L\angle - \theta$, $\overline{I}_b = I_L\angle - \theta - 120^{\circ}$, and $\overline{I}_c = I_L\angle - \theta + 120^{\circ}$. These are shown in the phasor diagram for a lagging PF case as in Figure 2.24(b). The line currents are the same as the phase currents in this case. The sum of the three currents is $\overline{I}_n = \overline{I}_a + \overline{I}_b + \overline{I}_c = 0$. Hence, no current flows through the neutral wire.

Wye-connected source and delta-connected load

Now consider the same wye-connected generator supplying a delta-connected load as in Figure 2.25(a). Since the three line-to-line voltages form a bal-

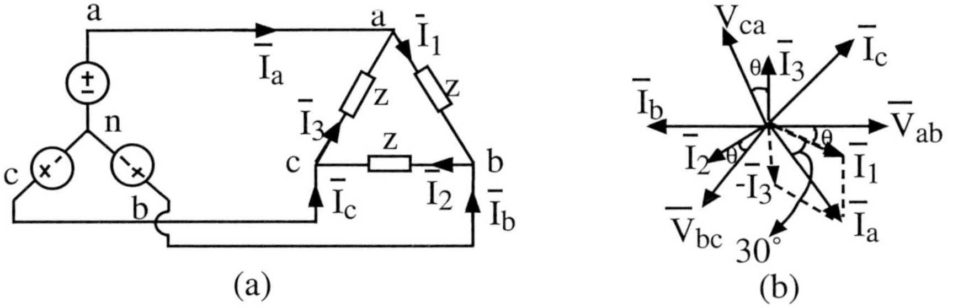

Figure 2.25: Phasor diagram for a delta-connected load

anced set, we may without loss of generality assume their phasor representation as (by taking $\overline{V}_{ab}$ as reference phasor)

$$\begin{aligned} \overline{V}_{ab} &= V_L\angle 0^\circ \\ \overline{V}_{bc} &= V_L\angle -120^\circ \\ \overline{V}_{ca} &= V_L\angle -240^\circ \end{aligned} \tag{2.44}$$

Assuming $Z = |Z|\angle\theta$, the currents $\overline{I}_1$, $\overline{I}_2$, and $\overline{I}_3$ in the three phases or legs of the delta connection lag the respective voltages by θ, as shown in the phasor diagram as in Figure 2.25(b). These currents, called *phase currents*, are all equal in magnitude denoted by I_ϕ. Since the voltages form a balanced set, the currents $\overline{I}_1$, $\overline{I}_2$, and $\overline{I}_3$ differ in phase by 120°. The line currents $\overline{I}_a$, $\overline{I}_b$, and $\overline{I}_c$ are obtained by applying Kirchhoff's current laws at the nodes of the delta-connected load as

$$\begin{aligned} \overline{I}_a &= \overline{I}_1 - \overline{I}_3 \\ \overline{I}_b &= \overline{I}_2 - \overline{I}_1 \\ \overline{I}_c &= \overline{I}_3 - \overline{I}_2 \end{aligned} \tag{2.45}$$

Note that $\overline{I}_a + \overline{I}_b + \overline{I}_c = 0$. From Figure 2.25(b), the magnitude of the phasor $\overline{I}_a$ is $2I_\phi \cos 30^\circ$ and the phase angle is $(-30 - \theta)$ degrees. Also using (2.45)

$$\begin{aligned} \overline{I}_a &= I_\phi\angle -\theta - I_\phi\angle -240^\circ - \theta \\ &= I_\phi\angle -\theta + I_\phi\angle -60^\circ - \theta \\ &= \sqrt{3}I_\phi\angle -30^\circ - \theta \\ \overline{I}_b &= \sqrt{3}I_\phi\angle -150^\circ - \theta \\ \overline{I}_c &= \sqrt{3}I_\phi\angle -270^\circ - \theta \end{aligned} \tag{2.46}$$

Geometrically, we can verify from Figure 2.25(b) that the magnitude of $\overline{I}_a$ is given by $I_L = 2I_\phi \cos 30^\circ = \sqrt{3}I_\phi$.

Thus, in a delta connection, the magnitude of all the three *line currents* is equal to $\sqrt{3}$ *times the phase currents*, and the line-to-line voltages have the same magnitude as the phase or the leg voltages and form a balanced set. To summarize, for a wye connection $V_L = \sqrt{3}V_\phi$ and $I_L = I_\phi$ and for a delta connection $V_L = V_\phi$ and $I_L = \sqrt{3}I_\phi$.

2.6.3 Power in a balanced three-phase circuit

Balanced Y-connected load

In a balanced three-phase load, whether wye or delta are connected, the power factor of the three-phase load is identical to the power factor of each phase or leg of the load.

Consider a wye-connected load with all three loads equal as shown in Figure 2.26 (balanced load). The phase angle difference between the voltage

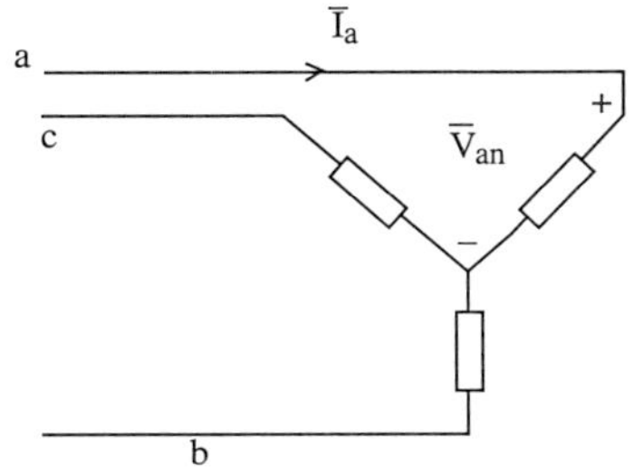

Figure 2.26: Balanced wye-connected load

across and current through each phase is the same and equal to a number $\theta(\theta > 0$ for lag, $\theta < 0$ for lead). If $\overline{V}_{an} = V_\phi \angle 0°$, then $\overline{I}_a = I_\phi \angle - \theta(\theta+$ for lag, $-$ for lead).

Since it is a balanced system, the magnitudes of voltages in all the three phases are the same and so also are the current magnitudes. Let these be V_ϕ and I_ϕ. Then the power per phase is

$$P_\phi = V_\phi I_\phi \cos\theta \tag{2.47}$$

The total power $P_T = 3P_\phi = 3V_\phi I_\phi \cos\theta$. It is also desirable to express P_T in terms of line-to-line voltage and line current. The line-to-line voltage magnitude is $V_L = \sqrt{3}V_\phi$ and the line current magnitude I_L is the same as I_ϕ. Thus,

$$\begin{aligned} P_T &= 3\left(\frac{V_L}{\sqrt{3}}\right) I_L \cos\theta \\ &= \sqrt{3} V_L I_L \cos\theta \end{aligned} \tag{2.48}$$

Note that θ is the phase angle between the phase voltage and phase current.

The complex power per phase is

$$\begin{aligned}\overline{S}_\phi &= \overline{V}_\phi \overline{I}_\phi^* \\ &= (V_\phi \angle 0^\circ)(I_\phi \angle \theta) \\ &= V_\phi I_\phi \cos\theta + jV_\phi I_\phi \sin\theta \\ &= P_\phi + jQ_\phi\end{aligned}$$

Thus the reactive power per phase is

$$Q_\phi = V_\phi I_\phi \sin\theta \tag{2.49}$$

The total reactive power is

$$Q_T = 3V_\phi I_\phi \sin\theta \tag{2.50}$$

The complex power per phase is

$$\begin{aligned}\overline{S}_\phi &= \overline{V}_\phi \overline{I}_\phi^* \\ &= V_\phi I_\phi \angle\theta \end{aligned} \tag{2.51}$$

and the total complex power is

$$\overline{S}_T = 3\overline{S}_\phi = 3V_\phi I_\phi \angle\theta = \sqrt{3} V_L I_L \angle\theta \ (\theta + \text{ for lag, } - \text{ for lead}) \tag{2.52}$$

since $V_\phi = V_L/\sqrt{3}$ and $I_\phi = I_L$.

Example 2.12 A three-phase wye-connected load of 24 kW at 0.8 PF lag is supplied by a three-phase wye-connected source of 480 V. Find the current and phase voltages. Assuming $\overline{V}_{an}$ as reference, express $\overline{I}_{an}$, $\overline{I}_{bn}$, $\overline{I}_{cn}$, $\overline{V}_{ab}$, $\overline{V}_{bc}$, and $\overline{V}_{ca}$ in phasor form. Also find total complex power.

Solution: Using per phase quantities, we have

$$\begin{aligned}V_\phi &= \frac{480}{\sqrt{3}} = 277.13 \text{ V} \\ \text{Power per phase} &= \frac{24}{3} = 8 \text{ kW} \\ V_\phi I_\phi \cos\theta &= 8 \times 10^3 \\ (277.13) I_\phi (0.8) &= 8 \times 10^3 \\ I_\phi &= 36.08 \text{ A} \\ \text{Assume } \overline{V}_{an} &= 277.13\angle 0^\circ \text{ V} \\ \text{PF angle } \theta &= \cos^{-1}(0.8) = 36.87^\circ\end{aligned}$$

Since it is a lagging PF and since $\overline{V}_{an} = 277.13\angle 0^\circ$ V,

$$\begin{aligned}
\overline{I}_{an} &= 36.08\angle -36.87^\circ \text{ A} \\
\overline{I}_{bn} &= 36.08\angle -156.87^\circ \text{ A} \\
\overline{I}_{cn} &= 36.08\angle 83.13^\circ \text{ A} \\
\overline{V}_{ab} &= \sqrt{3}\overline{V}_{an}\angle 30^\circ = 480\angle 30^\circ \text{ V} \\
\overline{V}_{bc} &= 480\angle -90^\circ \text{ V} \\
\overline{V}_{ca} &= 480\angle 150^\circ \text{ V}
\end{aligned}$$

Total complex power

$$\begin{aligned}
\overline{S}_T &= \sqrt{3}V_L I_L \angle\theta \text{ VA} \\
&= \sqrt{3}(480)(36.08)\angle 36.87^\circ \text{ VA} \\
&= 24,000 + j18,000 \text{ VA}
\end{aligned}$$

Verify

$$\begin{aligned}
\overline{S}_T &= 3\overline{V}_\phi \overline{I}_\phi^* \\
&= 3(277.13)(36.08)\angle 36.87^\circ \\
&= 24,000 + j18,000 \text{ VA}
\end{aligned}$$

Balanced delta-connected load

In Figure 2.27, we take $\overline{V}_{ab}$ as reference voltage, i.e., $\overline{V}_{ab} = V_\phi\angle 0^\circ = \overline{V}_\phi$.

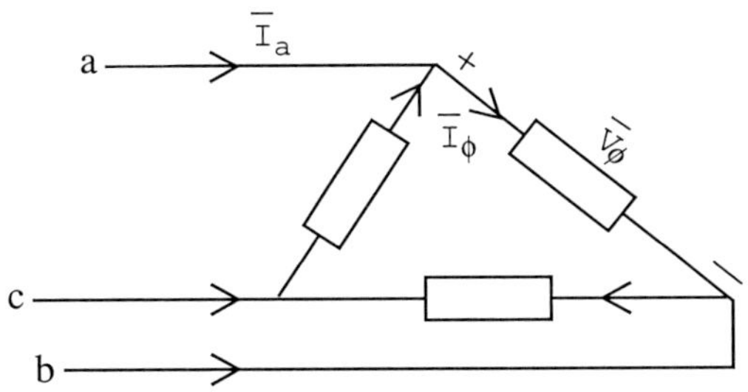

Figure 2.27: Balanced delta-connected load

The current $\overline{I}_{ab} = \overline{I}_\phi = I_\phi\angle -\theta$ where θ is the phase angle between $\overline{V}_\phi$ and $\overline{I}_\phi$, or impedance angle.

For a balanced load, the power per phase or leg of the delta connection is

$$\overline{S}_\phi = \overline{V}_\phi \overline{I}_\phi^* \tag{2.53}$$

Real power per phase is

$$P_\phi = V_\phi I_\phi \cos\theta \tag{2.54}$$

The current in each phase (leg) of the delta-connected load is $\frac{1}{\sqrt{3}}$ times the line current. The voltage per phase is the same as the line-line voltage. The total real power is

$$P_T = 3V_L\left(\frac{I_L}{\sqrt{3}}\right)\cos\theta = \sqrt{3}V_L I_L \cos\theta \tag{2.55}$$

The reactive power per phase is

$$Q_\phi = V_\phi I_\phi \sin\theta \tag{2.56}$$

The total reactive power is

$$Q_T = 3V_\phi I_\phi \sin\theta = \sqrt{3}V_L I_L \sin\theta \tag{2.57}$$

Finally the complex power per phase and the total complex power are given by

$$\overline{S}_\phi = P_\phi + jQ_\phi = \overline{V}_\phi \overline{I}_\phi^* \tag{2.58}$$

$$\overline{S}_T = 3\overline{V}_\phi \overline{I}_\phi^* = \sqrt{3}V_L I_L \angle\theta(\theta + \text{lag}, - \text{lead}) \tag{2.59}$$

From the above discussion, we see that for a balanced load, the expression for total complex power is the same both for wye and delta connections, provided line-line voltage and line currents are used. Hence, computation can be done on a three-phase or per-phase basis.

Example 2.13 A three-phase delta-connected load of 24 kW at 0.8 PF lag is supplied by a three-phase source with a line-line voltage of 480 V. Find the line and phase currents as well as the total complex power.

Solution: The real power per phase, with power factor angle θ, is given by

$$P_\phi = V_\phi I_\phi \cos\theta$$

Hence,

$$\begin{aligned}
\frac{24 \times 10^3}{3} &= (480) I_\phi (0.8) \\
I_\phi &= \frac{8 \times 10^3}{(480)(0.8)} \\
&= \frac{10^4}{480} \\
&= 20.8 \text{ A} \\
I_L &= \sqrt{3} I_\phi \\
&= 36.08 \text{ A} \\
\cos\theta &= 0.8 \\
\theta &= 36.87^\circ
\end{aligned}$$

Since it is a lagging PF, $\theta > 0$,

$$\begin{aligned}
\overline{S}_T &= (\sqrt{3}) V_L I_L \angle\theta \\
&= \sqrt{3}(480)(36.08)\angle 36.87^\circ \\
&= 30,000(0.8 + j0.6) \\
&= 24,000 + j18,000 \text{ VA}
\end{aligned}$$

The line current and total complex power is the same as in Example 2.12. Hence, in a balanced power case, the line current as well as the total complex power remain the same if the total real power load and the PF are unchanged.

2.7 Per-phase Equivalents

In balanced three-phase circuits, it is preferable to work with per-phase equivalents and then convert the variables to three-phase quantities.

$\Delta - Y$ conversion

We will show that, given a delta-connected circuit where impedance of each phase is Z_Δ, the equivalent wye circuit has a phase impedance of $Z_Y = Z_\Delta/3$, as shown in Figure 2.28. For terminal conditions to be the same in both the delta and wye connections, we equate the impedance across $a - b$

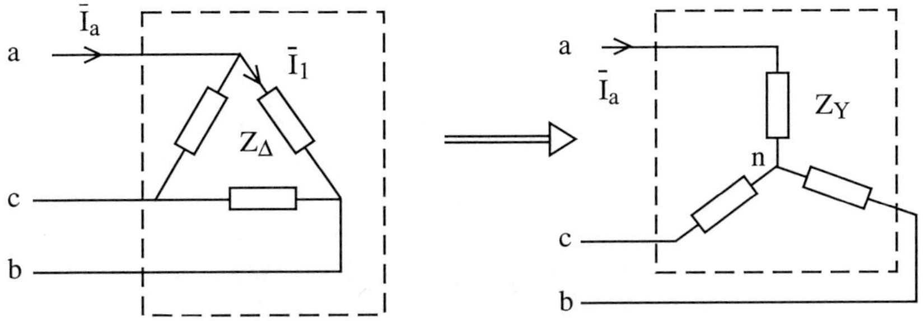

Figure 2.28: $\Delta - Y$ equivalence

in both cases. This yields

$$Z_\Delta \parallel 2Z_\Delta = 2Z_Y \tag{2.60}$$

i.e.,

$$\frac{(Z_\Delta)(2Z_\Delta)}{3Z_\Delta} = 2Z_Y \tag{2.61}$$

Hence $Z_Y = Z_\Delta/3$. Assume that a three-phase wye-connected source is connected across the loads. In terms of power, the complex power per phase in the delta circuit is

$$\overline{S}_\phi = \overline{V}_{ab}\overline{I}_1^* = \overline{V}_{ab}\left(\frac{\overline{V}_{ab}}{Z_\Delta}\right)^* = \frac{V_{ab}^2}{Z_\Delta^*} = \frac{3V_{an}^2}{Z_\Delta^*} = \frac{V_{an}^2}{Z_Y^*} \tag{2.62}$$

In the wye circuit, the complex power per phase is

$$\overline{S}_\phi = \overline{V}_{an}\overline{I}_a^* = \overline{V}_{an}\left(\frac{\overline{V}_{an}}{Z_Y}\right)^* = \frac{V_{an}^2}{Z_Y^*} \tag{2.63}$$

Thus, instead of analyzing the delta circuit, we can analyze the per-phase equivalent circuit after doing the delta-wye conversion as shown in Figure 2.29. The complex power across Z_Y is the same as power across Z_Δ in the delta circuit, i.e.,

$$\frac{V_{ab}^2}{Z_\Delta^*} = \frac{V_{an}^2}{Z_Y^*}$$

Example 2.14 Draw the per-phase equivalent circuit with respect to phase "a" for the circuit shown in Figure 2.30.

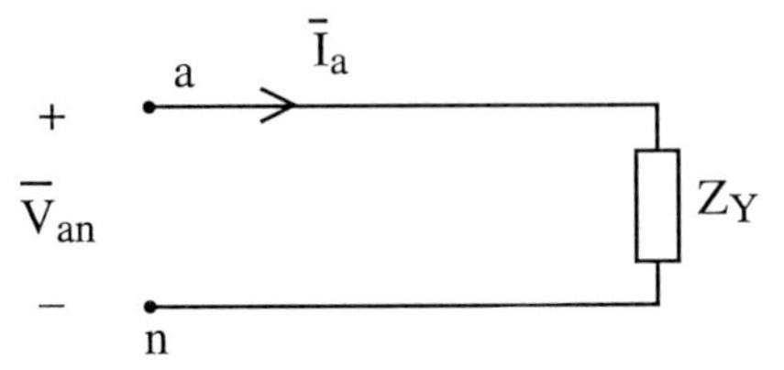

Figure 2.29: Per-phase equivalent circuit

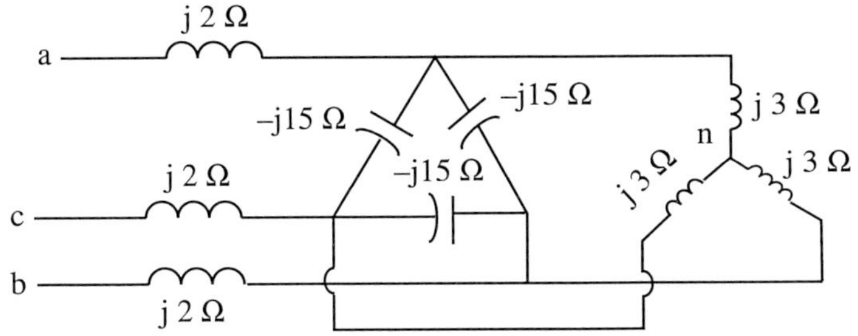

Figure 2.30: A three-phase circuit

Solution: The delta-connected capacitor "bank" is replaced by an equivalent wye-connected "bank" with phase impedance of $-j15/3 = -j5\ \Omega$. The per-phase equivalent is shown in Figure 2.31. The parallel combination of

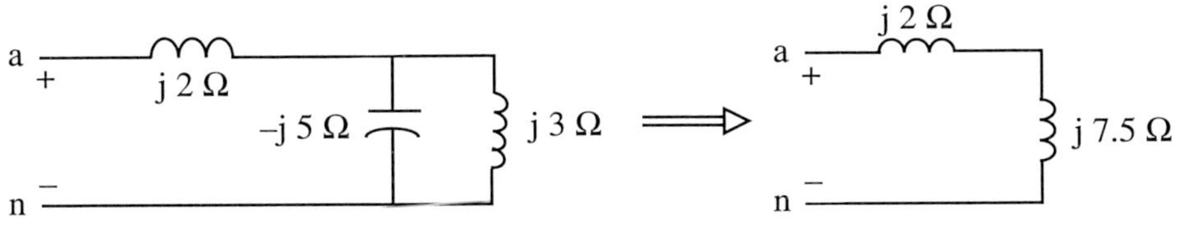

Figure 2.31: Per-phase equivalent circuit

$(-j5)$ and $j3$ gives $j7.5\ \Omega$.

Example 2.15 A three-phase industrial plant consists of 10 induction motors each drawing 30 kW at 0.6 PF lagging. Find the three-phase kVAR rating of a capacitor "bank" to improve the PF to unity.

Solution: Power triangle method The per-phase load is $30 \times 10/3 = 100$ kW at 0.6 PF lagging. The per-phase circuit as shown in Figure 2.32.

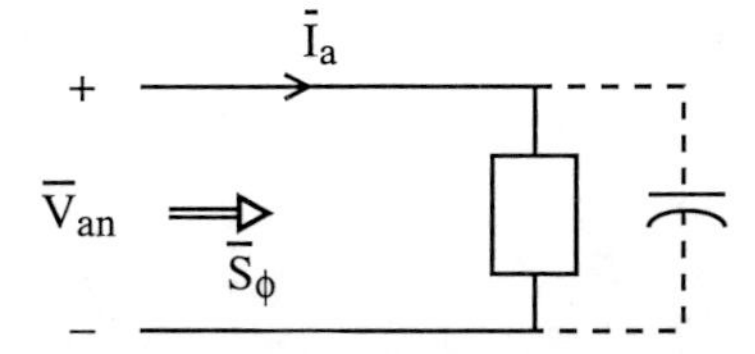

Figure 2.32: Per-phase equivalent circuit

The kVA per phase is equal to $\frac{100}{0.6}$. Therefore,

$$\begin{aligned}\overline{S}_\phi &= V_\phi I_\phi \angle \cos^{-1}(0.6) \\ &= \frac{100 \times 10^3}{0.6}(0.6 + j0.8) \text{ VA} \\ &= 100 + j133.33 \text{ kVA}\end{aligned}$$

Q_{cap} needed per phase to bring the PF to unity is $Q_{\text{cap}} = -133.33$ kVAR, and the three-phase $Q_{\text{cap}} = -3(133.33) = -400$ kVAR.

Example 2.16 Suppose in Example 2.15, the new PF is to be 0.9 lagging, what is the kVAR needed?

Solution: We use the power triangle method. $\overline{S}_\phi = 100 + j133.33$ kVA. New PF is 0.9 lag, i.e., $\cos\theta_{\text{new}} = 0.9$, and $\theta_{\text{new}} = 25.84°$. The power triangle is shown in Figure 2.33. New $Q_{\text{new}} = 100 \times \tan 25.84° = 48.4$ kVAR.

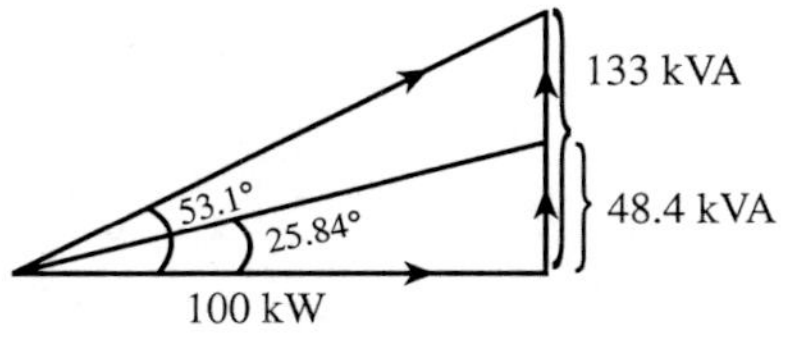

Figure 2.33: Power triangle

$Q_{\text{new}} = Q_{\text{old}} + Q_{\text{cap}}$. The capacitor should therefore supply $(48.4 - 133) = -84.93$ kVAR. The three-phase kVAR needed is $(-84.93) \times 3 = -254.79$ kVAR.

Example 2.17 The following two three-phase loads are connected in parallel across a three-phase 480 V wye-connected supply as shown in Figure 2.34.

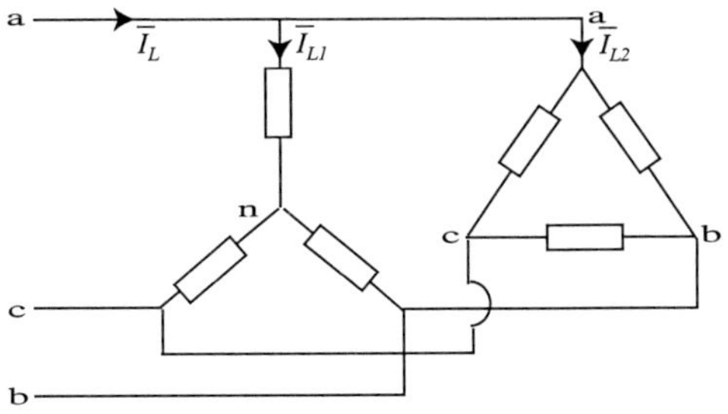

Figure 2.34: Two loads in parallel

Load 1: 24 kW at 0.8 PF lag (wye-connected)
Load 2: 30 kVA at 0.8 PF lead (delta-connected)
Find the line currents $\overline{I}_{L1}$ and $\overline{I}_{L2}$ for each of the two loads, total complex power $\overline{S}_T$ and total line current. Take $\overline{V}_{an}$ as reference.

Solution: Power triangle method Load 1

$$\begin{aligned}\overline{S}_{T1} &= \frac{24\times 10^3}{0.8}(0.8+j0.6)\text{ VA}\\ &= 24,000+j18,000\text{ VA}\end{aligned}$$

Load 2

$$\begin{aligned}\overline{S}_{T2} &= 30\times 10^3(0.8-j0.6)\\ &= 24,000-j18,000\text{ VA}\end{aligned}$$

Total load

$$\begin{aligned}\overline{S}_{T} &= \overline{S}_{T1}+\overline{S}_{T2}\\ &= 48,000\text{ W}\end{aligned}$$

Finding line current
For load 1

$$\begin{aligned}(\sqrt{3})V_L I_{L1}\cos\theta &= 24,000\\ I_{L1} &= \frac{24,000}{(480)(\sqrt{3})(0.8)}\\ &= 36.08\text{ A}\end{aligned}$$

Assuming $\overline{V}_{an}$ as reference,

$$\overline{I}_{L1} = 36.08\angle - 36.87^\circ \text{ since the current is lagging}$$

For load 2

$$\begin{aligned}(\sqrt{3})V_L I_{L2} &= 30,000 \\ I_{L2} &= \frac{30,000}{(\sqrt{3})480} \\ &= 36.08 \text{ A}\end{aligned}$$

Hence,

$$\overline{I}_{L2} = 36.08\angle 36.87^\circ \text{ since the current is leading.}$$

From the expression for total complex power, $Q_T = Q_{T1} + Q_{T2} = 18000 - 18000 = 0$. Hence the total complex power consists of real power only. Thus,

$$(\sqrt{3})V_L I_L = 48,000$$

Therefore,

$$\begin{aligned}I_L &= \frac{48,000}{(480)(\sqrt{3})} \\ &= 57.7 \text{ A}\end{aligned}$$

Per-phase equivalent method

The delta-connected load is equivalent to a wye-connected load with 10 kVA for each phase at 0.8 PF lead. Hence on a per-phase basis, the equivalent circuit is shown in Figure 2.35. Phase-to-neutral voltage is $\frac{480}{\sqrt{3}} = 277.13$ V.

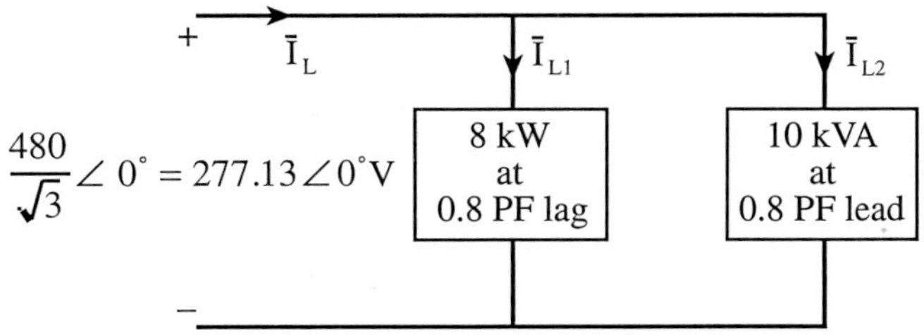

Figure 2.35: Two loads in parallel

For load 1

$$\begin{aligned} 8 \times 10^3 &= (277.13) I_{L1} (0.8) \\ I_{L1} &= 36.08 \text{ A} \\ \overline{I}_{L1} &= 36.08\angle - 36.87^\circ \text{ since the PF is lag} \end{aligned}$$

For load 2

$$\begin{aligned} (10 \times 10^3)(0.8) &= (277.13)(I_{L2})(0.8) \\ I_{L2} &= 36.08 \text{ A} \\ \overline{I}_{L2} &= 36.08\angle 36.87^\circ \text{ A since the PF is leading} \\ \overline{I}_L &= \overline{I}_{L1} + \overline{I}_{L2} \\ &= 36.08\angle - 36.87^\circ + 36.08\angle 36.87^\circ \\ &= 57.7\angle 0^\circ \text{ A} \end{aligned}$$

2.8 Conclusion

In this chapter, we have discussed the basic concepts of phasors and complex power, as well as real and reactive power. The notion of a phasor diagram is introduced. Balanced three-phase circuits are analyzed. Both delta and wye connections are discussed. Use of the power triangle and improvement of the power factor are discussed.

2.9 Problems

2.1 A voltage $v(t) = 100\cos(377t + 10^\circ)$ is applied to a two-terminal network and the current into the network is measured as $i(t) = \cos(377t + 55^\circ)$. Compute the average power into the network and power factor.

2.2

a) Find the average real power *into* the terminals as shown in Figure 2.36 for each of the following cases.

b) For cases (c), (d), (e), and (f), find the PF and state whether it is leading or lagging.

(a)

$$\begin{aligned} v(t) &= 100\cos(377t + 15^\circ) \\ i(t) &= -10\sin(377t + 45^\circ) \end{aligned}$$

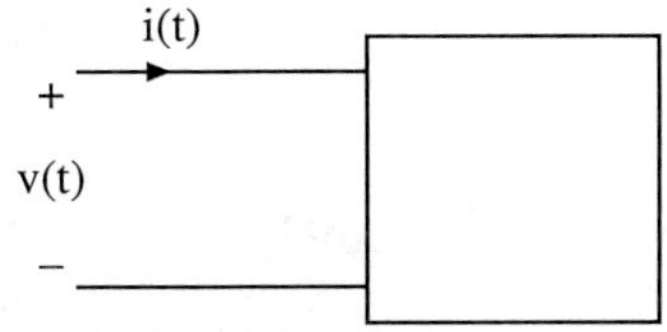

Figure 2.36: Two-terminal network

(b)

$$\begin{aligned} v(t) &= 100\cos(377t - 75°) \\ i(t) &= -10\sin(377t + 15°) \end{aligned}$$

(c)

$$\begin{aligned} v(t) &= \sqrt{2}(100)\sin(377t - 30°) \\ i(t) &= \sqrt{2}(10)\cos(377t - 60°) \end{aligned}$$

(d)

$$\begin{aligned} v(t) &= 100\sin(377t + 30°) \\ i(t) &= 10\sin(377t + 120°) \end{aligned}$$

(e)

$$\overline{V} = 10 + j10, \overline{I} = 5 + j10$$

(f)

$$\overline{V} = 100 - j90, \overline{I} = 10 - j10$$

(g)

$$\overline{V} = -10 - j10, \overline{I} = 5 + j10$$

2.3 In Problem 2.2, sketch the phasors in the complex plane and indicate the PF angle.

2.4 A single-phase source supplies a load of $R = 25\Omega$ and a reactance of $X_c = -15\Omega$, both of which are connected in parallel. The voltage across this combination is $v(t) = \sqrt{2}(120)\cos(\omega t - 45°)$ V.

a) Find $\overline{V}$, the total current $\overline{I}$, instantaneous current $i(t)$, and the complex power.

b) Repeat (a) if instead of the capacitive reactance we had inductive reactance of $X_L = 20\Omega$.

2.5 A parallel circuit consisting of two impedances $\overline{Z}_1 = 10\angle 45°\Omega$ and $\overline{Z}_2 = 25\angle -30°\Omega$ is connected across a source whose voltage is $\overline{V} = 100\angle 0°V$.

a) Find the current in each branch and the total current.

b) Find the complex power in each branch and the total complex power.

c) Find the overall power factor of the combined circuit.

d) Draw the power triangles for each branch and the combined power triangle.

2.6 A 5 hp, 220 V, single-phase motor operates at 90% efficiency while running at rated hp and has a PF of 0.707 PF lagging. (Hint: one horsepower (hp) is equivalent to 745.7 W)

a) Find the current drawn by the motor.

b) Find the real and reactive power absorbed by the motor plus its mechanical load.

2.7 A load of 10 kW at a PF of 0.7 lag is supplied at 230 V from a feeder whose impedance is $0.1 + j0.4\Omega$. Find the sending end voltage and the PF. (Partial answer: $PF = 0.66$ lag)

2.8 Two loads are connected in parallel and the currents are $\overline{I}_1 = 50 - j40$ and $\overline{I}_2 = 50 + j60$. The voltage across the combination is $\sqrt{2}(100)\angle 45°$. Find the total current, the complex power, and the overall PF. (Partial answer: $\overline{S} = 12 + j8$ kVA)

2.9 A 230 V source supplies two loads in parallel. Load 1 draws 5 kVA at a lagging PF of 0.8 and load 2 draws 3 kW at a lagging PF of 0.9. Take source voltage as the reference phasor $230\angle 0°$ V. Find the source current and the total complex power supplied by the source. (Partial answer: $\overline{S_T} = 7 + j4.45$ kVA)

2.10 Three loads are connected in parallel as follows: Load 1 absorbs an average power of 8 kW at a lagging power factor of 0.8. Load 2 absorbs 20 kVA at a leading power factor of 0.6. Load 3 is an impedance of $2.5 + j5.0\Omega$. The voltage across the parallel combination is $250\angle 0°$ V. The load is supplied through a feeder having an impedance of $0.1 + j1.0\Omega$. Find the source voltage $v_s(t)$ and $i_s(t)$ assuming the frequency is 60 Hz.

2.11 Three loads are connected in parallel as follows: Load 1 absorbs 250 kVA at 0.5 PF lag, load 2 absorbs 180 kW at 0.8 lead, load 3 absorbs $283 + j100$ kVA. Find the total complex power and the overall power factor of the combination. What is the kVAR of the capacitor needed to make the overall PF 0.8 lead? The capacitor is connected across the parallel combination. (Partial answer: 0.955 lag)

2.12 Two loads are connected in parallel across a source whose voltage is $1000\angle 60^\circ$ V. Load 1 is 100 kVA at 0.866 PF lagging and load 2 is 40 kW at 0.5 PF lagging. Find the total complex power, the total current $\overline{I}$ supplied by the source, and the currents in each of the loads. (Partial answer: $\overline{I} = 166.6 + j50$ A)

2.13 A transformer rated at 500 kVA is operating at full load (i.e., carrying rated kVA) at a PF of 0.6 lagging. Capacitors are added to improve the overall PF to become 0.9 lagging. Find kVAR of capacitors required. (Answer: -254 kVAR where negative sign confirms it is capacitive)

2.14 A single phase transformer rated at 100 kVA is operating at full load (i.e., carrying rated kVA) at a PF of 0.6 lagging. A capacitor is added to improve the overall PF to become 0.9 leading. Find the kVAR of the capacitor required.

2.15

a) In the circuit of Figure 2.37, write the nodal equations at

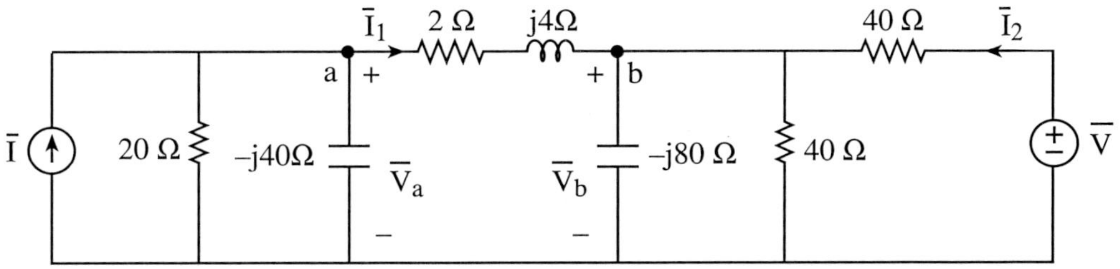

Figure 2.37: Circuit for Example 2.15

nodes a and b. Given $\overline{V} = 140\angle 0^\circ$ V and $\overline{I} = 7\angle -90^\circ$ A.

b) Arrange the equations in the form $Ax = y$ and use MATLAB or MATHEMATICA to solve for x, i.e., $\overline{V}_a$ and $\overline{V}_b$.

c) Find $\overline{I}_1$ and $\overline{I}_2$.

2.16 A 345 kV three-phase line supplies 750 MVA at 0.8 PF lagging to a three-phase load which is delta connected.

a) Find the complex impedance per phase.

b) Find the magnitudes of the line and phase currents.

c) Compute real and reactive power per phase.

d) Compute the total complex power.

2.17 The following three-phase loads are connected in *parallel* across a 2400 V (line-line) three-phase power supply.
Load 1: 120 kVA at 0.8 PF lead.
Load 2: 180 kW at 0.6 PF lag.
Load 3: 30 kW at unity power factor.

(a) Find the *total* complex power of the three loads. (Answer: $306 + j168$ kVA)

(b) Find the overall power factor. (Answer: 0.876 lag)

(c) Find line current in the supply line. (Answer: 84.07 A)

(d) Find the total kVAR of a capacitor needed to bring the PF to unity and the new line current. (Answer: −168 kVAR)

2.18 A three-phase wye-connected motor draws 40 kVA at 0.65 PF lagging from a 230 V (line-line) source. A wye-connected capacitor bank is connected across the motor terminals to improve the PF to 0.95 lagging.

a) Find the kVAR rating of the capacitor bank.

b) Find the line current before and after the capacitors are added.

(Answer: $Q_c = -21.86$ kVAR; $I_{\text{before}} = 100.4$ A; and $I_{\text{after}} =$ 68.7 A)

2.19 A three-phase wye-connected load draws 120 kW at a PF of 0.85 lagging from a 440 V (line-line) three-phase system. Three capacitors are connected in delta across the load. The kVAR of each capacitor is 50/3.

a) Find the line current before and after the capacitors are added.

b) The new PF.

(Answer: $I_{\text{new}} = 160.6$ A; PF = 0.98 lagging)

2.20 A three-phase power system consists of a wye-connected generator connected to a wye-connected load through a transmission line having an impedance of $Z_{\text{line}} = 1.5\angle 75°\Omega$ per phase. The load impedance is $Z_L = 20\angle 30°\Omega$ per phase. The voltage at the load is 4.16 kV (line-to-line).

a) Draw the per phase equivalent circuit for phase a. Taking phase a load voltage as reference, compute.

b) $\overline{I}_a$, $\overline{I}_b$, $\overline{I}_c$ and the voltages $\overline{V}_{an}$, $\overline{V}_{bn}$, $\overline{V}_{cn}$, $\overline{V}_{ab}$, $\overline{V}_{bc}$, $\overline{V}_{ca}$, at the sending end.

c) Total complex power of load.

d) Total complex power supplied by generator. Assume a-b-c sequence.

(Partial answer: $\overline{I}_c = 120\angle 90^\circ$ A, $\overline{V}_{ab} = 4382\angle 33^\circ$ A, and $\overline{S}_{\text{gen}} = 764 + j496$ kVA)

2.21 A three-phase load of 15 kVA with a PF of 0.8 lagging is connected in parallel with a three-phase load of 36 kW at 0.6 PF leading. The line-line voltage is 2000 V.

a) Find the total complex power and power factor.

b) How much kVAR is needed to make the PF unity?

2.22 A three-phase load of 15 kVA at 0.8 PF leading is connected in parallel with a three-phase 72 kW load at 0.8 PF lagging. The line-to-line voltage is 2000 V.

a) Find the line current and the PF of the combined total load.

b) Find the total kVAR needed to bring the PF to unity.

2.23 A 1000 kVA, three-phase, 60 Hz generator, 4160 volt (line-line) delivers full rated kVA at 0.8 PF lagging to the following two loads in parallel.

a) 200 kW of three-phase balanced lighting load (pure resistance) connected in wye.

b) A balanced three-phase impedance load connected in delta.

Find the total complex power in the three-phase impedance load and the impedance of each of the legs of the delta-connected load.

2.24 A balanced three-phase positive sequence wye-connected 60 Hz source voltage has line-neutral voltage of $V_\phi = 1000$ V. It is connected to a three-phase wye-connected load whose impedance is $\overline{Z} = 50 + j37.7\Omega$. Find line currents $\overline{I}_a$, $\overline{I}_b$, $\overline{I}_c$ and line-line voltages $\overline{V}_{ab}$, $\overline{V}_{bc}$, and $\overline{V}_{ca}$. Find the total complex power delivered to the load. (Partial answer: $\overline{S}_T = 38.25 + j28.86$ kVA)

2.25 An industrial plant consists of several three-phase induction motors and draws 300 kW at 0.6 PF lagging from the substations bus. Find the kVAR rating of a capacitor bank to improve PF to 0.9 lagging. ($Q_{\text{cap}} = -254.7$ kVAR)

Chapter 3

POWER ELECTRONICS

3.1 Control and Conversion of Electric Energy

A basic understanding of power electronics is important to understand modern energy systems and to enable you to further advance your studies in the area. First, what is power electronics? *Power electronics* is a subfield of the energy and power area focused on circuits designed for the control and conversion of electric energy. For example, power electronics are used to convert one type of voltage or current to another type of voltage or current. While most of this textbook covers AC systems, it is important to consider the role of DC systems, and the interfaces between DC and AC systems. Hence, power electronics plays a key role that cannot be neglected, and in fact, new grid edge and microgrid technologies are increasingly bringing in the need to understand the control of these power electronic systems together with the control of the AC power system. Hence, it it is a key building block of knowledge and cornerstone for next-generation energy systems.

Power electronics, as a core electrical engineering field and its technical knowledge base, is at the intersection of three other areas, as described in [10]: (1) electronics of devices, (2) systems and control, and (3) power and energy. We mean this context as a way of saying that power electronics is its own subject and textbook. The coverage provided in this book is however critical to link energy conversion topics and provide the end-to-end perspective of energy systems. It has been a missing piece that is increasingly critical to consider. Additional reading material to supplement this chapter exists in other resources, where we suggest the following works, by experts Krein, Hart, Sen, and Masters [10, ?, 11, 12] that provide greater depth into several of the key power electronics concepts that we describe in this chapter.

Consider a generic power system shown in Fig. 3.6 below. The source can be AC or DC, and the load can be AC or DC. The objective or role of the power electronics converter(s) in between is to match those two up efficiently by their voltages and currents.

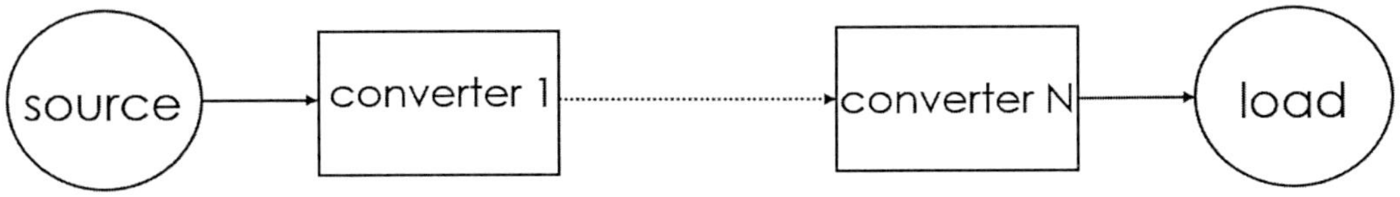

Figure 3.1: Energy conversion from source to converters to load

To achieve this objective, we can have (1) a rectifier that converts from AC to DC, (2) an inverter that converts from DC to AC, or (3) a power electronics converter that converts from DC to DC, or (4) a transformer that converts from AC to AC. In this chapter, we will cover (1)-(3); transformers are covered in Chapter 4.

The energy conversion step can be used for changing (1) voltage level, (2) frequency, (3) waveshape, and (4) number of phases.

3.2 Electronic Switches

Electronic switches play a key role in power electronics. The details of the seminconductor devices that form the switches is its own related technical area whose innovations contribute to improving electric energy conversion. Seminconductors are the basis of today's modern electronics: diodes, transistors, digital and analog integrated circuits. Power electronics usually utilize seminconductors for electronic switches that are characterized by on and off states.

The plots below (Fig. 3.2-3.4) show examples of three major types: a diode, thyristor, and transistor. Diodes are the simplest electronic switch. When a voltage is applied across the diode's terminals, the electric field formed in the diode excites electrons to form an electric current that is able to move in only one direction. Thyristors are three-terminal (anode, gate, cathode) devices that are often used when on/off control is needed. Transistors have both on-off and off-on controllability, and they also three terminals (base, collector, emitter). Each plot shows the device's basic configuration in a circuit, its i-v characteristics, and its idealized i-v characteristics.

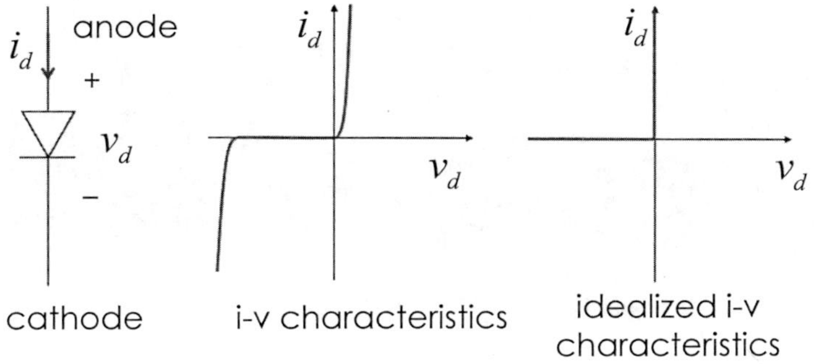

Figure 3.2: Diode as an electronic switch

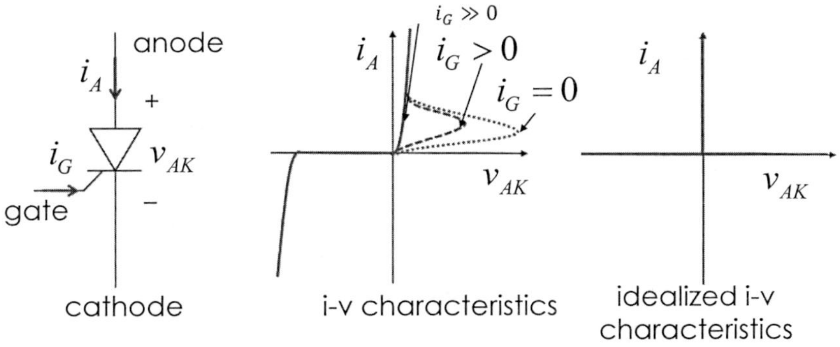

Figure 3.3: Thyristor as an electronic switch

3.3 DC-DC Converters

Why do we need DC-DC converters? To understand this, suppose we are interested in delivering power from a photovoltaic (PV) source to any given load, as in Fig. 3.6. In this case, the source is the PV panel which is DC. The "load" could mean the AC grid, a battery, a machine, etc. Now suppose the load is a resister R_L. With no converter, we would connect the load directly across PV terminals. Assuming R_L is fixed, the operating point of the resistor is always (V_L, I_L). Since R_L is fixed, it means that

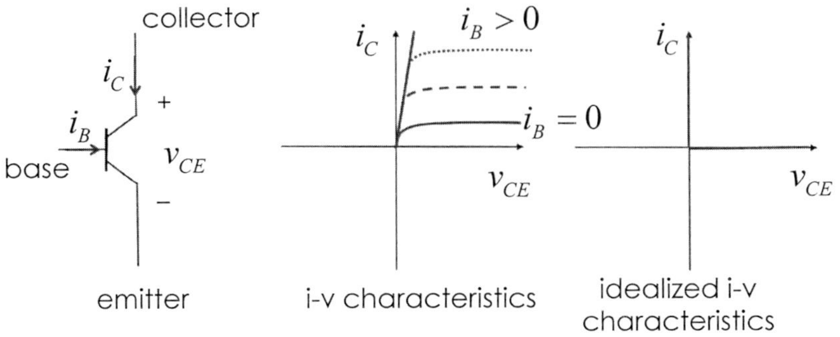

Figure 3.4: Bipolar junction transistor (BJT) as an electronic switch

the operating point (V_L, I_L) does not equal the point of maximum power transfer (V_M, I_M) : $(V_L, I_L) \neq (V_M, I_M)$.

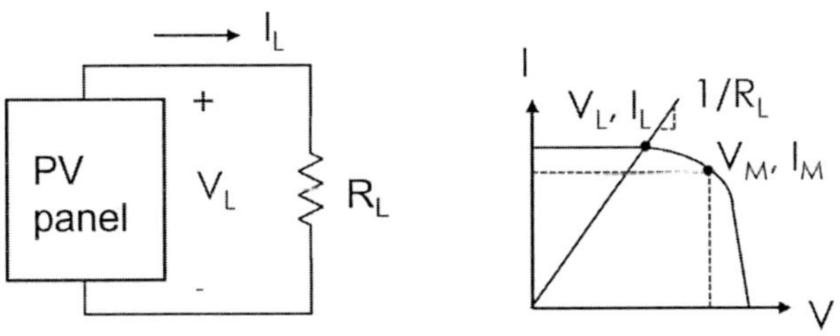

Figure 3.5: PV system directly connected to a load and its IV curve.

The $1/R_L$ line is where the circuit with only the resistor operates. You can see where it intersects the PV operating curve. The maximum power delievered corresponds to the largest square that fits inside this PV curve.

What does this mean? It means we could do better. This configuration (with no converter) does not deliver maximum available power to the load. We want to design some circuit that can do this. This is a key role of power electronics. That is, we want to change the way the load appears to the source to deliver maximum power.

An example solution that achieves this desired functionality is maximum power point tracking. Maximum power point trackers (MPPTs) are often a standard part of PV systems, especially grid-connected systems. The idea is to keep the operating point near the knee of the PV system's IV curve.

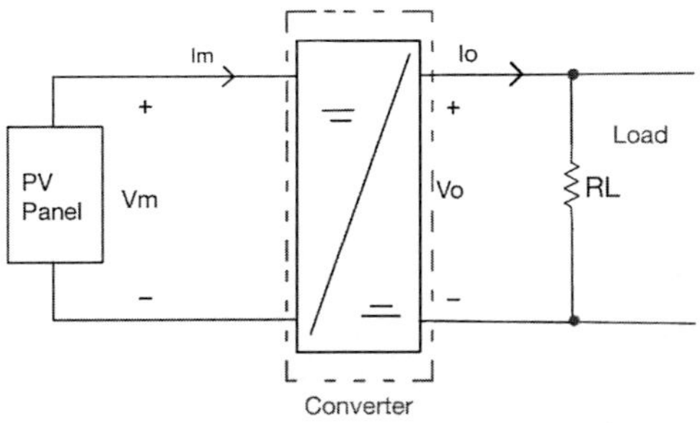

Figure 3.6: Maximum power point tracking in PV systems is a common example that uses power electronics converters to decouple a source from its load.

To do this, one might need to raise or to lower the DC voltage. A *buck boost converter* that we will describe in Section 3.3.3 is a DC to DC converter that can either "buck" (lower) or "boost" (raise) the voltage of the source to meet the needs of the load. Varying the duty cycle of a buck-boost conveter can be done such that the PV system will delvier the maximum power to the load. Solar PV systems and other applications are covered more in Chapter on distributed energy resources.

DC-DC Converter Topologies

In this chapter, we will cover three typical power electronics DC-DC converter technologies:

- Buck converter - lowers DC voltage.
- Boost converter - raises DC voltage.
- Buck-boost converter - does both.

With these topologies, the goal is to present a primer on how power electronics work in the context of energy conversion, from which you will be able to extend your knowledge more deeply. This is not meant to be an exhaustive list, and there are other topologies.

The essential components that comprise power electronics converters are their (1) energy storage (inductors and capacitors), (2) switches (Chapter 3.2), and (3) control (digital and/or analog).

To understand how energy storage devices play a key role in the design and function of power electronics converters, it is important to recall the basic principles of inductors and capacitors. These principles will be used to analyze how power electroncis converters work, especially in an average sense, to determine what they do. Total energy stored in an inductor is

$$E_L = \frac{1}{2}LI^2 \tag{3.1}$$

and the total energy stored in a capacitor is

$$E_C = \frac{1}{2}CV^2 \tag{3.2}$$

Inductors in power electronics

Current through an inductor cannot change instantaneously.

$$v = L\frac{di}{dt} \tag{3.3}$$

$$i(t) = \frac{1}{L}\int v(t)\,dt \tag{3.4}$$

If the current could change instantaneously, $\frac{di}{dt}$ would be infinite, and the voltage would rise uncontrollably (Eq. 3.3). Another way to look at this is that it takes some time for current to build up in an inductor (Eq. 3.4).

The *steady-state inudctor principle* says that the average voltage across an inductor is zero. If we look at Eq. 3.4 on average, it means that voltage across the inductor V_L is zero, or it is a short circuit. In periodic steady-state, it means that the values of current are always repeated one cycle later (Eq. 3.5).

$$i(t_0 + T) = i(t_0) + \frac{1}{L}\int_{t_0}^{t_0+T} v(t)\,dt \tag{3.5}$$

This voltage integral in Eq. 3.5 MUST be zero, or the current does not end up back where it started, and hence it is not operating in periodic steady state.

Capacitors in power electronics

The capacitors exhibit the dual relationships to those described for the inductors. The voltage across a capacitor cannot change instantanously. If it could, $\frac{dv}{dt}$ would be infinite, and the current would rise uncontrollably

(Eq. 3.6). It takes some time for its voltage to build up across the capacitor (Eq. 3.7).

$$i = C\frac{dv}{dt} \tag{3.6}$$

$$v(t) = \frac{1}{C}\int i(t)\,dt \tag{3.7}$$

The *steady-state capacitor principle* says that the average current through a capacitor is zero. If we look at Eq. 3.7 on average, it means that current through the capacitor i_C is zero, or the capacitor is an open circuit. In periodic steady-state, it means that the value of capacitor voltage must be repeated one cycle later (Eq. 3.8).

$$v(t_0 + T) = v(t_0) + \frac{1}{C}\int_{t_0}^{t_0+T} i(t)\,dt \tag{3.8}$$

The best way to fully appreciate the operation of power electronics is to take a power electronics lab course and to design and build a converter yourself. If you do this, the fundamentals will take on a concrete physical meaning that will serve you well throughout your career.

The way to connect inductors and capacitors safely is a function of their characteristics, as discussed above. The penalty of connecting these elements the wrong way is that bad things will happen, as the laws of physics will still apply to enforce KVL, KCL, and conservation of power. Components may explode, and you probably will not be allowed back into the lab! So, it is important to proceed carefully and get the fundamentals right. We will illustrate what interconnections would be safe by doing a thought experiment.

Example 3.1 In Fig. 3.7, are these two circuits showing connections that are safe or unsafe, and why?

Solution:

The answer is that these are both unsafe connections.

First, you should not ever connect a DC voltage source across an inductor, as shown in the left figure of Fig. 3.7. This is a bad idea! Why? Well, in this topology, it forces V_L=V_{DC}, and V_{DC} is a constant. However, $V_L = L\frac{di}{dt}$. So, if you connect a constant voltage across an

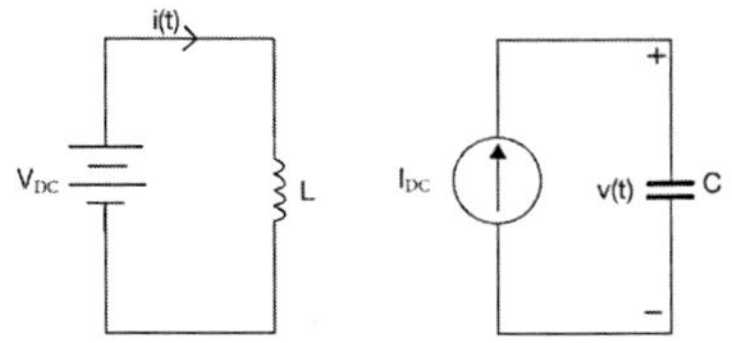

Figure 3.7: Unsafe connections example: never connect an inductor across a DC voltage source or a capacitor in series with a DC current source.

inductor, you constrain its increase in current to be a positive constant $\frac{di}{dt} = \frac{V_{DC}}{L} = constant$. This means that the current will continue to rise! Recall that for periodic steady state we need the voltage across the inductor to be zero on average. This bad configuration would never allow that. Continually rising current in a circuit is a very bad idea!

Second, you should not ever connect a DC current source across a capacitor, as shown in the right figure of Fig. 3.7. Similarly, the problem is that it forces $I_C{=}I_{DC}$, where I_{DC} is a constant. This is again a bad idea! Why? Following the dual of the discussion above for the inductor, if you hold the current through the capacitor fixed $I_C = C\frac{dv}{dt}$ or $\frac{dv}{dt} = \frac{I_{DC}}{C} = constant$, it will cause the voltage to rise uncontrollably. This is also a very bad idea in a circuit!

Because DC-DC converters are composed of inductors and capacitors, it is important to make sure to recall and follow their guiding principles in your circuit design. In the following topologies, we will judiciously use these principles that we restate using the following shorthand from the fundamentals above.

The average voltage across an inductor over a cycle is zero:

$$< v_L >= 0 \tag{3.9}$$

The average current through a capactior over a cycle is zero:

$$< i_C >= 0 \tag{3.10}$$

Switching Waveforms

The switching waveform or switching signal is used to control the switches in a DC-DC converter. An example is shown in Fig. 3.8, where the signal $q(t)$ is the switching signal that controls the switch. It has a value of 1 when the switch is closed and 0 when it is open. The switch is closed for DT seconds.

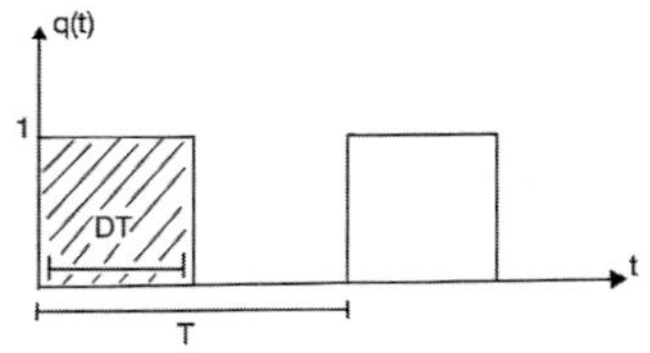

Figure 3.8: Switching signal

The average of $q(t)$ gives us the proportion of time that the switch is "on" or closed, which is also called the duty cycle, D. D is between 0 and 1 and tells us how often the switch is on. Understanding the average behavior is important to understanding DC-DC converter operation.

DC-DC Converter Design Goals

Converter design should allow the input to be specified independently from the load. That is, a main functionality that we want to achieve by using power electronics in energy conversion is electrical decoupling between the load and the source. We also want efficiency to be as close to 100% as possible and maximum control flexibility.

The source may be any DC source. The load may be any DC load. To keep with the above example, let us say the source is a PV (solar) panel and the load is a resistor. Now let's look at the three topologies.

3.3.1 Buck Converter

When presented a topology, the first step is to ask yourself what happens when the switch is open, and what happens when the switch is closed? This reveals its operating principle.

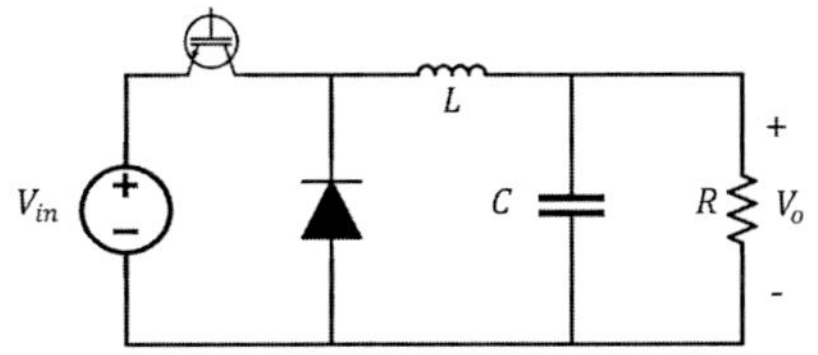

Figure 3.9: Buck converter topology

In the buck converter topology above, let us look first at what happens when the switch is closed, so DT is 1. We will assume the diode is ideal and the capacitor is large.

Switch Closed

When the switch is closed (Fig. 3.10), the diode is reverse biased, so it does not conduct. This sends the current through the inductor, the capacitor, and the resistor. Then, we can write the following equation for V_L from basic circuit rules, where its multiplication with $q(t)$ tells us that this equation only applies when the switch is on:

$$V_L(t) = q(t)[V_{IN} - V_{OUT}(t)] \tag{3.11}$$

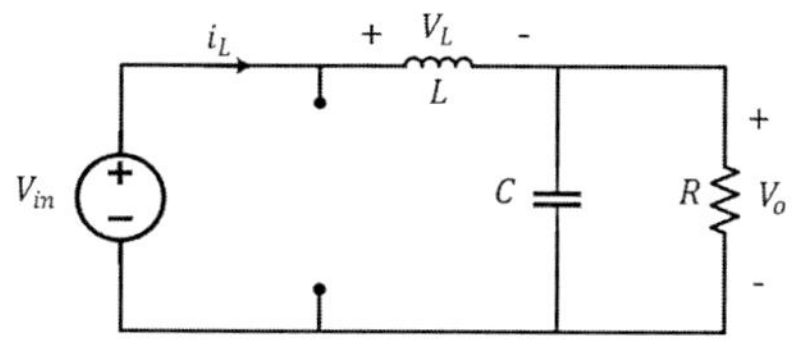

Figure 3.10: Buck with switch closed, with $V_L = V_{IN} - V_{OUT}$.

Switch Open

When the switch is open (Fig. 3.12), the current cannot change instantaneously in an inductor, so the current needs somewhere to go to complete the circuit. Hence, the diode turns on (conducts). Like for the "switch closed" case above, we can now write the following equation for V_L from basic circuit rules, where its multiplication with $[1 - q(t)]$ tells us that this equation only applies when the switch is off:

$$V_L(t) = [1 - q(t)](-V_{OUT}(t)) \tag{3.12}$$

Putting it all Together: Buck Converter Operation

Using these insights about these two states, and recalling that the average voltage across an inductor over a cycle is zero, we will analyze the circuit.

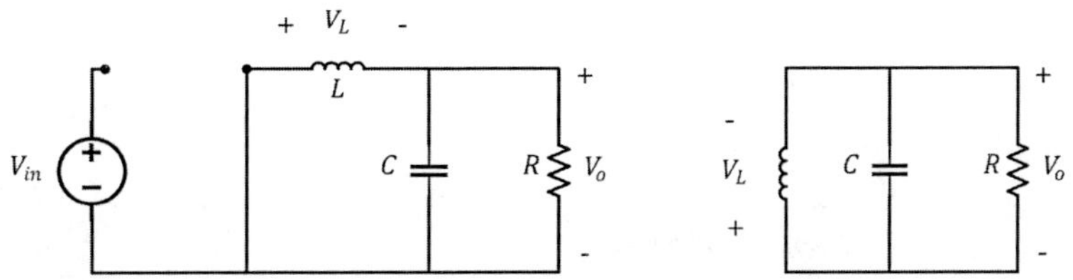

Figure 3.11: Buck with switch open, with $V_L = -V_{OUT}$.

Specifically, we obtain the following equation for the voltage across the inductor, which covers the behavior over the entire cycle, both when the switch is closed and when the switch is open:

$$V_L(t) = q(t)[V_{IN} - V_{OUT}(t)] + [1 - q(t)](-V_{OUT}(t)) \tag{3.13}$$

Then, we apply the fact that average voltage across an inductor over a cycle is zero. When we do so, we obtain the following principle of operation of the DC-DC buck converter:

$$< V_L(t) >= 0 \tag{3.14}$$

$$< q(t) > [V_{IN} - V_{OUT}(t)] + < 1 - q(t) > (-V_{OUT}(t)) = 0 \tag{3.15}$$

$$D[V_{IN} - V_{OUT}] + (1 - D)(-V_{OUT}) = 0 \tag{3.16}$$

$$DV_{IN} - DV_{OUT} - V_{OUT} + DV_{OUT} = 0 \tag{3.17}$$

$$\boxed{V_{OUT} = DV_{IN}} \tag{3.18}$$

where $(0 \leq D \leq 1)$, and D is tunable. Hence, this topology represents a tunable, controllable DC-DC converter that can deliver output voltage that is always less than input voltage: $V_{OUT} < V_{IN}$. If the capacitor C is large, V_{OUT} is approximately constant.

What does the current do? When the switch is closed, i_L increases because the voltage across it $v_L(t) = V_{IN} - V_{OUT} > 0$. When the switch is open, $v_L(t) = -V_{OUT} < 0$, so i_L decreases. Then, recall on average that we know the following:

$$< V_L >= 0 \tag{3.19}$$

$$V_{OUT} = DV_{IN} \tag{3.20}$$

Then, recall that the average current in a capacitor is zero over a cycle:

$$< i_C >= 0 \tag{3.21}$$

This means that on average the same current flows through the resistor and inductor with the following relationship:

$$I_L = V_{OUT}/R \tag{3.22}$$

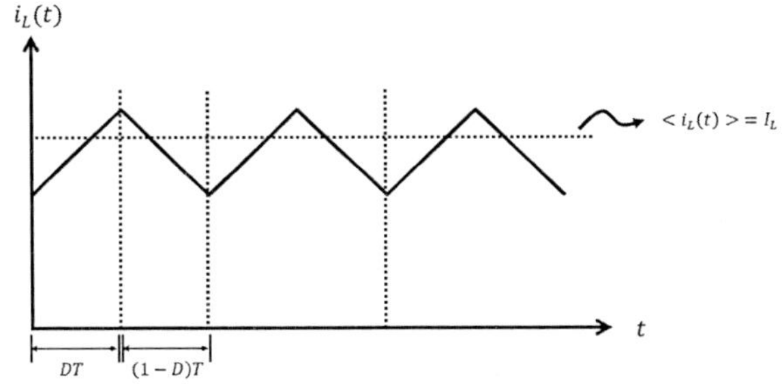

Figure 3.12: Buck inductor current with ripple.

In summary, in a buck converter, output voltage is controlled by changing duty cycle, D, of the switch which operates at high frequency. The output and input voltages are shown in Fig. 3.17 for various D values. Duty cycle D is the fraction of time the switch is closed. Example switches include an insulated-gate bipolar transistor (IGBT) or a silicon controlled rectifier (SCR). When the switch closed, the current in the inductor increases. When the switch is open, the current in the inductor decreases.

Buck Converter Comments:

- The buck converter always decreases the voltage.
- A large L effectively acts as a DC current source, with some ripple.
- A large C effectively acts as a DC voltage source, with some ripple.

3.3.2 Boost Converter

In the above section, we see that a buck converter can only lower the voltage. Now, what if we want to raise the voltage? A boost converter does this. The

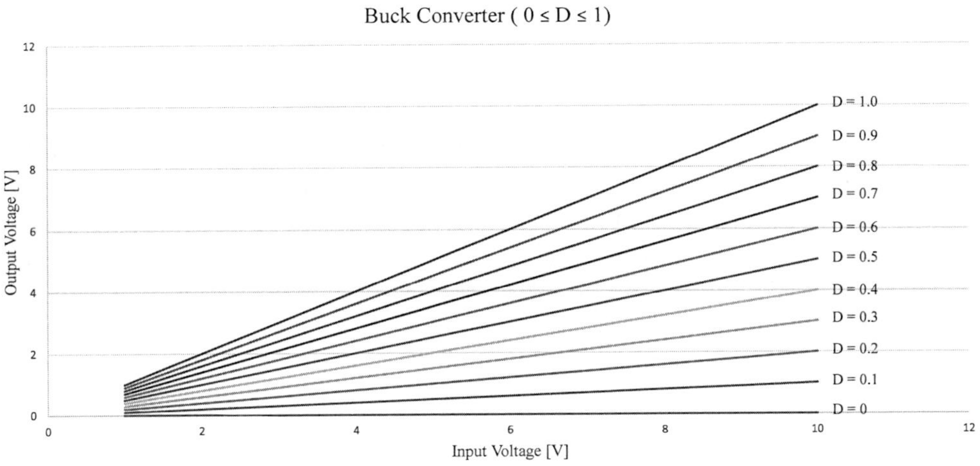

Figure 3.13: Buck converter output vs. input voltage, as D is varied.

process to analyze its topology is the same as for the buck converter. We start by looking at the circuit with the switch closed, and then looking at the cirucit with the switch open. Taking the average behavior over a cycle will tell us how the boost converter behaves. As with the buck converter, we will assume that the diode is ideal and that L and C are large.

Switch Closed

When the switch is closed, the diode is reversed biased. The current from the source flows through the inductor. The output is essentially disconnected from the input, as shown in Fig. 3.15 below. Then, the voltage across the inductor is equal to the voltage across the input (Eq. 3.23), where the multiplication of V_{IN} with $q(t)$ tells us that this equation only applies when the switch is closed:

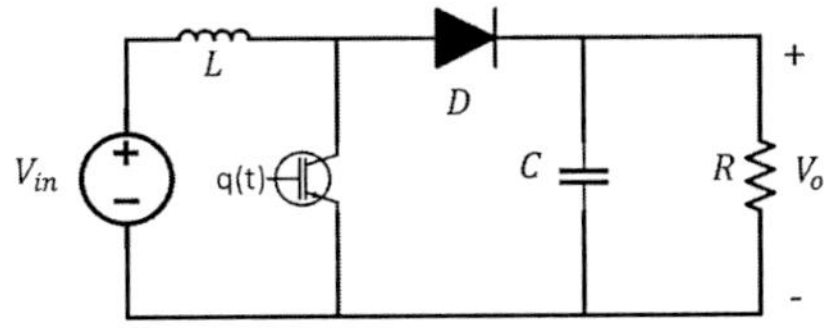

Figure 3.14: Boost converter topology.

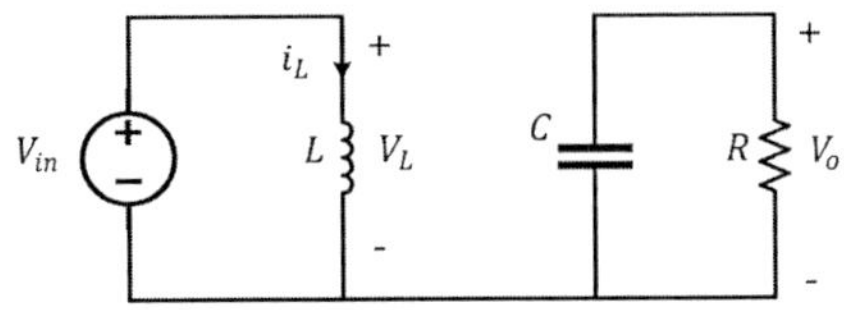

Figure 3.15: Boost converter - switch closed.

$$v_L(t) = q(t)[V_{IN}] \tag{3.23}$$

Switch Open

When the switch is open, the inductor current needs a path, so the diode is conducting, as shown in Fig. 3.16 below. Then, the voltage across the inductor is equal to the voltage difference between the input and output voltages (Eq. 3.24), where the multiplication by $[1 - q(t)]$ tells us that this equation only applies when the switch is open.

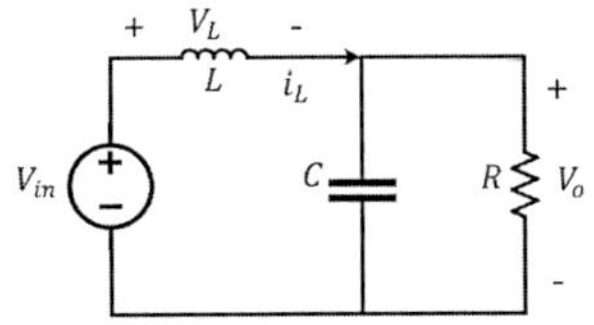

Figure 3.16: Boost converter - switch open.

$$v_L(t) = [1 - q(t)](V_{IN} - V_{OUT}(t)) \tag{3.24}$$

Putting it all Together: Boost Converter Operation

Now, we will apply the fact that the average voltage across an inductor over a cycle is zero, in a similar manner to how we analyzed the buck converter.

$$v_L(t) = q(t)(V_{IN}) + [1 - q(t)](V_{IN} - V_{OUT}(t)) \tag{3.25}$$

$$< v_L(t) >= 0 \tag{3.26}$$

In doing so, we obtain the following, which describes the behavior of the boost converter over a cycle:

$$< q(t) > V_{IN} + < 1 - q(t) > (V_{IN} - V_{OUT}) = 0 \tag{3.27}$$

$$DV_{IN} + (1 - D)(V_{IN} - V_{OUT}) = 0 \tag{3.28}$$

$$DV_{IN} + V_{IN} - V_{OUT} - DV_{IN} + DV_{OUT} = 0 \tag{3.29}$$

$$\boxed{V_{OUT} = \frac{V_{IN}}{(1 - D)}} \tag{3.30}$$

where $(0 \leq D \leq 1)$, and D is tunable. This is a controllable DC-DC converter that can deliver output voltage greater than input voltage, $V_{OUT} > V_{IN}$, as shown in Fig. 3.17 for various D values.

In summary, the boost converter is used to raise output voltage above input voltage. Assuming L and C are sufficiently large, we can treat L as a current source, and C as a voltage source. When the switch is closed, diode is reverse biased, so the inductor current increases. When the switch is open, the inductor drives current into the diode.

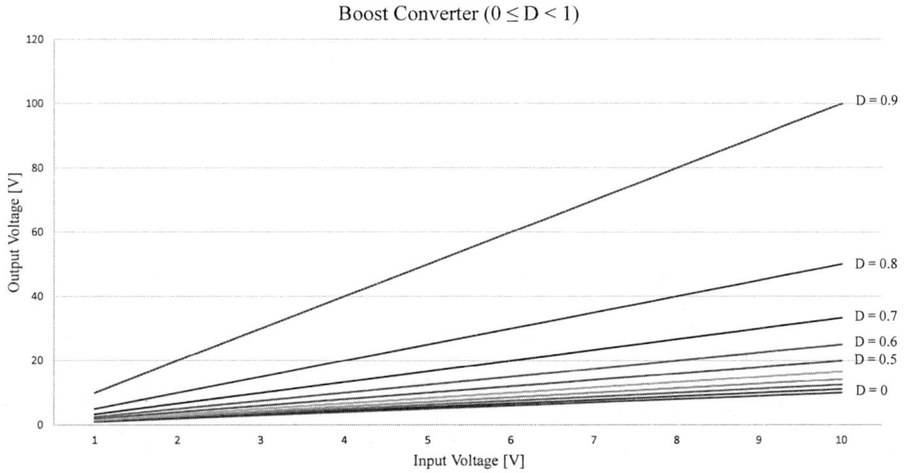

Figure 3.17: Boost converter output vs. input voltage, as D is varied.

Analysis of Other Converter Topologies

Many DC-DC converter topologies have been proposed and continue to be proposed (this is not all). Examples include the flyback converter and the buck-boost converter. In any case, the approach suggested here is that to analyze a circuit, one should start by examining its fundamental operating principles. When inductors and capacitors are involved, and when electronic switches are involved, the following tips provide a good starting point:

- Recall that $< v_L >= 0$.
- Recall that $< i_C >= 0$.
- Draw the circuit with its [switch closed] vs. [switch open]. How does the current flow?

3.3.3 Buck-Boost Converter

With the above general suggestions now to consider, we will present one more topology, the buck-boost converter, that can both lower and raise the output voltage. Again, we will assume that L and C are sufficiently large, so that we can treat L as a current source and C as a voltage source.

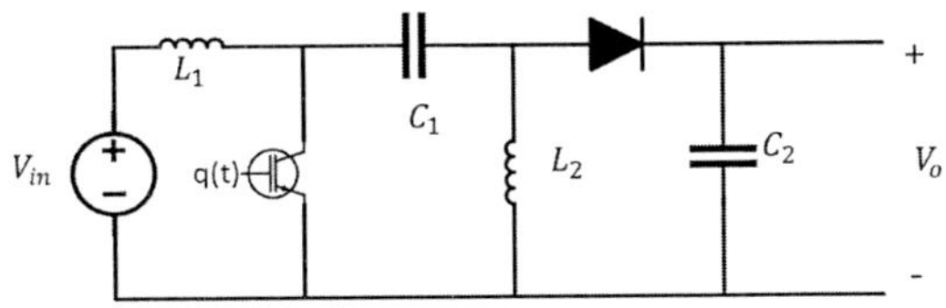

Figure 3.18: Buck-boost converter topology.

We will look at the average voltage across the second inductor L_2.

Switch Closed

When the switch is closed, the branch where the switch is located becomes a short, isolating the source and first inductor from the rest of the curciut. Additionally, the diode is reversed biased and open, isolating the middle portion of the circuit from the output. Then, the voltage across L_2 is equal to the negative of the input voltage (Eq. 3.31).

$$V_{L2} = -V_{IN} \tag{3.31}$$

To see why this is true, it helps to look at what happens to the capacitor voltage on C_1. Recall that the voltage across a capacitor cannot change instantaneously. Hence, when the switch is open (see circuit below), the

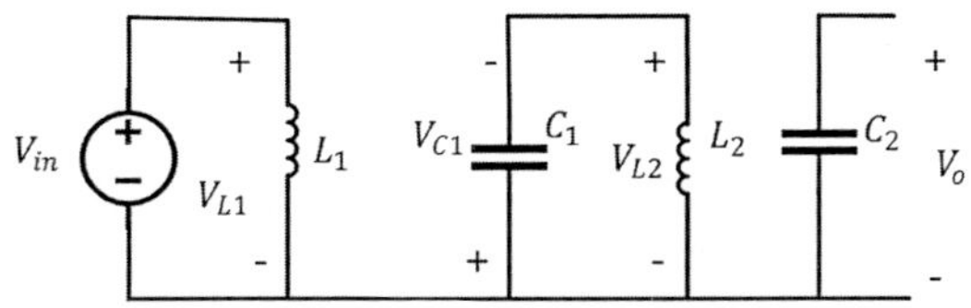

Figure 3.19: Buck-boost - switch closed.

voltage across the capacitor C_1 must be V_{IN}, polarized as shown, due to KVL (the inductor L_2 would act as a short). When the switch opens, since the voltage $V_{C1} = V_{IN}$ cannot change instantanously, the voltage across inductor L_2 has to be the same as that of the capacitor to satisfy KVL. Since we are assuming C_1 and C_2 are very large, it means that the capacitor maintains essentially the same voltage the entire time the switch is closed.

Switch Open

When the switch is open, the current in the inductor needs a path, so the diode is conducting.

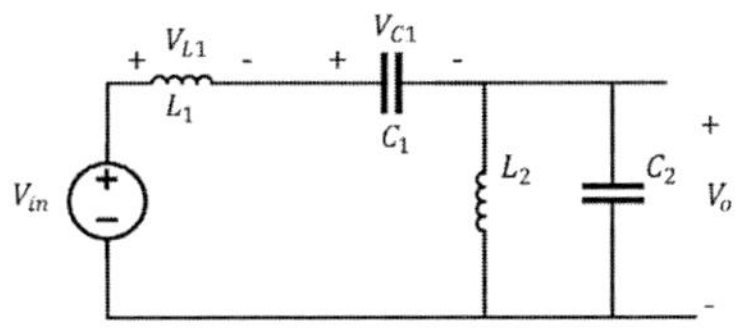

Figure 3.20: Buck-boost - switch open.

$$V_{L2} = V_{OUT} \tag{3.32}$$

Putting it all Together: Buck-Boost Converter Operation

We apply the fact that the average voltage across an inductor over a cycle is zero, in a similar manner to how we analyzed the buck and boost converters, and analyze the buck-bookst circuit (Eq. 3.33). Hence, the following equation describes for the voltage across the inductor, over the entire cycle, when the switch is closed and open:

$$V_{L2} = q[-V_{IN}] + (1 - q)[V_{OUT}] \tag{3.33}$$

Now, we set it equal to zero to analyze its operation.

$$< V_{L2}(t) >= 0 \tag{3.34}$$

$$< q(t) > [-V_{IN}] + < 1 - q(t) > (V_{OUT}) = 0 \tag{3.35}$$

$$-DV_{IN} + (1 - D)(V_{OUT}) = 0 \tag{3.36}$$

$$\boxed{V_{OUT} = \frac{D}{(1 - D)} V_{IN}} \tag{3.37}$$

This is a controllable DC-DC converter that can raise or lower output voltage, based on whether the duty cycle $D > 0.5$ or $D < 0.5$, respectively, as shown in Fig. 3.17 for various D values.

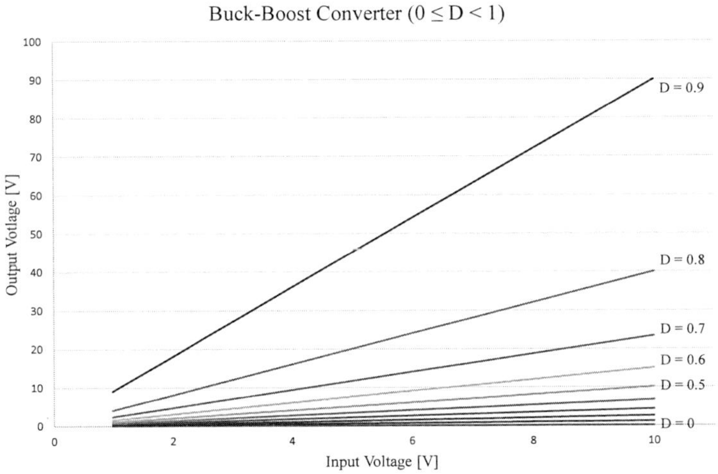

Figure 3.21: Buck-boost converter output vs. input voltage, as D is varied.

3.4 AC-DC and DC-AC Converters

AC to DC conversion is done using rectifiers, while DC to AC conversion is done using inverters. Both types can also be referred to as converters.

3.4.1 Rectifiers

Rectifiers have a wide range of applications and are prevalent in converting the AC power delivered by the grid to the DC power required by many appliances: laptop chargers (wall outlet to battery), appliances such as vacuum, telephone, rice cooker, etc., as in Fig. 3.22. Motor drives often have a front-end AC source to DC. High voltage DC (HVDC) transmission lines require converters with high power ratings.

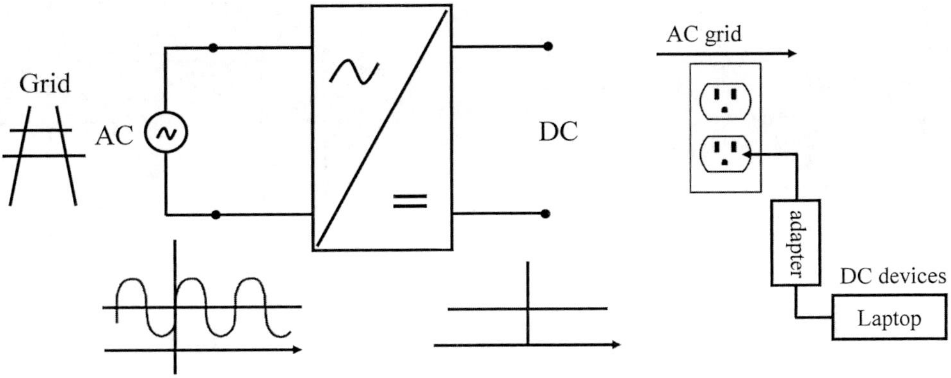

Figure 3.22: AC-DC converter (rectifier) applications

The purpose is to convert AC to DC, with either a pure DC output, or with a DC component in the output. Types include half-wave rectifiers, full-wave rectifiers, single-phase rectifiers, and three-phase rectifiers.

Half-Wave Rectifier

A half-wave rectifier is ideal to begin with, to build intuition on the basics that also govern how even more complicated converters work. Hence, we first examine a half wave rectifier. A half wave rectifier with an output capacitor and a resistive load is given in Fig. 3.23.

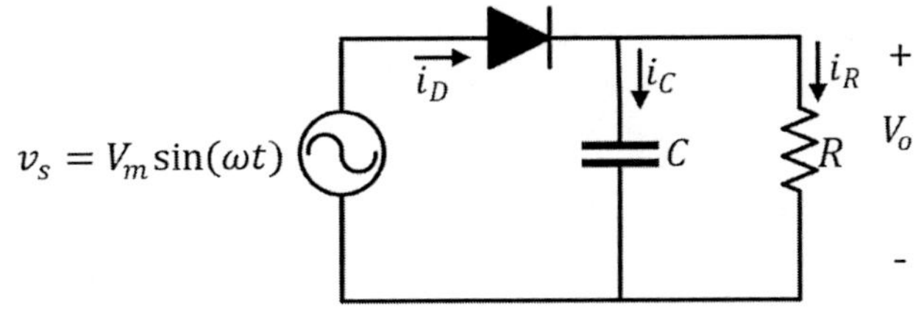

Figure 3.23: Half-wave rectifier circuit example

To build intuition about converter operation, we need to examine the

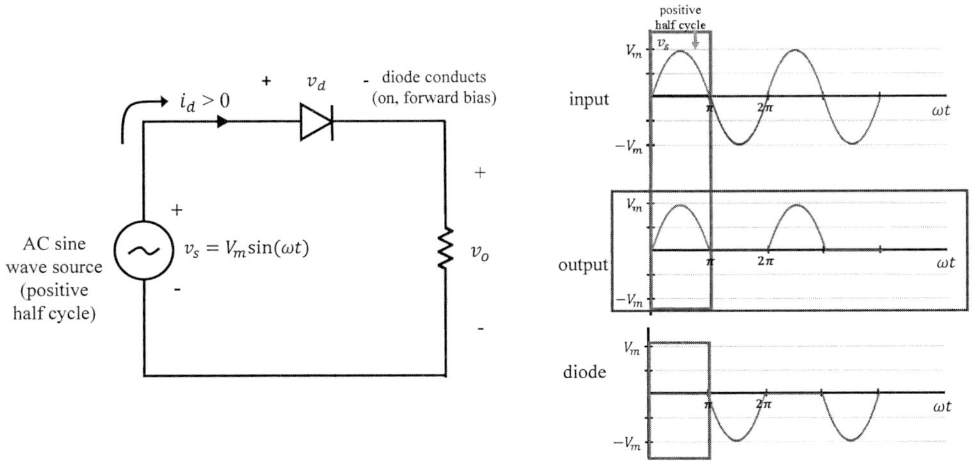

Figure 3.24: Half-wave rectifier circuit operation for the positive half cycle.

circuit's states. The input voltage is a sine wave. The diode is the electronic switch that changes the circuit's behavior when it is on or off. The interesting feature of rectifiers is that AC source itself is what causes the diode to turn on or off, and we will show how this works next. It means that to understand what the circuit is doing, it will be essential to consider both the positive and negative half cycles of the AC input voltage.

Suppose the AC voltage is polarized in the direction shown in Fig. 3.24 for the first half-cycle. During that time, the current flows as shown by the arrow, allowing the diode to conduct. Fig. 3.24 also shows the input, output, and diode voltage waveforms.

During the negative half-cycle, the diode is off (reverse biased). The output looks like the second waveform. The rectified value or the DC value is the average of this output over one cycle.

In the half-wave rectifier, an output capacitor would usually be added. The capacitor acts like a filter. Rather than the output voltage dipping all the way down to zero at each negative half-cycle, it only dips down to the terminal voltage of the discharging capacitor, which is governed by the capacitor's equation, V_0, where V_θ is the output voltage at the time θ the switch is opened.

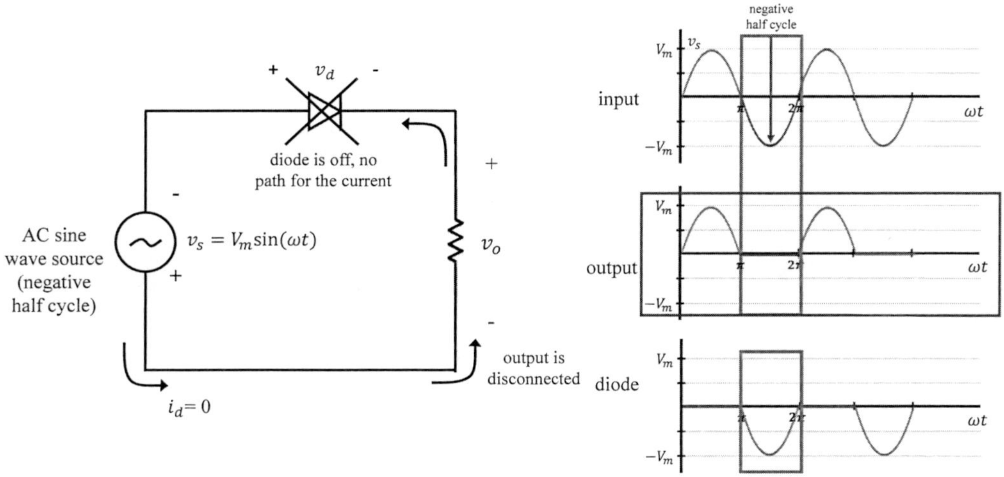

Figure 3.25: Half-wave rectifier circuit operation for the negative half cycle.

$$V_0(\omega t) = \begin{cases} V_m \sin(\omega t), & \text{diode on} \\ V_\theta e^{(-(\omega-\theta)/\omega RC)}, & \text{diode off} \end{cases} \tag{3.38}$$

A larger capacitor will hence result in a smaller voltage ripple, ΔV_0. A typical design question would be to find how much capacitance C is required to achieve a certain output voltage ripple.

Full-Wave Rectifier

From the analysis of the half-wave rectifier above, we can conclude that the output does not look flat like a DC voltage, although it does have a DC component. Extending on the concepts introduced above, we now present the full-wave rectifier. The advantages of a full-wave rectifier over a half-wave rectifier include that it produces less ripple and has improved power utilization, as the source power is drawn during both positive and negative AC half-cycles. Thus, it achieves higher performance.

During the positive half-cycle, diodes $D1$ and $D2$ conduct (Fig. 3.27). During the negative half-cycle, diodes $D3$ and $D4$ conduct (Fig. 3.28). In both of these states, the polarization of current into and voltage across the load stays the same, and the values stay the same as the input values.

The result is that the diode bridge rectifier output rectifies the negative parts of the input AC voltage waveform and makes them all positive. The

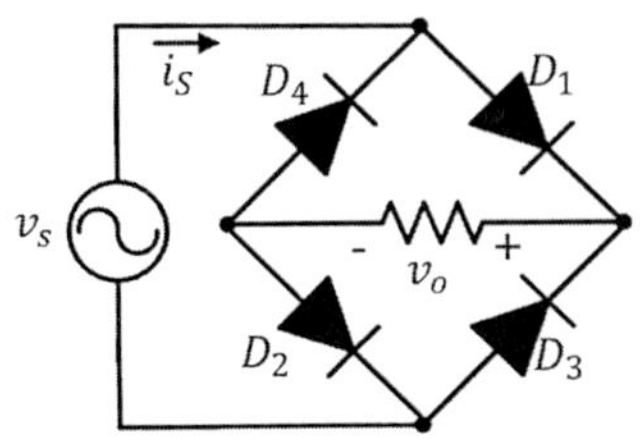

Figure 3.26: Full-wave diode bridge rectifier circuit

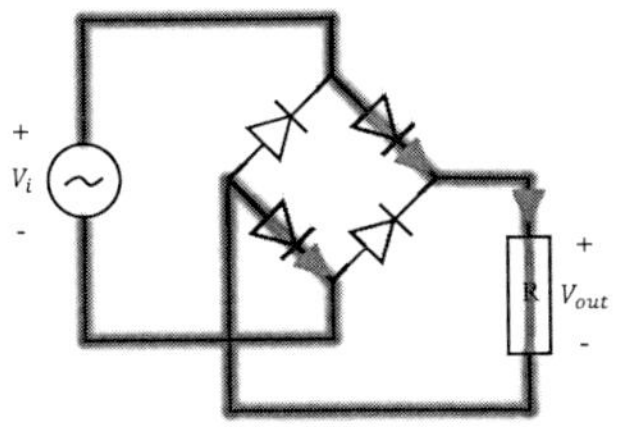

Figure 3.27: Full-wave diode bridge rectifier circuit - positive half cycle

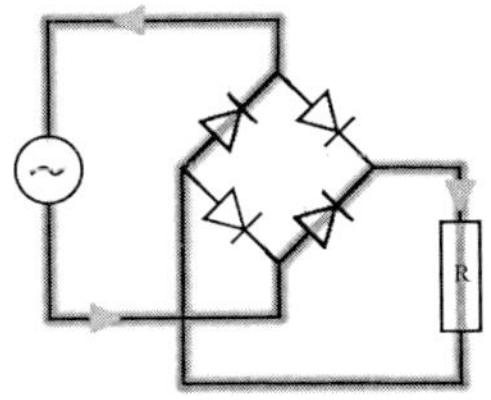

Figure 3.28: Full-wave diode bridge rectifier circuit - negative half cycle

output voltage is the absolute value of the input voltage.

Full-Wave Rectifier with Output Capacitor

When an output capacitor is added to the full-wave rectifier, as in Fig. 3.29, we can consider four states. We have the positive and negative half cycles, described above, and for each of these states, the capacitor can be either charging or discharging. It is charging if $|V_i|$ greater than V_O and discharging if $|V_i|$ is less than than V_O.

The output capacitor has the same function as for the half-wave rectifier; it keeps the output voltage high and reduces the ripple as shown in Fig. 3.30. The capacitor keeps V_0 high and steady. This is used in many wall powered

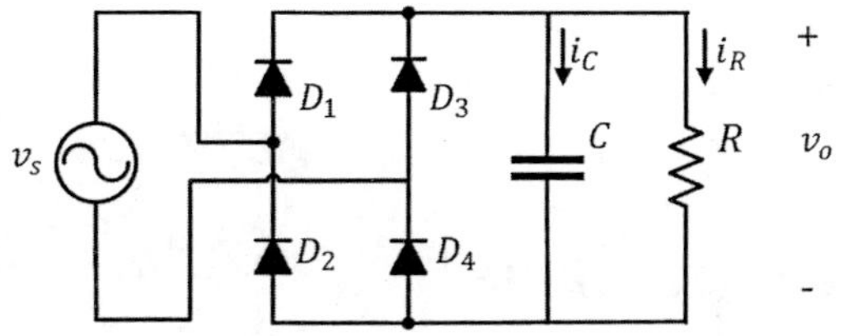

Figure 3.29: Full-wave diode bridge rectifier circuit with output capacitor

devices.

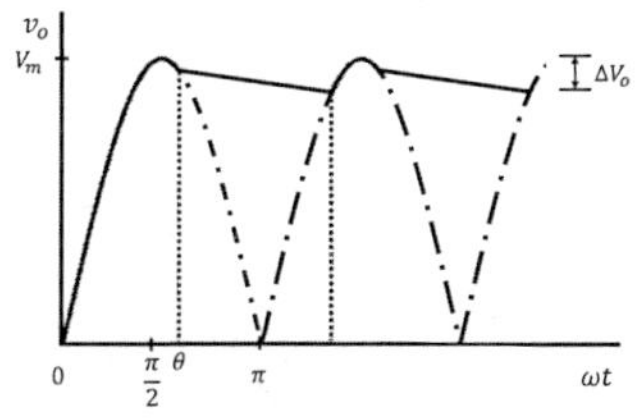

Figure 3.30: Output voltage of diode bridge rectifier with output capacitor

Three-Phase Rectifier

For three-phase power systems, a three-phase rectifier takes advantage of the balanced three-phase AC sources to deliver a larger amount of power to the load (Fig. 3.31). It also achieves a smaller ripple size (less or no need for a filter) and higher efficiency. It is used in industry for large loads. Its operation follows the same principles discussed above, where its first two states are shown in Fig. 3.32.

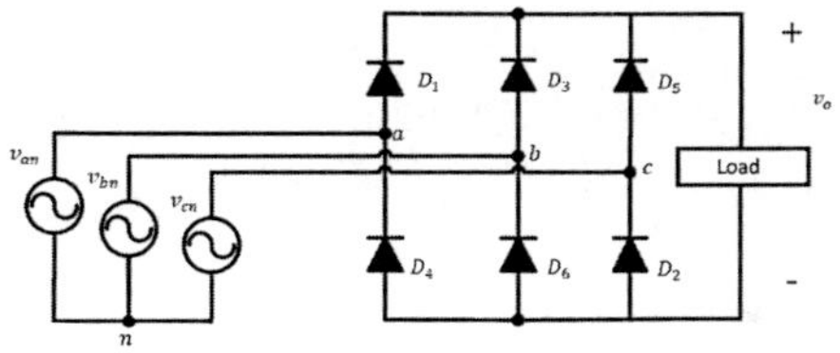

Figure 3.31: Three-phase rectifier circuit

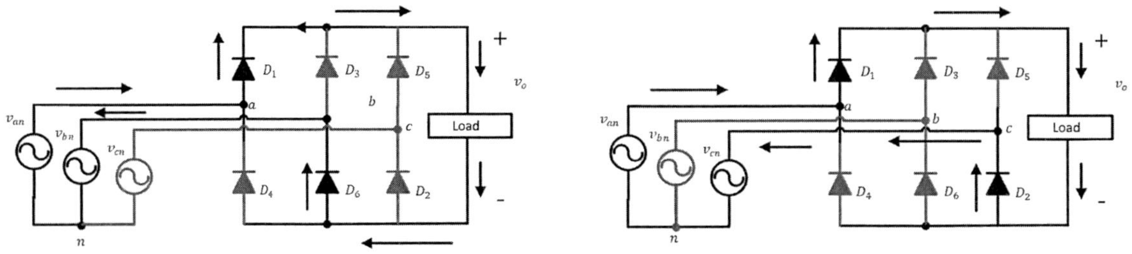

Figure 3.32: Three-phase rectifier operation - current flow direction follows the polarity and orientation of the balanced three-phase phasors, so current flows through the diodes in a direction from a to b, then from a to c, etc., producing a rectified output voltage

3.4.2 Total Harmonic Distortion

Harmonics are an important issue to get familiar with when working with power electronics. If a sinusoidal voltage source is applied to a nonlinear load, the current waveform will not be sinusoidal. This produces harmonics at multiples of the fundamental frequency. An area of power systems deals more in depth in analyzing and mitigating harmonics, called *power quality*. Harmonics can cause serious transformer overheating. Losses are higher due to the higher RMS current that increases as total harmonic distortion increases. Digital clocks that work by counting zero crossings gain time. Telephone interference, motor heating, and overloaded building neutral conductors are additional symptoms [13].

Harmonics are described by Fourier analysis, which you will cover in depth in other electrical engineering courses. The Fourier theorem says that any waveform can be written as a summation of sinusoids at multiples of a fundamental frequency, as given in (Eq. 3.39),

$$\begin{aligned} f_t = \frac{a_0}{2} + a_1 cos(\omega t) + a_2 cos(2\omega t) + a_3 cos(3\omega t) + \\ ... + b_1 sin(\omega t) + b_2 sin(2\omega t) + b_3 sin(3\omega t) + ... \end{aligned} \tag{3.39}$$

where $\omega = 2\pi f = \frac{2\pi}{T}$ and

$$a_n = \frac{2}{T}\int_0^T f(t) cos(n\omega t)\, dt, n = 0, 1, 2, ... \tag{3.40}$$

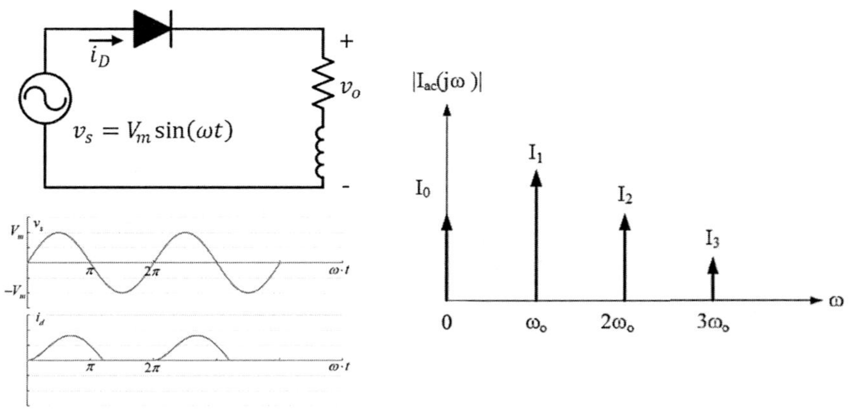

Figure 3.33: Harmonics example for a half wave diode bridge rectifier circuit

$$b_n = \frac{2}{T}\int_0^T f(t)sin(n\omega t)\,dt, n = 0, 1, 2, ... \tag{3.41}$$

Harmonics can appear in both voltage and current waveforms, and they can be caused by power electronics and other nonlinear loads– anything connected to the grid that is not a perfect sinusoid. Fig. 3.33 shows an example. To get a first-hand experience with harmonics, you can do an interesting experiment (only safely and with permission/help from your professor!): connect an oscilloscope to the wall outlet in your classroom, and you will see that indeed the voltage waveform that you view has some sharp edges, and it does not look at all like a perfect sinusoid. This is because your building has many computers connected to the wall outlet, as well as other nonlinear loads. It is interesting to think about the implications of changing load profiles on power quality. From this experiment with the wall outlet, we can see that if we do not consider harmonics, and we design something (an electronic device or power electronics converter) that is grid-connected, we might contribute to the problem rather than doing our part to make it better. We want to be good citizens of the grid and make it better. Hence, the goal is to minimize the non-fundamental frequency components while maximizing the fundamental component that delivers power to the load. This can be done in practice by utilizing the characteristics of inductors and capacitors in the design of passive and active filters, which provide a low-impedance path to ground for the harmonics currents, and hence act like a drain to pull them out of the system. One common source of harmonics is a compact flourescent lamp (CFL) light bulb. CFLs have electronic ballasts that draw current waveforms that contain significant non-fundamental frequency components.

In summary, harmonics are multiples of the fundamental frequency, and they occur due to non-linear loads and cause power quality issues. THD is a term to quantify the non-sinusoidal property of a waveform. It is the ratio of the RMS value of all the non-fundamental frequency terms to that of the fundamental frequency term. It is always greater than or equal to zero. The goal is to have THD as low as possible– no more than 1-3% max.

We can calculate the I_{RMS} and THD for a device's waveforms, as in Eq. 3.44 and Eq. 3.43, respectively.

$$I_{RMS} = \sqrt{I_1^2 + I_2^2 + I_2^2 + ... + I_k^2} \tag{3.42}$$

$$THD = \frac{\sqrt{I_2^2 + I_2^2 + ... + I_k^2}}{I_1} \tag{3.43}$$

Similarly, THD can also be calculated for a voltage waveform based on V_{RMS} components:

$$V_{RMS} = \sqrt{V_1^2 + V_2^2 + V_2^2 + ... + V_k^2} \tag{3.44}$$

Example 3.1 Consider a residential 2-phase system with the following phase currents:

$$i_A = 6\sqrt{2}cos(377t) + 2\sqrt{2}cos(2(377t)) + \sqrt{2}cos(3(377t))$$

$$i_B = 6\sqrt{2}cos(377t + \pi) + 2\sqrt{2}cos(2(377t + \pi)) + \sqrt{2}cos(3(377t + \pi))$$

(a) What is the total RMS current in each phase?

(b) What is the THD in each phase's current?

(c) What is the RMS current in the neutral?

Solution:

(a) For phase a, the total RMS current is the square root of the sum of the squares of the three current components at each of the three frequencies:

$$I_A, RMS = \sqrt{6^2 + 2^2 + 1^2} = \sqrt{41} = 6.403A$$

For phase b, the RMS current is calculated the same way, and we can observe that it has the same value:

$$I_B, RMS = \sqrt{6^2 + 2^2 + 1^2} = \sqrt{41} = 6.403A$$

(b) THD is also the same for phases a and b, since RMS and THD only depend on the current magnitudes which are the same for both phases:

$$THD_A = THD_B = \sqrt{2^2 + 1^2}/6 = 0.373$$

(c) Neutral wires, which usually carry zero or nearly zero currents, are a place where problems from harmonics tend to show up. Even in this simple example, we can see this, as the phase differences of the currents due to the harmonics causes some cancellations and some additions. The first and third harmonics cancel here, but the second harmonics add. The neutral current is given by.

$$i_n = i_a + i_b$$

Recall that $cos(x + \pi) = cos(x)$. Now, let us look at one harmonic at a time.

At the fundamental frequency, we have the following neutral current contribution:

$$i_{n,1} = 6\sqrt{2}cos(377t) + \sqrt{2}cos(377t + \pi) = 6\sqrt{2}cos(377t) - \sqrt{2}cos(377t) = 0$$

Then, for the second harmonic, $cos(x + 2\pi) = cos(x + 0)$, so the phase difference disappears, and the current magnitudes will sum up:

$$i_{n,2} = 2\sqrt{2}cos(2(377t)) + 2\sqrt{2}cos(2(377t + \pi))$$

$$= 2\sqrt{2}cos(2(377t)) + 2\sqrt{2}cos(2(377t) + 2\pi))$$

$$= 2\sqrt{2}cos(2(377t)) + 2\sqrt{2}cos(2(377t) + 0))$$

$$= 4\sqrt{2}cos(2(377t))$$

For the third harmonic, $cos(x+3\pi) = cos(x+\pi) = -cos(x)$, so we have the same situation as for $i_{n,1}$, where the current magnitudes cancel:

$$i_{n,3} = 6\sqrt{2}cos(3(377t)) + \sqrt{2}cos(3(377t+\pi)) = 6\sqrt{2}cos(377t) - \sqrt{2}cos(377t) = 0$$

Finally, to answer the question about the RMS current in the neutral, we need to sum up the neutral current contributions at each frequency:

$$i_n = i_a + i_b = i_{n,1} + i_{n,2} + i_{n,3}$$

Then, we find its RMS value:

$$I_{n,RMS} = \sqrt{I_{n,1}^2 + I_{n,2}^2 + I_{n,3}^2}$$

In this case:

$$I_{n,RMS} = \sqrt{0 + 4^2 + 0} = 4A$$

Problems with Harmonics

A key problem with the third harmonic in three-phase systems is that it will not cancel out. This creates a neutral current in an otherwise balanced three-phase system that would not usually have neutral current. This is also true for other triplen harmonics. This is because the 120 degree phase shift becomes 360 degrees when multiplied by 3– meaning both b and c phase 3rd harmonic phasor components will be in phase with the a phase, and hence will add up instead of cancel out. This feature can be seen by constructing a small example.

Example 3.2 A three-phase system has the following currents:

$$i_A = 120\sqrt{2}cos(\omega t + 0°) + 10\sqrt{2}cos(2(\omega t + 0°)) + 10\sqrt{2}cos(3(\omega t + 0°))$$

$$i_B = 120\sqrt{2}cos(\omega t - 120°) + 10\sqrt{2}cos(2(\omega t - 120°)) + 10\sqrt{2}cos(3(\omega t - 120°))$$

$$i_C = 120\sqrt{2}cos(\omega t + 120°) + 10\sqrt{2}cos(3(\omega t + 120°)) + 10\sqrt{2}cos(3(\omega t + 120°))$$

What is the RMS current in the neutral?

Solution:

As in Example 3.1, we first look each harmonic separately to find its contribution to the neutral current:

$$i_{n,1} = 120\sqrt{2}cos(\omega t{+}0^\circ){+}120\sqrt{2}cos(\omega t{-}120^\circ){+}120\sqrt{2}cos(\omega t{+}120^\circ) = 0A$$

$$i_{n,2} = 10\sqrt{2}cos(2(\omega t{+}0^\circ)){+}10\sqrt{2}cos(2(\omega t{-}120^\circ)){+}10\sqrt{2}cos(2(\omega t{+}120^\circ))) = 0A$$

$$i_{n,3} = 10\sqrt{2}cos(3(\omega t{+}0^\circ)){+}10\sqrt{2}cos(3(\omega t{-}120^\circ)){+}10\sqrt{2}cos(2(\omega t{+}120^\circ))$$

$$= 10\sqrt{2}cos(3(\omega t + 0^\circ)) + 10\sqrt{2}cos(3\omega t - 360^\circ)) + 10\sqrt{2}cos(3\omega t + 360^\circ)$$

$$= 10\sqrt{2}cos(3(\omega t)) + 10\sqrt{2}cos(3\omega t)) + 10\sqrt{2}cos(3\omega t)$$

$$= 30\sqrt{2}cos(3\omega t)A$$

Then, the RMS of the neutral current can be found:

$$I_{n,RMS} = \sqrt{I_{n,1}^2 + I_{n,2}^2 + I_{n,3}^2}$$

In this case:

$$I_{n,RMS} = \sqrt{0 + 0 + 30^2} = 30A$$

Harmonics also cause transformer overheating since core losses are proportional to frequency. This effect can be significant during certain events like geomagnetic disturbances. Resonance is also an issue, particularly with shunt capacitors; these tend to occur around the 5th or 7th harmonic. Negative impacts of total harmonic distortion also include (1) interference with communications, (2) distortion of source voltage, and (3) heating up of transformers, motors, generators, etc.

Minimizing the Impact of Harmonics

Substation design and protection systems must consider these factors. Good engineering selects transformers that can eliminate or minimize the impact of harmonics. Selection of the protection systems for equipment also should consider harmonics and their potential impacts on the equipment. For example, delta to grounded-wye transformers prevent triplen harmonic currents from flowing back into the grid. There can be no triplen harmonics in a delta. The grounded wye also provides a path to ground. Essentially, what we try to do is provide a low-impedance path to ground for the harmonics so that they flow out, instead of into the rest of the network. We can reduce THD by modifying the system frequency response by filters, inductors, and/or capacitors. We can also reduce the harmonic currents drawn by the load.

Because of the wide-ranging negative effects of harmonics in power systems, understanding the impact of power electronics converters on harmonics is essential and must be considered during converter design and testing.

3.4.3 Inverters

An inverter is a circuit that converts DC input voltage to AC output voltage, with a desired magnitude and frequency.

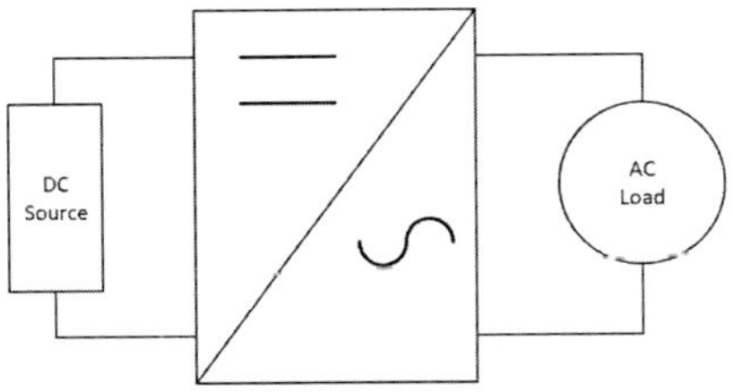

Figure 3.34: DC to AC conversion: Inverter operation

There are many common applications of inverters. Some of these include (1) connecting uninterruptable power supplies (UPS) to a building, (2) adjustable speed AC motor drives, (3) running AC appliances from an automotive electrical outlet, (4) connecting photovoltaic solar panels to the grid or to the house, and (5) connecting electric vehicle batters to the grid. Essentially, inverters appear anywhere you need to convert a DC power source to an AC power load (or to the AC grid).

First, to generate a conceptual understanding of inverter operation, let us do a thought experiement. Think about the inverter as having two DC rails at the input, $\frac{V_{DC}}{2}$ and $\frac{-V_{DC}}{2}$. Then, the inverter has a control mechanism and its electronic switches, used to switch the output between the two rails (i.e., between $\frac{V_{DC}}{2}$ and $\frac{-V_{DC}}{2}$). This is done at a fast rate, and in such a way as to produce an AC-like output: see the example with S_1 and S_2 switching signals and output voltage V_0 in Figure 3.36 below. The output in this example is actually a square wave, with mean value 0, where its value switches between $\frac{V_{DC}}{2}$ and $\frac{-V_{DC}}{2}$.

Half-Bridge Inverter

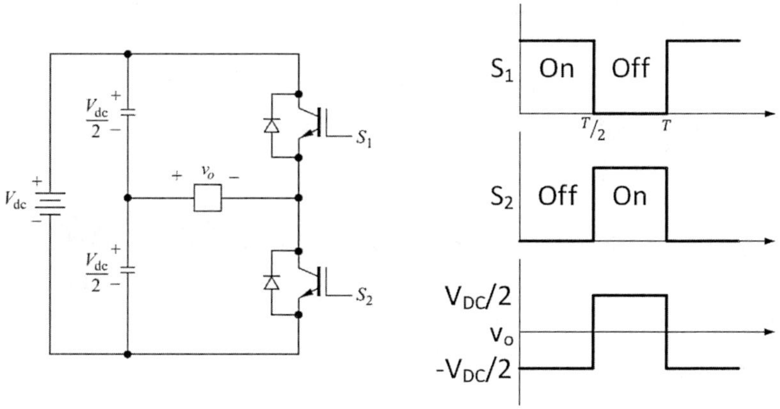

Figure 3.35: Half-bridge inverter circuit with switching signals

The example in Figure 3.36 for the thought experiment is called a half-bridge inverter, with two switches, S_1 and S_2. It is important to note that S_1 and S_2 cannot be on at the same time; if they were, it would create a short across the voltage source. Feedback diodes are required to provide continuity of current required of an inductive load.

Full-Bridge Inverter

The concept of operation is the same for a full-bridge inverter. Now there are four switches. S_1 and S_4 cannot be on at the same time. S_2 and S_3 cannot be on at the same time. Four switch combinations are possible.

The output V_0 looks more sinusoidal than for the half-bridge inverter. It still switches between the $\frac{V_{DC}}{2}$ and $\frac{-V_{DC}}{2}$ rails, but it now also has an

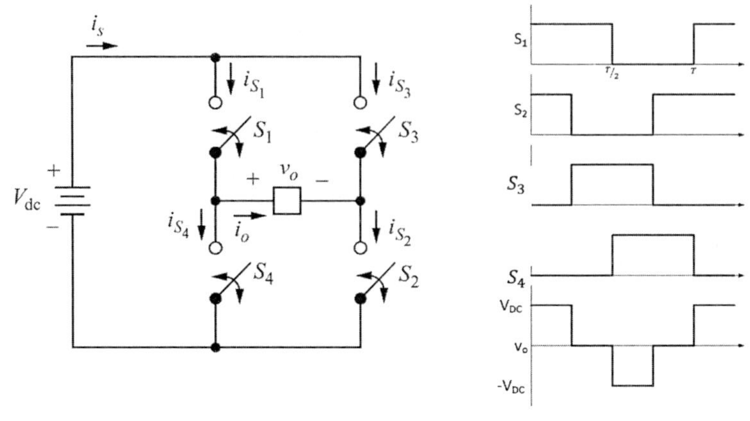

Figure 3.36: Full-bridge inverter circuit and switching signals

intermediate output position at 0. In an average sense, we can see that the output is starting to look more sinusoidal.

The power electronics circuit design element appears here in several places to achieve a sinusoidal output, including the design of the controller that operates the switches and the design of output filters (inductors and capacitors) to help produce a smooth AC waveform. In reality, real switches cannot turn on and off instantaneously. They are also not ideal and thus have a little bit of voltage drop. While overlap of switch 'on' times should be avoided, in reality, it can occur, and it is called *shoot through faults.* This creates a short circuit across the DC voltage short. To avoid this, the real switching transition times need to be considered in the controller design. This is called blanking time or dead time, and it is a time delay between switching actions to prevent shoot-through faults.

Full-Bridge Inverter with Feedback Diodes

Switches must carry both positive and negative currents for RL loads, and adding feedback diodes to the design supports this, while not adversely affecting the inverter's function. Feedback diodes are placed in parallel with each switch (Fig. 3.37). During the time when current in switches is negative, feedback diodes carry current.

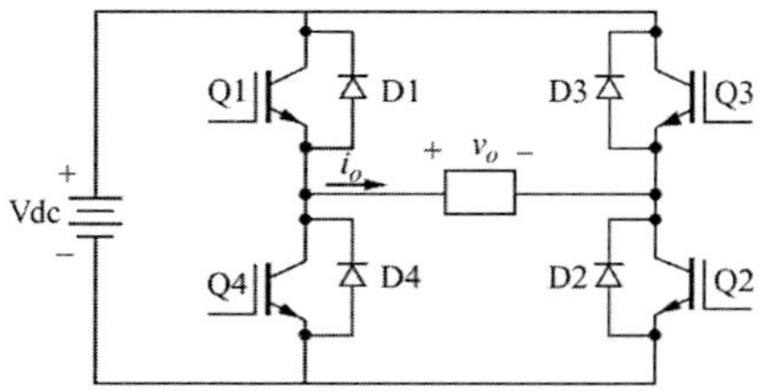

Figure 3.37: Full-bridge inverter circuit with feedback diodes shown

Multi-level Inverters

We can see from our discussion of half-bridge and full-bridge inverters that adding more switches in a strategic way can help achieve our goal of making the output more sinusoidal. Multi-level inverters do this further: they produce a more sinusoidal output voltage, and this of course also reduces the THD. These are a good solution for connecting renewable energy sources (e.g., connecting PV) to the grid. In this example, we increase the number of positions by stacking two inverters, fed by two DC sources; we can then have switching positions for our output at $2V_{DC}$, V_{DC}, 0 $-V_{DC}$, $-2V_{DC}$. The key observation is that the output is more sinusoidal.

Inverter with Pulse Width Modulation (PWM)

The idea is to vary timing of switches to modulate pulse width (duty cycle). With higher duty cycle, average output voltage is larger, as shown in Fig. 3.38, where the dotted line is the average value. This further helps produce a sinusoidal output.

The way PWM works is that it controls switching by comparing a sinusoidal reference signal and a triangular carrier signal. There are different ways to do this, and different chips exist can be added to a power electronics circuit to provide the desired functionality. An example is bipolar switching. This causes the output voltage V_0 to alternate between high and low rails (between V_{DC} and $-V_{DC}$), as shown in Fig. 3.39. The logic is as follows, considering the triangle wave as the carrier wave and the sine wave as the reference:

$$V_0 = \begin{cases} +V_{DC}, & \text{if } V_{sin} > V_{tri} \text{ turn on } S_1 \text{ and } S_2 \\ -V_{DC}, & \text{if } V_{sin} < V_{tri} \text{ turn on } S_3 \text{ and } S_4 \end{cases}$$

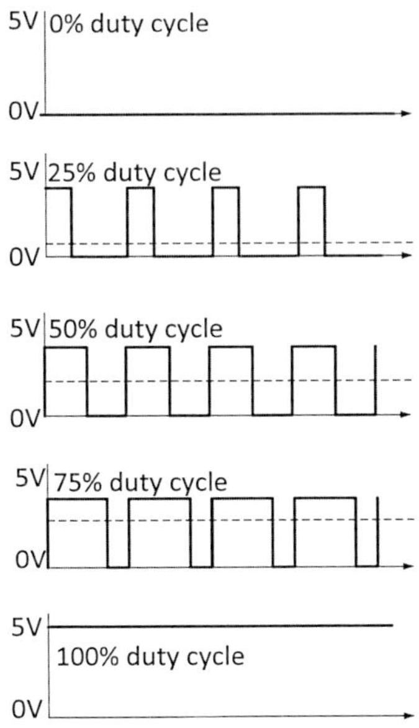

Figure 3.38: PWM example

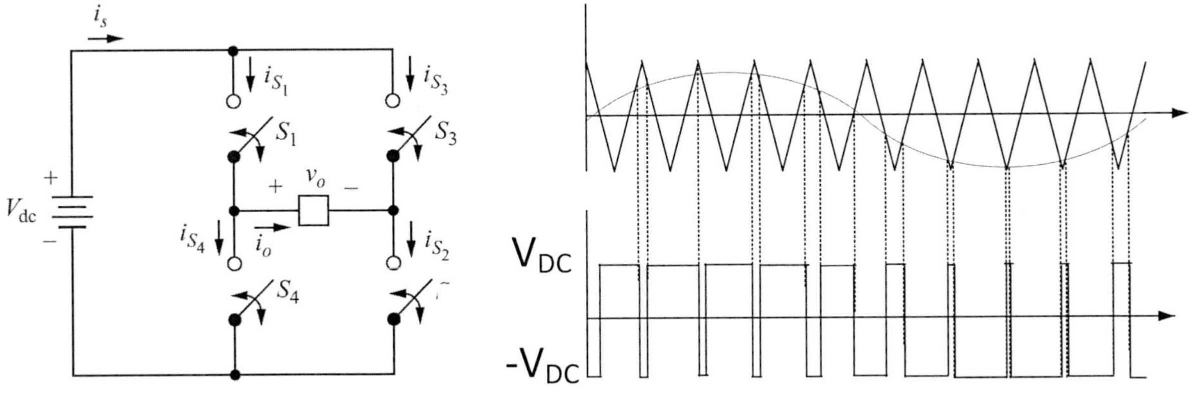

Figure 3.39: Inverter example with PWM operation for bipolar switching

For a three-phase inverter with PWM, the PWM uses three-phase voltage signals (with the 120 degree angle shifts) as the reference, and compares the voltage signals with a triangle carrier signal to control the switches to achieve an output that is much more sinusoidal.

3.5 Problems

3.1 Suppose that we have a PV module running at its maximum power point with V_m=40 V and I_m=5 A. We wish to deliver power to a 12-volt battery to charge it. We want to use a DC-DC converter to accomplish this, while extracting maximum power from the source.

(a) Propose a converter topology and specify the reasons for that.

(b) Solve for the duty cycle (D).

3.2 Consider the converter in Figure 3.40 below. You may assume that the inductor is large enough that the current through it can be considered constant.

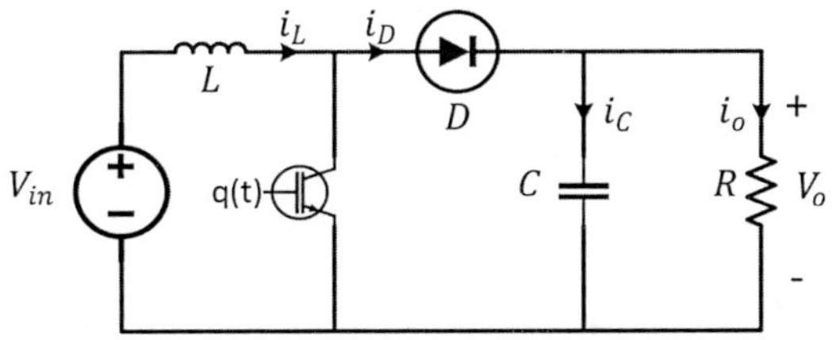

Figure 3.40: Converter topology for Problem 3.2

(a) Draw the circuit when the active switch is on and off.

(b) Draw the circuit when the active switch is off.

(c) Based on your answer to (a) and (b), derive an equation in terms of D that relates the input voltage V_{in} to the output voltage V_{out}.

(d) What does this circuit do?

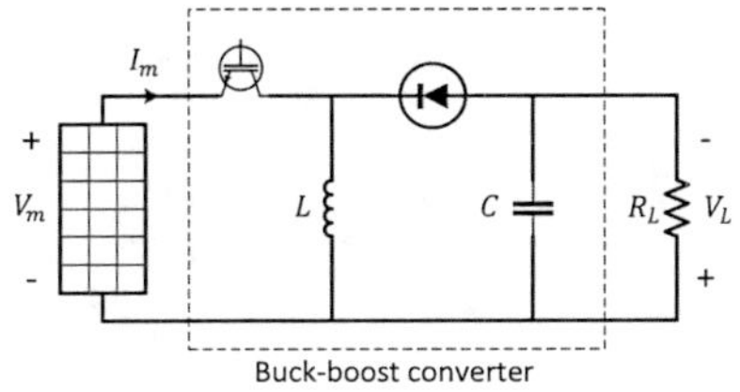

Figure 3.41: Circuit for problem 3.3

3.3 A PV module is connected to a 70 resistive load through a maximum power point tracker (MPPT), which consists of a buck-boost converter.

The PV has its maximum power point at $V_m = 120V$ and $I_m = 10A$. Suppose that we want to add a power electronics converter such that the power to the load is equal to the maximum power. Hence, we want $P_m = (V_L^2)/R_L$.

(a) What is the duty cycle D needed for the buck-boost converter to transfer the maximum power to the load?

(b) In reality the converter operates with some efficiency η less than 100% (although efficiencies can be quite high), so $\eta \cdot P_m = V_L^2/R_L$. If the MPPT has an efficiency of 94%, what output voltage would be needed at the converter?

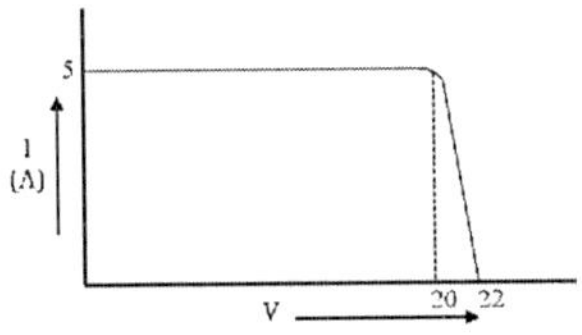

Figure 3.42: IV curve for PV module

3.4 Figure 3.42 shows the IV curve for a PV panel at noon.

(a) What is the load resistance R that would result in maximum power delivered by the panel? Draw what its load line looks like on top of the PV panel's IV curve.

(b) How much power is that in W?

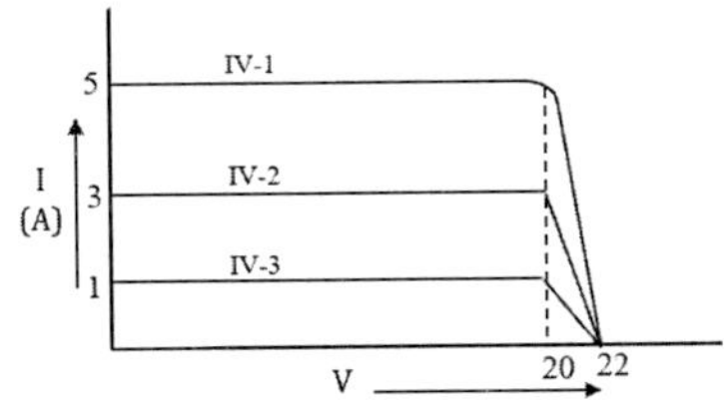

Figure 3.43: Three IV curves of the PV module for different levels of insolation throughout the day

3.5 As shown in Figure 3.43, the amount of insolation (solar irradiation) changes throughout the day, causing the IV curves to shift. With less insolation, the current from the PV panel drops. As the PV output changes, suppose it has three IV curves labeled IV-1, IV-2, and IV-3. The problem is that for a fixed load, power will be lost as the energy from the sun changes. The new operating points can be found from intersections of the load line with the PV panel's IV curve.

(a) Draw and label the new operating points from intersections of the load line, with each of the PV panel's IV curves, assuming the resistance found n Problem 3.3 (a).

(b) Now, there are 24 hours in the day. Suppose that the panel operates at IV-3 for 12 hours, IV-2 for 9 hours, and IV-1 for 3 hours. Calculate the amount of energy that would be delivered to the load in one day.

(c) If the module has an MPPT, what type of convert topology would you use, and how would it adjust the voltage and current to the load? Draw this on the curves in Figure 3.43.

3.6 If you use the MPPT for the scenario in Problem 3.5, what is the amount and the percentage improvement you will get (in terms of energy delivered to the load in one day), compared to if it was always directly connected to the resistance found in Problem 3.5 (a)? As in Problem 3.5, suppose the panel operates at IV-3 for 12 hours, IV-2 for 9 hours, and IV-1 for 3 hours.

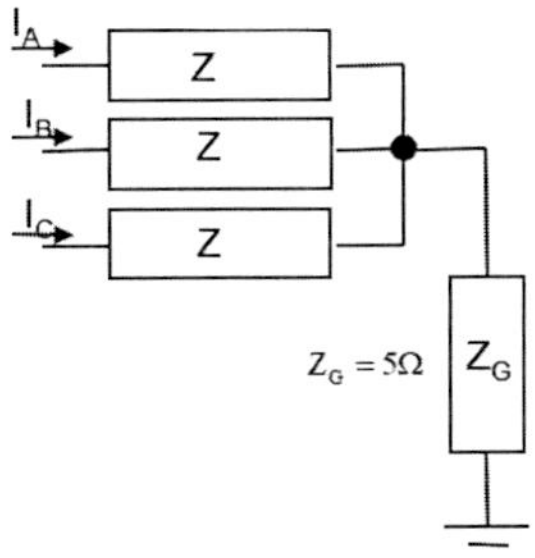

Figure 3.44: Grounded-wye connected circuit for 3.7 and 3.8

3.7 For the system in Fig. 3.44, with the three-phase currents given below, find the real power losses in Z_G.

$$i_A = 5\sqrt{2}cos(\omega t + 0°) + 4\sqrt{2}cos2(\omega t + 0°)$$

$$i_B = 5\sqrt{2}cos(\omega t - 120°) + 4\sqrt{2}cos2(\omega t - 120°)$$
$$i_C = 5\sqrt{2}cos(\omega t + 120°) + 4\sqrt{2}cos2(\omega t + 120°)$$

3.8 For the system in Fig. 3.44, with the three-phase currents given below, find the real power losses in Z_G.

$$i_A = 5\sqrt{2}cos(\omega t + 0°) + 4\sqrt{2}cos3(\omega t + 0°) + 3\sqrt{2}cos9(\omega t + 0°)$$
$$i_B = 5\sqrt{2}cos(\omega t - 120°) + 4\sqrt{2}cos3(\omega t - 120°) + 3\sqrt{2}cos9(\omega t - 120°)$$
$$i_C = 5\sqrt{2}cos(\omega t + 120°) + 4\sqrt{2}cos3(\omega t + 120°) + 3\sqrt{2}cos9(\omega t + 120°)$$

Chapter 4

MAGNETIC CIRCUITS, MUTUAL INDUCTANCE, AND TRANSFORMERS

4.1 Introduction

Electromagnetic theory is the basis for explaining the operation of all electrical and electromechanical systems. There are two types of fields, magnetic field systems and electric field systems. Since the bulk of devices fall in the magnetic category, we will restrict our discussion to magnetic field systems. A discussion of Maxwell's equations and the resulting quasi-static approximation useful in analyzing electromechanical systems is contained in Reference [4]. These laws can be in either differential or integral form. From the point of view of applications, the integral form is more useful. The integral forms of equations are

$$\begin{aligned}
\oint_c \underline{H} \bullet d\underline{\ell} &= \int_s \underline{J}_f \bullet \underline{n}\, da && \text{Ampere's law} \\
\oint_c \underline{E} \bullet d\underline{\ell} &= -\int_s \frac{d\underline{B}}{dt} \bullet \underline{n}\, da && \text{Faraday's law} \\
\oint_s \underline{J}_f \bullet \underline{n}\, da &= 0 && \text{Conservation of charge} \\
\oint_s \underline{B} \bullet \underline{n}\, da &= 0 && \text{Gauss's law}
\end{aligned}$$

We will assume that $\underline{B}$ is a linear function of $\underline{H}$. In the last two expressions, s represents a *closed* surface. In the first two expressions, s represents the surface defined by the contour c.

A large class of electromagnetic systems are static devices such as inductors, transformers, or simple magnetic circuits. We discuss these in this chapter. Only Ampere's law and Gauss's law are needed to compute the flux linkages of the electrical system. Faraday's law is then used to compute the voltage across the electrical coils.

4.2 Static Magnetic Circuits

In static magnetic circuits, there are no moving members. The most important device in this category is the transformer. It also gives us an insight into the more useful dynamic magnetic circuits in the next chapter where mechanical motion is involved. We use Ampere's current law (ACL) and Gauss's law (GL) for magnetic fields to derive useful flux-current terminal relations. The analysis is helpful in the design of inductors, and study of transformers, and provides a useful conceptual basis for analyzing electromechanical systems.

4.2.1 Toroid

Consider the toroid in Figure 4.1 wound uniformly with N turns. Figure 4.2

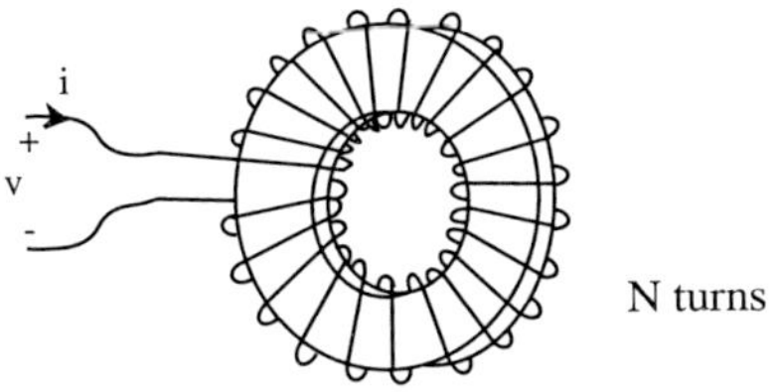

Figure 4.1: Toroid

represents a vertical cross section. The inner and outer radii are r_0 and r_1,

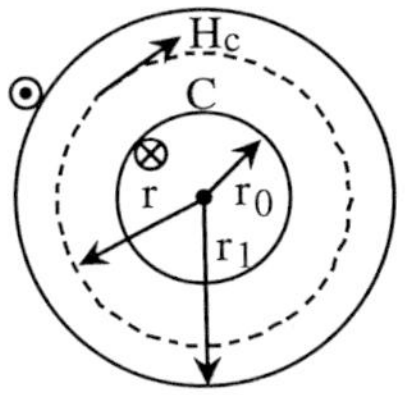

Figure 4.2: Cross section of the toroid

respectively. Choose a mean radius inside the toroid $r = \frac{r_0+r_1}{2}$. Magnetic field intensity is assumed to be uniform inside the core. We choose the contour corresponding to the mean radius as shown in Figure 4.2. The magnetic field intensity H_c is the same around the contour C and is tangential to the contour. Using Ampere's circuit law (ACL) around the contour, we have $\oint_c \underline{H} \bullet d\underline{\ell} = \int_s \underline{J}_f \bullet \underline{n}\, da$ resulting in $H_c(2\pi r) = Ni$. Ni is the magnetomotive force (mmf), with units of ampere-turns. Therefore,

$$H_c = \frac{Ni}{2\pi r} \text{ (ampere-turns/meter)} \tag{4.1}$$

Let $2\pi r = \ell_c$ (mean length in core). Therefore, (4.1) becomes

$$H_c \ell_c \;=\; Ni \tag{4.2}$$

Assuming a linear relationship between B and H in the core, the flux density in the core is

$$B_c = \mu H_c = \mu \frac{Ni}{\ell_c} \tag{4.3}$$

where μ is the magnetic permeability of the material. If the cross-sectional area of the core is A_c, then the flux ϕ_c is given by $\phi_c = B_c A_c = \frac{\mu Ni}{\ell_c} A_c = \frac{Ni}{\ell_c/\mu A_c}$. ϕ_c has units of webers (Wb) and B_c has (Wb)/m^2.

Define $Ni \triangleq$ mmf. Then, we define *reluctance* $\mathcal{R}$ as

$$\frac{Ni}{\phi_c} = \frac{\text{mmf}}{\text{flux}} = \frac{\ell_c}{\mu A_c} \triangleq \mathcal{R} \tag{4.4}$$

$\mathcal{R}$ has units of ampere-turns/weber. $\frac{1}{\mathcal{R}} \triangleq \mathcal{P}$ is called *permeance*, which has Wb/ampere-turn. In terms of permeance $\mathcal{P}$, we can express $\phi_c = \mathcal{P}Ni$. Now the flux linkage of the coil is defined as $\lambda = N\phi_c = \mathcal{P}N^2 i$. By definition, $\frac{\lambda}{i} \triangleq L = \mathcal{P}N^2 = \frac{N^2}{\mathcal{R}}$. Thus, self inductance L of a coil is given by $N^2\mathcal{P} = \frac{N^2}{\mathcal{R}}$ and has units of Henries. For the simple toroid, the magnetic equivalent circuit defined by (4.4) is given in Figure 4.3. It has similarities to a resistive circuit, with the following equivalences.

$$\begin{aligned} \text{mmf} &\Rightarrow \text{voltage} \\ \text{flux} &\Rightarrow \text{current} \\ \text{reluctance} &\Rightarrow \text{resistance} \\ \text{permeance} &\Rightarrow \text{conductance} \end{aligned}$$

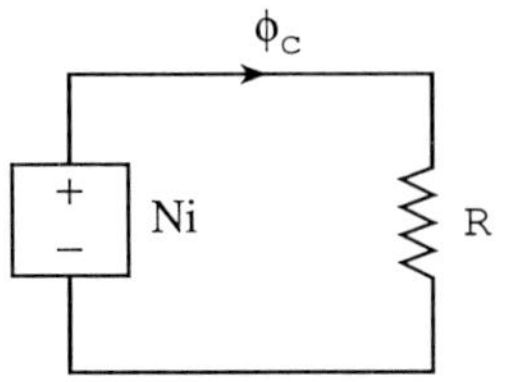

Figure 4.3: Equivalent magnetic circuit

4.2.2 Toroid with air gap (no fringing)

Consider again the toroid with an air gap whose vertical cross section is shown in Figure 4.4. Now there is magnetic field intensity H in both the air

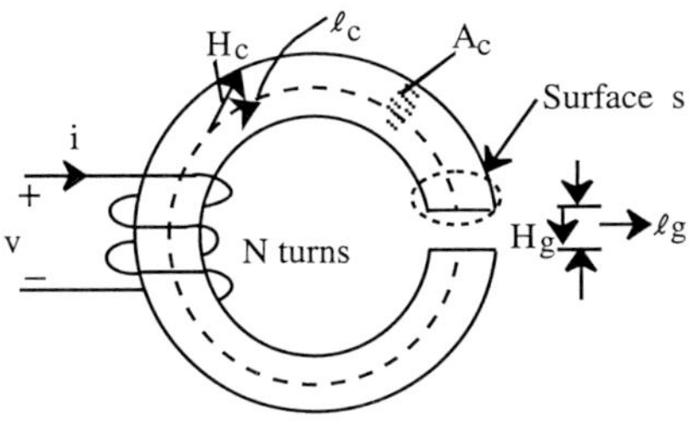

Figure 4.4: Toroid with an air gap

gap and the iron portion as shown. The length of the air gap is ℓ_g, and the mean length in the iron portion is ℓ_c. Applying ACL around the contour c

$$\oint_c \underline{H} \bullet d\underline{\ell} \quad = \quad \int_s \underline{J}_f \bullet \underline{n}\ da \tag{4.5}$$

i.e.,

$$Ni = H_g \ell_g + H_c \ell_c$$

In general, assuming linearity in the core,

$$\begin{aligned} B &= \mu H \text{ where } \mu \triangleq \text{permeability of the material} \\ \mu &= \mu_o \mu_r \text{ where } \mu_r \text{ is relative permeability of the material} \\ \mu &= \mu_o = 4\pi \times 10^{-7}\text{H/m for air} \end{aligned}$$

Equation (4.5) now becomes

$$Ni = H_g \ell_g + H_c \ell_c = \frac{B_g}{\mu_o}\ell_g + \frac{B_c}{\mu}\ell_c \tag{4.6}$$

where B_g is flux density in the gap and B_c is the flux density in the core. Also $\mu \stackrel{\Delta}{=} \mu_r\mu_o$ where μ_r is called the *relative permeability.* For ferromagnetic materials, μ_r is of the order of $10^3 - 10^5$. Therefore (4.6) becomes

$$Ni = \underbrace{\frac{B_g\ell_g}{\mu_o}}_{\text{Air gap}} + \underbrace{\frac{B_c\ell_c}{\mu_o\mu_r}}_{\text{Iron}} \tag{4.7}$$

For high μ_r and hence, higher μ, it is clear that the air gap accounts for most of the ampere-turns (mmf).

Gauss's law indicates that flux must be conserved. Hence, application of Gauss's law on the closed surface s as shown in Figure 4.4 establishes the continuity of flux ϕ, namely, $\oint_s \underline{B} \bullet \underline{n}\ da = 0$, and we get $B_gA_g = B_cA_c$. If there is no fringing of the flux, then $A_g = A_c$. Therefore, $B_g = B_c$. Hence, flux $\phi = B_gA_g = B_cA_c$. Dividing (4.7) by the flux $\phi = B_gA_g = B_cA_c$, we get

$$\frac{Ni}{\phi} = \frac{\ell_g}{\mu_oA_g} + \frac{\ell_c}{\mu A_c} \tag{4.8}$$

As in the previous section, we can define a quantity called *reluctance* $\mathcal{R} = \ell/\mu A$. Thus, from (4.8)

$$\frac{Ni}{\phi} = \mathcal{R}_g + \mathcal{R}_c \tag{4.9}$$

where $\mathcal{R}_g$ and $\mathcal{R}_c$ are reluctances of the air gap and the core, respectively. From (4.9),

$$\phi = \frac{Ni}{\mathcal{R}_g + \mathcal{R}_c} \tag{4.10}$$

Therefore, (4.10) can be represented by the magnetic equivalent circuit of Figure 4.5. The reluctances $\mathcal{R}_g$ and $\mathcal{R}_c$ are in series.

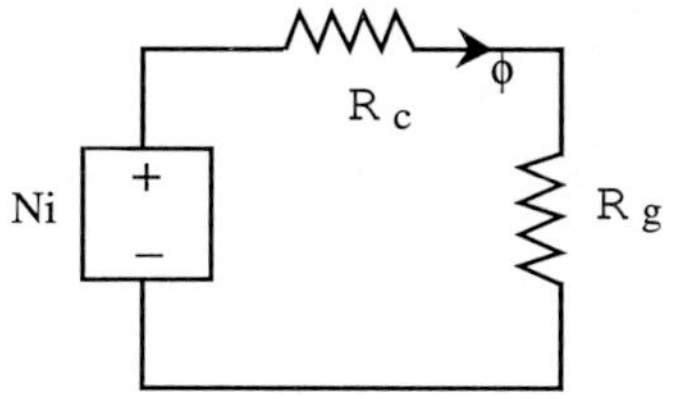

Figure 4.5: Equivalent magnetic circuit

We have assumed so far that there is no fringing of the flux. Suppose there is "fringing," i.e., not all the flux is confined to the area between the two faces of iron portion, as in Figure 4.6. With fringing present, the flux bulges at the air gap, such that there is also flux on the sides of the air gap

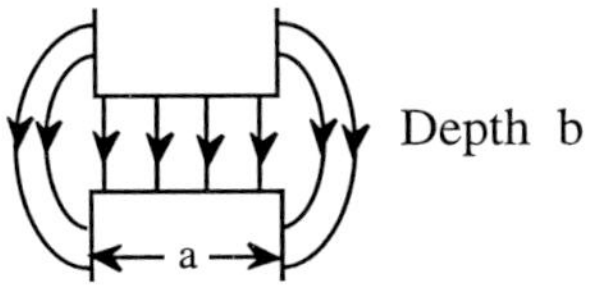

Figure 4.6: Fringing

as shown. This makes the effective area of the air gap larger.

In the presence of fringing, $A_g > A_c$, i.e., the effective air gap area A_g increases to be larger than the area of the core A_c. We account for this empirically as follows. Add the length of the air gap ℓ_g to each of the two dimensions to compute the area; i.e., if the core has width a and depth b, then

$$A_c = ab, A_g = (a + \ell_g)(b + \ell_g) \tag{4.11}$$

Sometimes A_g is given as a percentage times A_c, e.g., $A_g = 1.1A_c$. (4.9) is still applicable.

Example 4.1 In Figure 4.4, find the mmf required to produce a flux density of 0.5 Wb/m^2 in the air gap $\ell_g = 0.1$ cm, $A_c = 1 \text{ cm}^2$, $\ell_c = 6$ cm, $\mu_r = 10^4$. Assume $A_g = 1.1A_c$.

Solution:

$$\begin{aligned}
\mathcal{R}_c &= \frac{0.06}{(10^4)(4\pi \times 10^{-7})(10^{-4})} = 47.7 \times 10^3 \text{ At/Wb} \\
\mathcal{R}_g &= \frac{.001}{(4\pi \times 10^{-7})(1.1 \times 10^{-4})} = 7.23 \times 10^6 \text{ At/Wb} \\
\phi &= B_g A_g = (0.5)(1.1 \times 10^{-4}) = 5.5 \times 10^{-5} \text{ Wb}
\end{aligned}$$

Applying (4.9),

$$Ni = (\mathcal{R}_c + \mathcal{R}_g)\phi = 400 \text{ At}$$

Example 4.2 In Figure 4.7, each air gap length is $g = 0.1$ cm and cross-sectional area $A_g = 4 \text{ cm}^2$. $\mu = \infty$ in iron. Find the flux through the coils and their direction. Ignore fringing.

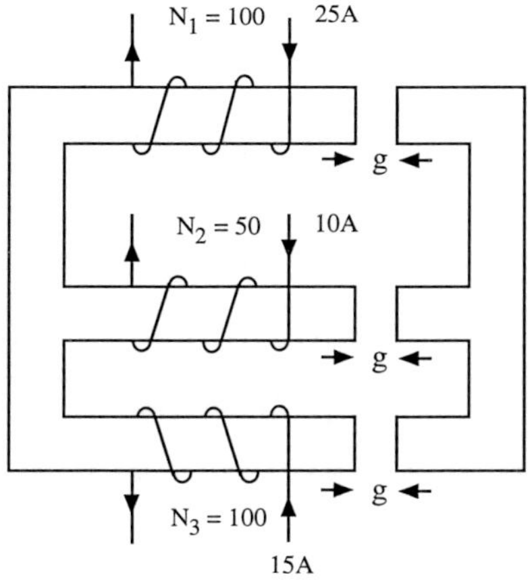

Figure 4.7: Magnetic circuit

Solution: Since all air gaps are equal in length,

$$\mathcal{R}_1 = \mathcal{R}_2 = \mathcal{R}_3 = \mathcal{R} = \frac{(0.1)10^{-2}}{4\pi \times 10^{-7} \times 4 \times 10^{-4}} \approx 2 \times 10^6 At/Wb$$

The equivalent circuit is given in Figure 4.8. Assign positive directions for

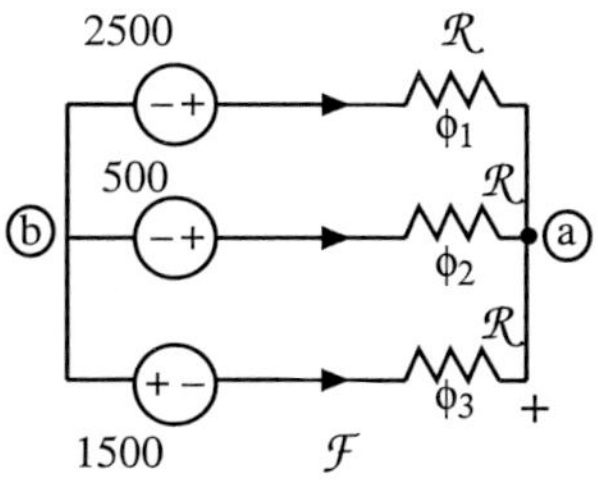

Figure 4.8: Equivalent magnetic circuit

ϕ_1, ϕ_2, and ϕ_3 as shown. The algebraic sum of the fluxes at magnetic node (a) is equal to zero.

$$\phi_1 + \phi_2 + \phi_3 = 0$$

Let the mmf of node (a) with respect to (b) be $\mathcal{F}$. Then

$$\frac{2500 - \mathcal{F}}{\mathcal{R}} + \frac{500 - \mathcal{F}}{\mathcal{R}} - \frac{(\mathcal{F} + 1500)}{\mathcal{R}} = 0$$

Solving the equation, we get $\mathcal{F} = 500At$. Therefore,

$$\phi_1 = \frac{2500 - \mathcal{F}}{\mathcal{R}} = 10^{-3} \text{ Wb}$$

Similarly,

$$\phi_2 = 0, \phi_3 = -10^{-3} \text{ Wb}$$

Example 4.3 In Figure 4.9, the closed magnetic path lengths from 1 to 2

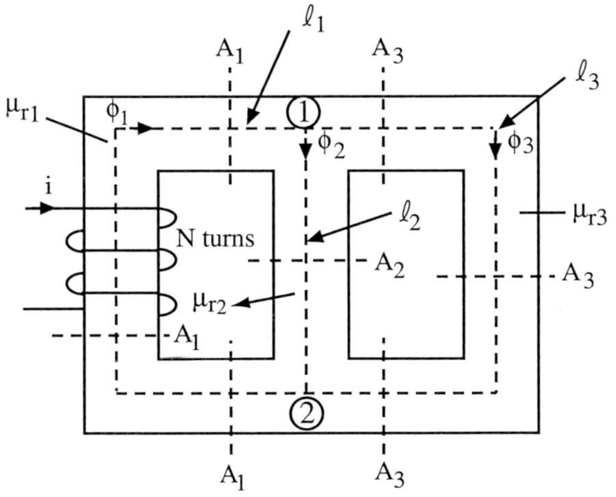

Figure 4.9: Magnetic circuit

are

$$\ell_1 = \ell_3 = 30 \text{ cm } \ell_2 = 10 \text{ cm}$$

Cross-sectional areas are

$$A_1 = A_3 = 2 \text{ cm}^2, A_2 = 4 \text{ cm}^2$$

The relative permeabilities are

$$\mu_{r1} = \mu_{r3} = 2250, \mu_{r2} = 1350$$

For $N = 25$ turns and $i = 0.5$ A find the flux densities B_1, B_2, and B_3.

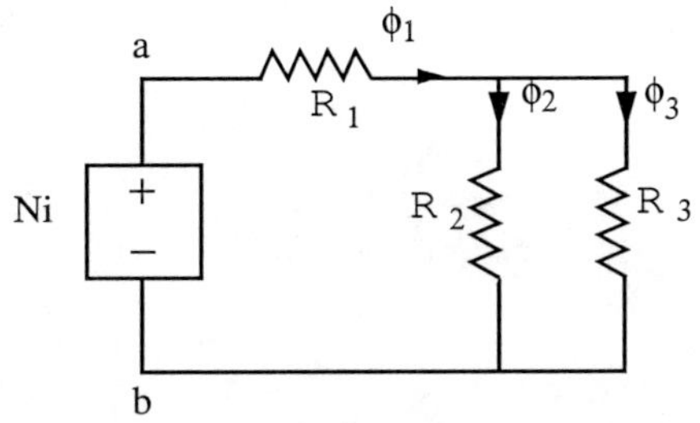

Figure 4.10: Equivalent magnetic circuit

Solution: The magnetic equivalent circuit is drawn in Figure 4.10.

$$\begin{aligned}
\mathcal{R}_1 &= \mathcal{R}_3 = \frac{30 \times 10^{-2}}{4\pi \times 10^{-7} \times 2250 \times 2 \times 10^{-4}} = 0.531 \times 10^6 At/Wb \\
\mathcal{R}_2 &= 0.148 \times 10^6 At/Wb \text{ (verify)} \\
\mathcal{R}_2 \| \mathcal{R}_3 &= 0.116 \times 10^6 At/Wb \text{ (verify)}
\end{aligned}$$

Total reluctance at $a - b = \mathcal{R}_1 + (\mathcal{R}_2 \| \mathcal{R}_3) = 0.647 \times 10^6 At/Wb$

$$\begin{aligned}
\phi_1 &= \frac{Ni}{0.647 \times 10^6} = \frac{25 \times 0.5}{0.647 \times 10^6} = 19.32 \times 10^{-6} \text{ Wb} \\
B_1 &= \frac{\phi_1}{A_1} = 0.0966 \text{ Wb/m}^2 \\
\phi_2 &= \frac{\phi_1 \mathcal{R}_3}{\mathcal{R}_2 + \mathcal{R}_3} = 15.11 \times 10^{-6} \text{ Wb} \\
B_2 &= \frac{\phi_2}{A_2} = 0.0377 \text{ Wb/m}^2 \\
\phi_3 &= \phi_1 - \phi_2 = 4.21 \times 10^{-6} \text{ Wb} \\
B_3 &= \frac{\phi_3}{A_3} = 0.021 \text{ Wb/m}^2
\end{aligned}$$

4.2.3 Flux linkage and inductance

Suppose that a coil of N turns has a flux ϕ passing through each of those turns. Then the total flux linked by the N turns of the coil is $\lambda = N\phi$. This total flux is commonly referred to as *flux linkage.* In Example 3.3, the flux linkage of the coil is $N\phi_1 = (25 \times 19.32) = 483 \times 10^{-6}$ *weber turns.* In a linear inductor, the flux linkage is directly proportional to the current.

Hence, $\lambda = Li$ where L is called the self inductance. In Example 3.3,

$$\begin{aligned} L(0.5) &= (25)(19.32 \times 10^{-6}) \\ L &= 50 \times 19.32 \times 10^{-6} \text{ H} \\ &= 0.966 \text{ mH} \end{aligned}$$

4.3 Mutual Inductance

Mutual inductance is the circuit parameter that relates the voltage induced in one coil due to a time varying current in another coil. It forms the basis for the study of transformers. Consider Figure 4.11 where there are two

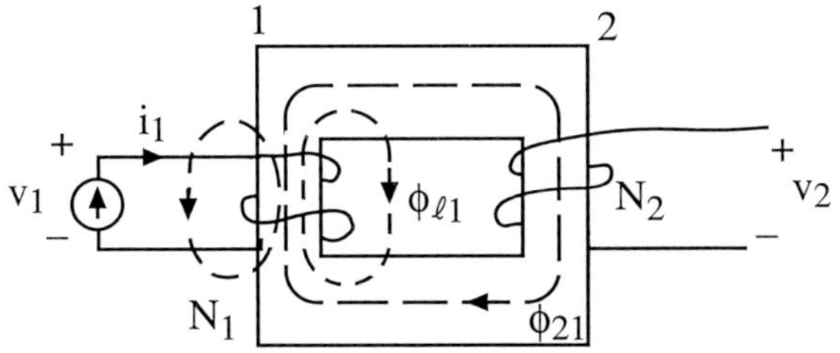

Figure 4.11: Two mutually coupled coils

coils wound on the same magnetic core. Coil 1 is excited while coil 2 is open circuited. The total flux linking coil 1 is

$$\phi_{11} = \phi_{\ell 1} + \phi_{21}$$

$\phi_{\ell 1}$ (called the *leakage flux*) is linking coil 1 only; whereas, ϕ_{21} is the mutual flux linking both the coils. It is also the flux linking coil 2 due to current in coil 1. The order of subscripts is to be noted.

Since coil 2 is open circuited, no current flows in the coil. The flux linkage of this coil is

$$\lambda_2 = N_2\phi_{21}$$

The voltage v_2 is the rate of change of flux linkage and is given by

$$v_2 = \frac{d\lambda_2}{dt}$$

(We shall justify the polarity as shown later.) The flux ϕ_{21} is linearly proportional to the current i_1. Hence, we may write

$$\lambda_2 = N_2\phi_{21} = M_{21}i_1 \tag{4.12}$$

Thus,

$$v_2 = \frac{d\lambda_2}{dt} = M_{21}\frac{di_1}{dt}$$

M_{21} is called the *mutual inductance* between the coils. The voltage in coil 1 is

$$v_1 = \frac{d\lambda_1}{dt}$$

where λ_1 is the flux linkage given by $\lambda_1 = N_1\phi_{11}$. As in the case of the linear inductor, ϕ_{11} is also proportional to i_1. Hence,

$$N_1\phi_{11} = L_1 i_1 \tag{4.13}$$

where L_1 is called the self inductance. Hence,

$$v_1 = \frac{d\lambda_1}{dt} = L_1\frac{di_1}{dt}$$

Now consider the case where coil 2 is excited and coil 1 is open circuited, as shown in Figure 4.12. Now ϕ_{12} is the flux linking coil 1 due to current in

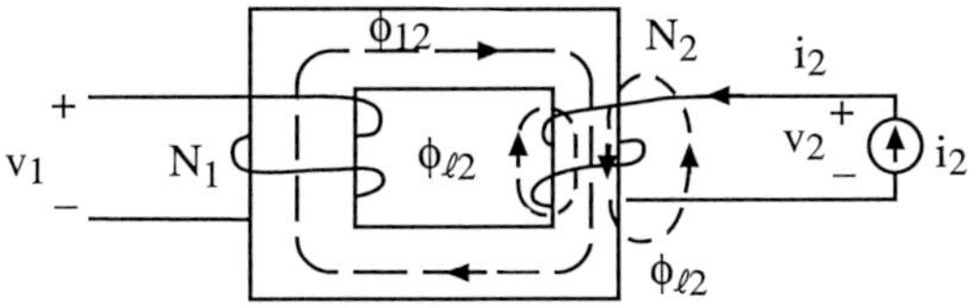

Figure 4.12: Magnetic equivalent circuit of Figure 3.11

coil 2. Now $\phi_{22} = \phi_{\ell 2} + \phi_{12}$. Flux linkage of coil 1 is

$$\lambda_1 = N_1\phi_{12} = M_{12}i_2 \tag{4.14}$$

since ϕ_{12} is proportional to i_2. Voltage across the coil 1 is

$$v_1 = \frac{d\lambda_1}{dt} = N_1\frac{d\phi_{12}}{dt} = M_{12}\frac{di_2}{dt}$$

It can be shown from energy considerations that $M_{12} = M_{21} = M$. We refer to M as the mutual inductance. The voltage across coil 2 is given by

$$v_2 = \frac{d\lambda_2}{dt}$$

where

$$\lambda_2 = N_2\phi_{22} \tag{4.15}$$

Defining

$$N_2\phi_{22} = L_2 i_2$$

we have

$$v_2 = L_2\frac{di_2}{dt}$$

where L_2 is the self inductance of coil 2.

Now consider the general case where both i_1 and i_2 are non-zero. The fluxes linking coils 1 and 2 are as shown in Figure 4.13.

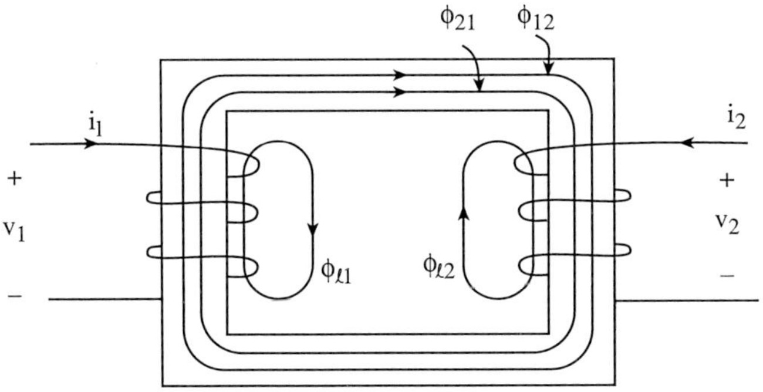

Figure 4.13: Two mutually coupled coils

$$\begin{aligned}\phi_1 &= \phi_{\ell 1} + \phi_{21} + \phi_{12} = \phi_{11} + \phi_{12}\\ \phi_2 &= \phi_{21} + \phi_{\ell 2} + \phi_{12} = \phi_{21} + \phi_{22}\end{aligned}$$

The flux linkages are, therefore,

$$\begin{aligned}\lambda_1 &= N_1\phi_1 = N_1\phi_{11} + N_1\phi_{12}\\ \lambda_2 &= N_2\phi_2 = N_2\phi_{21} + N_2\phi_{22}\end{aligned} \tag{4.16}$$

Substituting (4.12–4.15) into (4.16) and noting $M_{12} = M_{21} = M$, we get the flux linkages as

$$\begin{aligned}\lambda_1 &= L_1 i_1 + M i_2\\ \lambda_2 &= M i_1 + L_2 i_2\end{aligned} \tag{4.17}$$

Differentiating (4.17), we get

$$\begin{aligned} v_1 &= L_1\frac{di_1}{dt} + M\frac{di_2}{dt} \\ v_2 &= M\frac{di_1}{dt} + L_2\frac{di_2}{dt} \end{aligned} \tag{4.18}$$

The coefficient of coupling between the two coils indicates the amount of coupling and is defined by

$$k = \frac{M}{\sqrt{L_1 L_2}} \tag{4.19}$$

Clearly, $k = 0$ if no coupling exists between the coils and, hence, $M = 0$. We can find the upper limit of k by substituting for M_{21}, M_{12}, L_1, and L_2 from (4.12)–(4.15) in (4.19) as (noting $M = \sqrt{M_{21}M_{12}}$)

$$\begin{aligned} k &= \left(\sqrt{\frac{M_{21}}{L_1}}\right)\left(\sqrt{\frac{M_{12}}{L_2}}\right) \\ &= \sqrt{\frac{\phi_{21}}{\phi_{11}}}\sqrt{\frac{\phi_{12}}{\phi_{22}}} \end{aligned} \tag{4.20}$$

Since

$$\phi_{11} = \phi_{\ell 1} + \phi_{21} \text{ and } \phi_{22} = \phi_{\ell 2} + \phi_{12}$$

we have $\phi_{11} \geq \phi_{21}$ and $\phi_{22} \geq \phi_{12}$. Thus from (4.20), we must have

$$0 \leq k \leq 1 \tag{4.21}$$

or equivalently

$$0 \leq M \leq \sqrt{L_1 L_2} \tag{4.22}$$

$k = 1$ indicates perfect coupling and $M = \sqrt{L_1 L_2}$. The value of k (and hence, M) depends on the physical parameters, i.e., number of turns and dimensions of the coil. Coils are loosely coupled if $k < 0.5$ and tightly coupled if $k > 0.5$. Most air core transformers are loosely coupled; whereas, iron core transformers are tightly coupled, in which case, k can approach 1.

Example 4.4 The reluctances of the three air gaps shown in Figure 4.14 are $\mathcal{R}_1 = 3 \times 10^6$, $\mathcal{R}_2 = 2 \times 10^6$, and $\mathcal{R}_3 = 2 \times 10^6$, all in At/Wb.

a) Draw the magnetic equivalent circuit and write the loop equations.

b) Compute the flux linkages λ_1 and λ_2 and find L_1, L_2, and M.

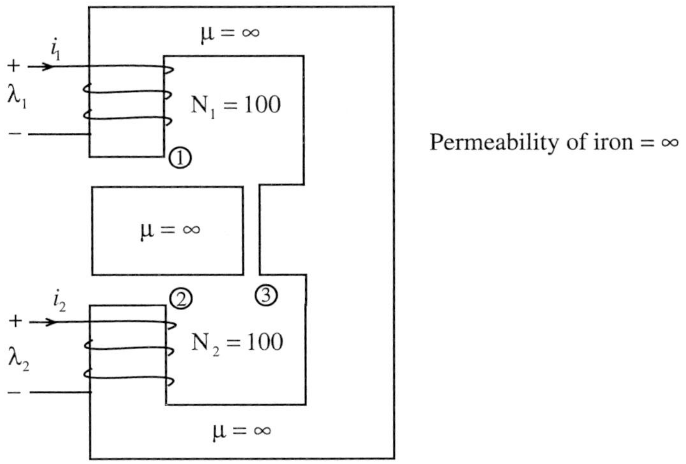

Figure 4.14: A coupled magnetic circuit for Example 3.4

Solution:

a) The equivalent circuit is shown in Figure 4.15.

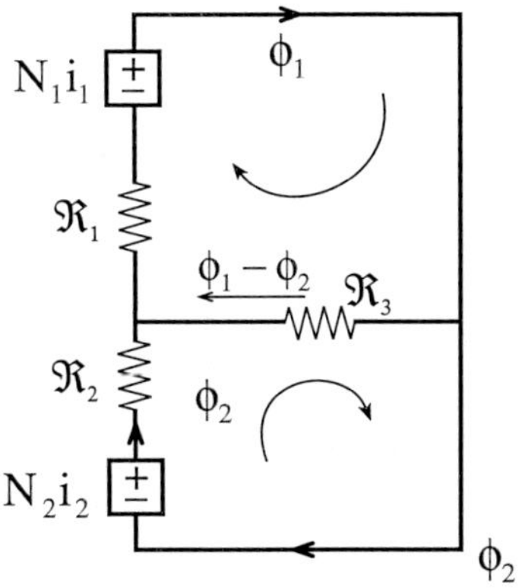

Figure 4.15: Magnetic equivalent circuit for Example 3.4

b) Assigning loop directions as shown, the loop equations are

$$N_1 i_1 = \mathcal{R}_3(\phi_1 - \phi_2) + \mathcal{R}_1 \phi_1$$
$$N_2 i_2 = \mathcal{R}_2 \phi_2 - \mathcal{R}_3(\phi_1 - \phi_2)$$

Substituting the values for N_1, N_2, and the reluctances

$$\begin{aligned} 100i_1 &= (5\phi_1 - 2\phi_2)10^6 \\ 100i_2 &= (-2\phi_1 + 4\phi_2)10^6 \end{aligned}$$

Solving this system of equation for ϕ_1, ϕ_2, we get

$$\phi_1 = (25i_1 + 12.5i_2)10^{-6}, \qquad \phi_2 = (12.5i_1 + 31.25i_2)10^{-6}$$

The flux linkages are

$$\begin{aligned} \lambda_1 &= N_1\phi_1 = (25i_1 + 12.5i_2)10^{-4} \\ \lambda_2 &= N_2\phi_2 = (12.5i_1 + 31.25i_2)10^{-4} \end{aligned}$$

Hence,

$$\begin{aligned} L_1 &= 25 \times 10^{-4}H = 2.5 \text{ mH} \\ L_2 &= 31.25 \times 10^{-4}H = 3.125 \text{ mH} \\ M &= 12.5 \times 10^{-4}H = 1.25 \text{ mH} \end{aligned}$$

4.3.1 Polarity markings (the dot convention)

When time varying flux links a coil, there is an induced voltage. For example, in Figure 4.11 when i_1 increases in the reference direction, the polarity of v_2 is positive on the upper terminal. This follows from the so-called Lenz's law, which states that the voltage induced is in such a direction that the current due to it produces a flux that opposes the change in flux. Thus, a positive v_2 would create a current that will oppose ϕ_{21}.

After establishing the polarity of the induced voltage, we can write v_2 as $M\frac{di_1}{dt}$. On the other hand, if v_2 is given a + sign at the lower terminal, then $v_2 = -M\frac{di_1}{dt}$. Similarly, for the case when coil 2 is excited as in Figure 4.12, $v_1 = \pm M\frac{di_2}{dt}$ where a + sign is when the top terminal of coil 1 is given a positive sign and a − sign is when a lower terminal of coil 1 is given a positive sign.

In general, it is a tedious process of showing the details of the mutually coupled windings. We monitor the signs of mutually induced voltages through a *dot marking* convention. These dots carry information about the signs of the mutually induced voltages. For example, for the coils in Figure 4.11, either of the dot conventions shown in Figure 4.16 is valid.

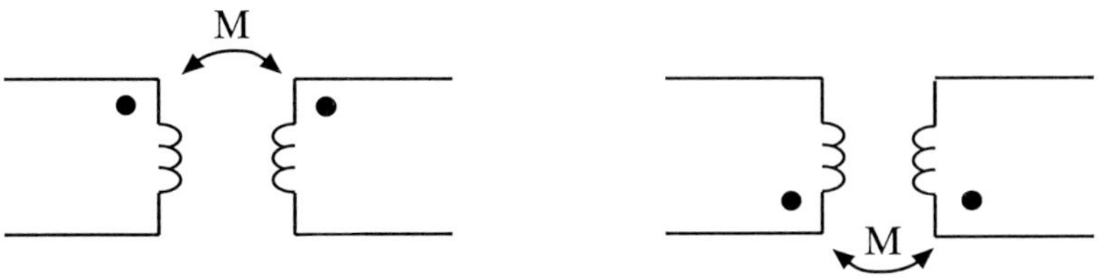

Figure 4.16: Dot markings for coil in Figure 3.11

The polarity markings are assigned such that a positively increasing current in the dotted (undotted) terminal in one winding induces a positive voltage at the dotted (undotted) terminal of the other winding. The following rule may be stated: A current i entering a dotted (undotted) terminal in one winding induces a voltage $M\frac{di}{dt}$ with positive polarity at the dotted (undotted) terminal of the other winding.

There are two problems. (1) Given the coil configuration, determine the dot markings. (2) Given the dot markings, how they are used in writing equations? We discuss both these questions next.

4.3.2 Determining dot markings

We will illustrate the dot marking technique through Figure 4.17

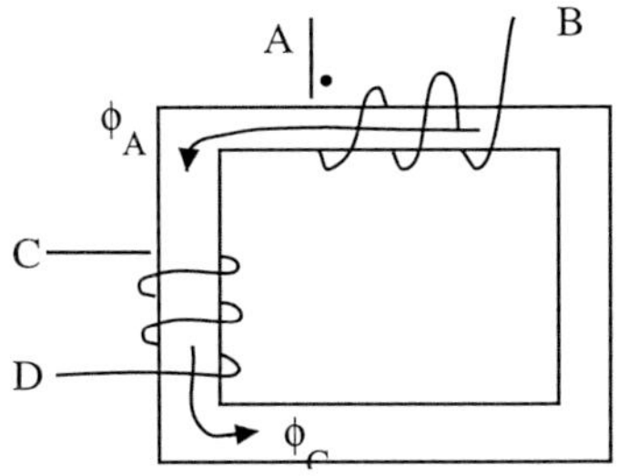

Figure 4.17: Illustrating the dot marking method

1. Arbitrarily select one terminal of a coil and assign a dot in one coil. We select terminal A.

2. Assume a current flowing into the selected dotted terminal and using the right-hand rule, see how the flux is flowing in the core. We determine ϕ_A as shown.

3. We select an arbitrary terminal of the second coil and assign a positive test current to it. We choose terminal C.

4. Use the right-hand rule to determine the direction of the flux due to this current. We determine it as flux ϕ_C.

5. We now compare directions of ϕ_A and ϕ_C. If both are additive, then a dot is placed in the second coil where the test current enters. This is point C.

6. If the fluxes are in opposite directions, then a dot is placed in the second coil at the terminal where the current leaves.

7. In the case of Figure 4.17, a dot is therefore put at terminal C.

Practical way of determining dot markings

When a device such as a transformer is given, there is no way of knowing how the coils are wound. Hence, we use a DC voltage to excite one coil of the transformer, as shown in Figure 4.18. Place the dot on the terminal to which

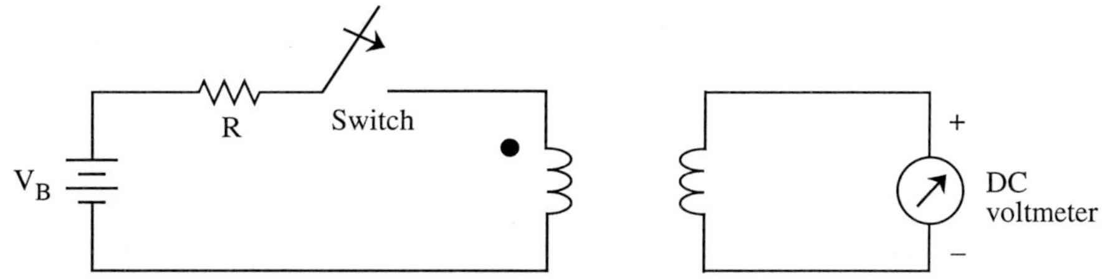

Figure 4.18: Method to determine dot markings

the + side of the DC voltage is connected. A DC voltmeter is connected on the other coil. When the switch is closed, if the voltmeter gives an up-scale kick, then the dot on the other coil is placed on the terminal to which the positive of the voltmeter is connected. If the momentary deflection is down-scale, then the dot is placed on the terminal to which the negative side of voltmeter is connected.

Example 4.5 Put the dot markings on the two coils as shown in Figure 4.19. Following the rules stated above, the dots are placed at terminals 1 and 2′.

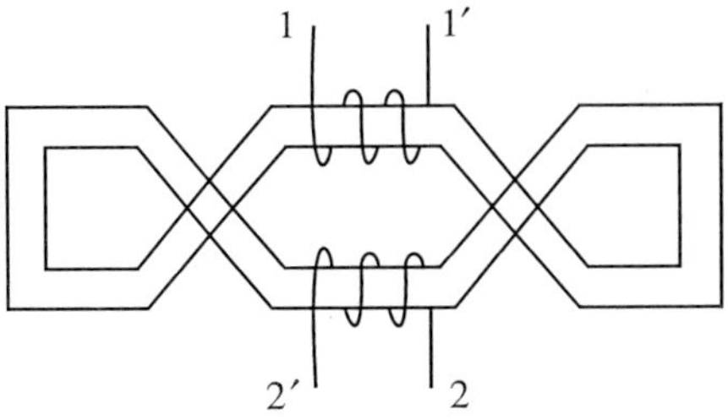

Figure 4.19: Example illustrating dot markings

4.3.3 Writing equations with mutually coupled coils

Suppose we have two mutually coupled coils and the dot markings are as shown in Figure 4.20. It is easy to write the loop equations. We can choose a reference direction for the currents in any arbitrary way. Once this is done, the equations have to be written in a *consistent* manner. Choosing i_1 and i_2 as indicated, there is an induced voltage in each coil due to the current in the other coil. The self induced voltage due to the self inductance is in the direction of the current and is a voltage drop. The polarity of the mutually induced voltage depends on the dot marking and is determined as follows.

If the reference current enters a dotted (undotted) terminal of a coil, the polarity of the induced voltage in the other coil is positive at the dotted (undotted) terminal (equal to $M\frac{di}{dt}$ *where* i *is the current into the dot of the other coil). If the reference current in a coil leaves the dotted (undotted) terminal, then the voltage induced at the dotted (undotted) terminal of the other coil has a negative sign.* Practicing using this rule to write loop equations for different circuit configurations will help. These rules are illustrated in Figure 4.20, where the reference directions are shown. The equations for Figure 4.20(a) are

$$\begin{aligned} v_1 &= i_1 R_1 + L_1\frac{di_1}{dt} + M\frac{di_2}{dt} \\ v_2 &= i_2 R_2 + L_2\frac{di_2}{dt} + M\frac{di_1}{dt} \end{aligned} \tag{4.23}$$

The equations for Figure 4.20(b) are

$$\begin{aligned} v_1 &= i_1 R_1 + L_1\frac{di_1}{dt} - M\frac{di_2}{dt} \\ v_2 &= i_2 R_2 + L_2\frac{di_2}{dt} - M\frac{di_1}{dt} \end{aligned} \tag{4.24}$$

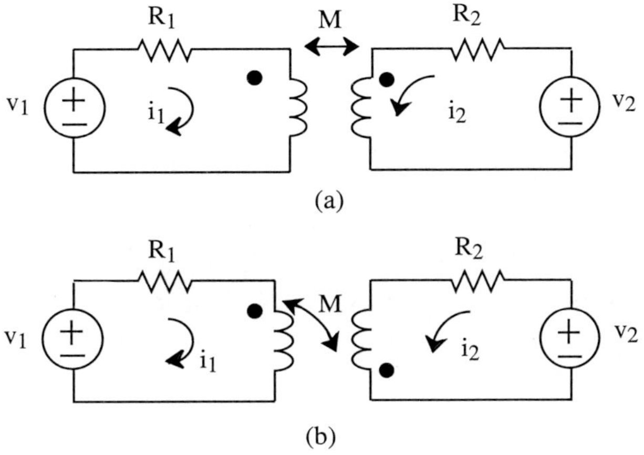

Figure 4.20: Loop equations for mutually coupled coils

Example 4.6 Write the loop equations for the circuit in Figure 4.21. As-

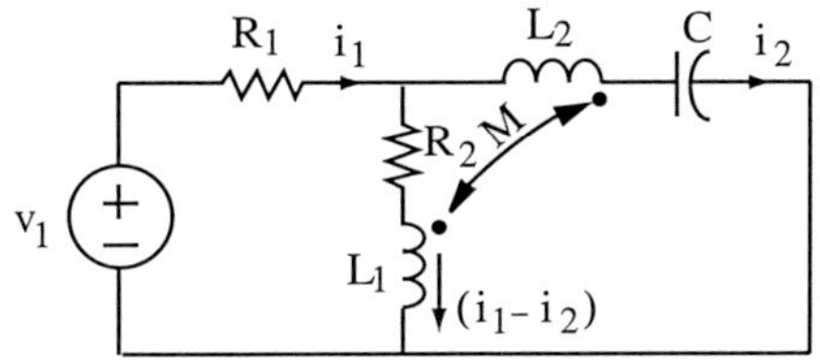

Figure 4.21: Circuit with mutual inductance

sume zero initial conditions. The two coils have mutual inductance M with the polarity as indicated. The currents in the left and right branches are shown in Figure 4.21. The current in the middle branch is $(i_1 - i_2)$. The two loop equations are written as (assuming zero initial voltages on capacitors)

$$\begin{aligned} v_1 &= i_1 R_1 + (i_1 - i_2)R_2 + L_1 \frac{d}{dt}(i_1 - i_2) - M\frac{di_2}{dt} \\ 0 &= \frac{1}{C}\int_0^t i_2 dt + L_2 \frac{di_2}{dt} - M\frac{d}{dt}(i_1 - i_2) + L_1\frac{d}{dt}(i_2 - i_1) + M\frac{di_2}{dt} \\ &\quad +(i_2 - i_1)R_2 \end{aligned} \tag{4.25}$$

Note that in (4.25) the mutually induced voltage appears twice, once for each coil. It may be remarked here that the writing of node equations is not so straightforward, but it can be done if we define a so-called reciprocal inductance matrix.

4.4 Transformers

The transformer transfers electrical energy from one circuit to another through the changing magnetic field that links both circuits. It has many applications in both the power and communications fields. In the transmission, distribution, and utilization of electrical energy, the transformer changes voltage levels at a fixed frequency, such as 50 Hz or 60 Hz, and the power to be handled can vary from a few hundred watts to several hundred megawatts. On the other hand, in the communications field, the use of the transformer is based on altogether different considerations. Impedance matching, DC isolation, and changing voltage levels with power handling capacities of the order of a few watts but with the capability of operating satisfactorily over a very wide frequency range, are some of the applications. It is therefore apparent that design considerations and, consequently, modeling issues are quite different. In this section, we restrict ourselves mainly to power transformers.

4.4.1 Ideal transformer

Consider a magnetic core with two coils wound as shown in Figure 4.22.

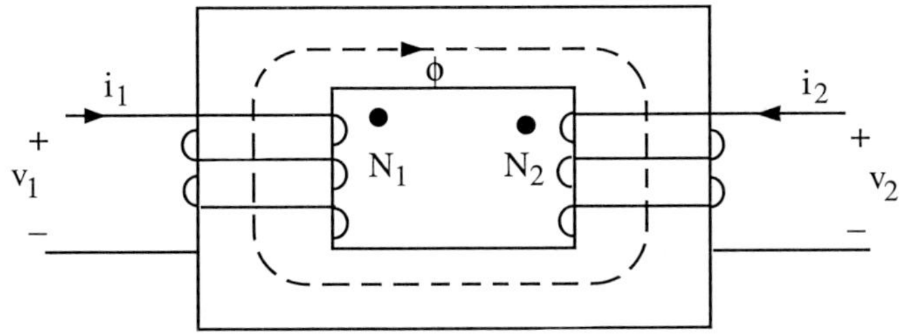

Figure 4.22: Two-winding transformer

From the orientations of the coils, dots are placed as shown. The positive orientations for the currents and voltages are as shown. Note that this assignment of orientation is arbitrary. The following assumptions are made:

1. The coils have neither losses nor stray capacitance.

2. The $\phi - i$ (flux versus mmf) curve for the magnetic material has infinite slope and is linear. This is equivalent to saying that magnetic permeability is infinite.

3. There is no leakage flux, i.e., all the flux that links coil 1 also links coil 2.

Under the foregoing assumptions, let the common flux be denoted by ϕ. Then

$$v_1(t) = N_1 \frac{d\phi}{dt} \tag{4.26}$$

$$v_2(t) = N_2 \frac{d\phi}{dt} \tag{4.27}$$

We therefore have

$$\frac{v_1(t)}{v_2(t)} = \frac{N_1}{N_2} = a \tag{4.28}$$

for all t. a is called the *turns ratio*. Now, ϕ can be expressed in terms of the mmf and reluctance of the circuit. The net mmf corresponding to the choice of reference directions of currents in Figure 4.22 is given by

$$\text{mmf} = N_1 i_1 + N_2 i_2 \tag{4.29}$$

Since the mmf equals reluctance times flux, we have

$$N_1 i_1 + N_2 i_2 = \mathcal{R}\phi \tag{4.30}$$

Now, if permeability μ is infinite, then the reluctance becomes zero since it is inversely proportional to μ. Hence, we have

$$N_1 i_1 + N_2 i_2 = 0$$

i.e.,

$$\frac{i_1(t)}{i_2(t)} = -\frac{N_2}{N_1} = -\frac{1}{a} \tag{4.31}$$

for all t. Equations (4.28) and (4.31) constitute the terminal equations for the two-winding ideal transformer. The ideal transformer is defined by these terminal equations and the symbol shown in Figure 4.23.

If the reference direction for the current i_2 were to be reversed (as shown in Figure 4.24), then the negative sign in the terminal (4.31) gets eliminated.

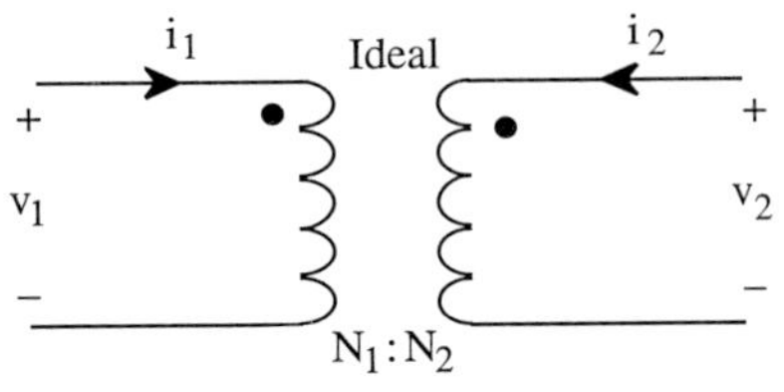

Figure 4.23: Ideal transformer

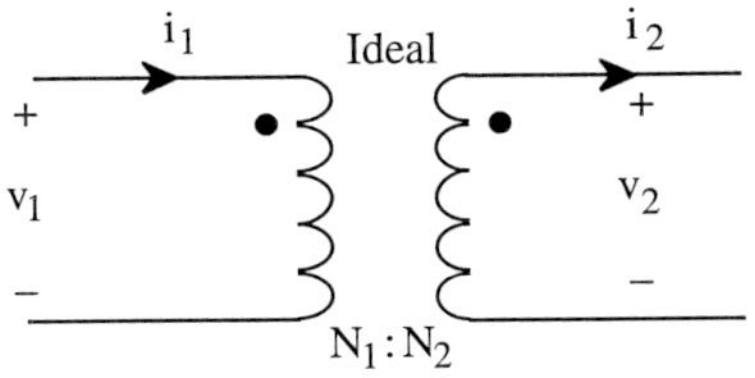

Figure 4.24: Alternative representation of an ideal transformer

This can be easily verified. The terminal relations for this representation are

$$\begin{aligned} \frac{v_1}{v_2} &= \frac{N_1}{N_2} = a \\ \frac{i_1}{i_2} &= \frac{N_2}{N_1} = \frac{1}{a} \end{aligned} \tag{4.32}$$

Figure 4.24 for the ideal transformer is more popular in dealing with power transformers. Some interesting consequences arise out of the definition of the ideal transformer. Following the convention in Figure 4.23, first, we verify from (4.28) and (4.31) that

$$v_1(t)i_1(t) + v_2(t)i_2(t) = 0 \tag{4.33}$$

For the convention in Figure 4.24 we have

$$v_1(t)i_1(t) = v_2(t)i_2(t) \tag{4.34}$$

This implies that instantaneous power flowing into one pair of terminals flows out of the other pair. Hence, there is no energy storage. Contrast this with the capacitors, inductors, and the pair of mutually coupled coils that are non-dissipative but capable of energy storage.

The self inductance of each coil of the ideal transformer is infinite. This can be shown as follows: Consider coil 2 open circuited, i.e., $i_2 = 0$. Hence, i_1 must be equal to zero in view of (4.31). Since the flux linkage $\lambda_1 = L_1 i_1$ is finite, it follows that the self inductance of coil 1 is infinite.

Finally, we look at the ideal transformer from a purely circuit point of view as a set of coupled coils. The energy stored in the two coils may be expressed as

$$\begin{aligned} W &= \frac{1}{2}L_1 i_1^2 + \frac{1}{2}L_2 i_2^2 + M i_1 i_2 \\ &= \frac{1}{2}(\sqrt{L_1}\, i_1 + \sqrt{L_2}\, i_2)^2 + (k-1)\sqrt{L_1\, L_2}\;\, i_1 i_2 \end{aligned} \tag{4.35}$$

where $k = M/\sqrt{L_1 L_2}$ is the coefficient of coupling. Note that i_1 and i_2 have opposite signs for the convention in Figure 4.23. Since the stored energy W must be identically zero, we must therefore have:

$$\sqrt{L_1}\, i_1 = -\sqrt{L_2}\, i_2$$

and

$$k = 1 \tag{4.36}$$

This implies that the coefficient of coupling is unity and

$$\frac{i_1}{i_2} = -\frac{\sqrt{L_2}}{\sqrt{L_1}} = -\frac{v_2}{v_1} = -\frac{1}{a} \tag{4.37}$$

This, together with the fact that the self inductances are infinite, leads us to the following interesting conclusion. If the ideal transformer is viewed as a set of coupled coils with zero energy storage, their self inductances are infinite, yet the ratio of these self inductances is finite.

To summarize, the ideal transformer consisting of two mutually coupled coils having N_1 and N_2 turns, respectively, is characterized by the following three properties.

1. The coefficient of coupling k is unity ($k = 1$).

2. The self inductance of each coil is infinite ($L_1 = L_2 = \infty$).

3. The coil losses are negligible.

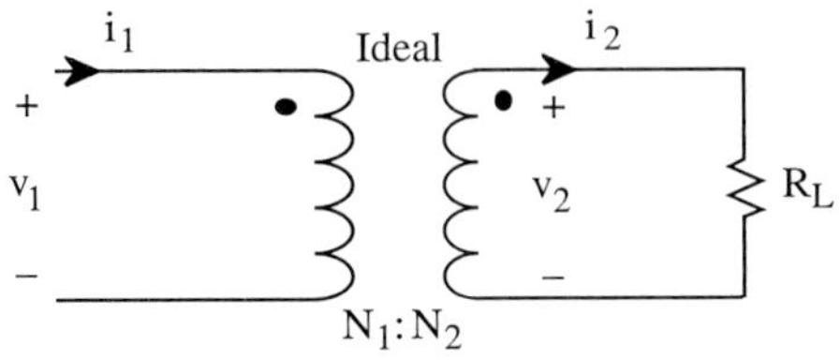

Figure 4.25: Ideal transformer with resistive load

4.4.2 Impedance-changing properties of the ideal transformer

Consider a resistive load with resistance R_L connected across the terminals of winding 2, as shown in Figure 4.25. It is convenient to adopt the convention as in Figure 4.24 since we are interested in the transfer of power from one winding to another. By Ohm's law

$$\frac{v_2}{i_2} = R_L \tag{4.38}$$

Using the relations in (4.32), $v_2 = \frac{v_1}{a}$, and $i_2 = ai_1$. Substituting in (4.38), we get

$$\frac{v_1}{i_1} = a^2 R_L = \left(\frac{N_1}{N_2}\right)^2 R_L \tag{4.39}$$

But v_1/i_1 is the resistance looking into the terminals of winding 1. Hence, the apparent resistance viewed from the terminals of winding 1 gets modified, as given by (4.39). The reader can easily verify that if R_L is connected across winding 1, then the apparent resistance viewed from winding 2 is given by $R_L(N_2/N_1)^2$.

This discussion is easily extended to the sinusoidal steady-state case. Let the phasors corresponding to the currents and voltage variables be denoted as shown in Figure 4.26. An impedance Z_L is connected across winding 2. It can be verified that

$$\begin{aligned}\frac{\overline{V}_1}{\overline{I}_1} &= \left(\frac{N_1}{N_2}\right)^2 \frac{\overline{V}_2}{\overline{I}_2} \\ &= \left(\frac{N_1}{N_2}\right)^2 Z_L = a^2 Z_L \end{aligned} \tag{4.40}$$

The impedance viewed from winding 1 is the load impedance across winding 2 multiplied by the square of the turns ratio, as given by (4.40).

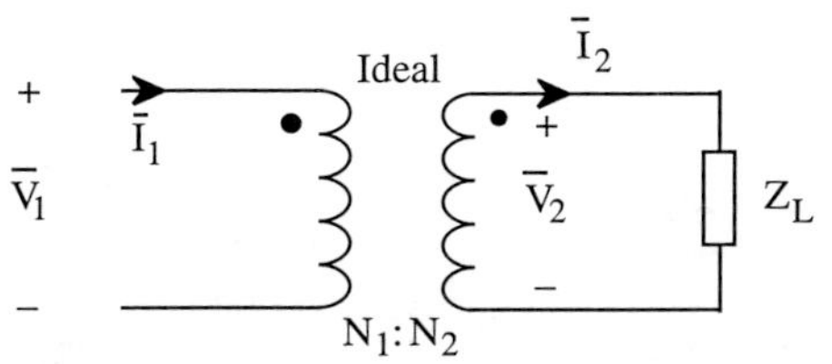

Figure 4.26: Ideal transformer with impedance load

This illustrates the impedance-changing properties of the transformer, which can be exploited in situations where the matching of circuits with different impedances is required or where maximum power is to be transferred from a source with a given source impedance to a fixed load. We shall now illustrate this property.

Impedance matching

Suppose that, in Figure 4.27(a), it is desired to transfer maximum power to

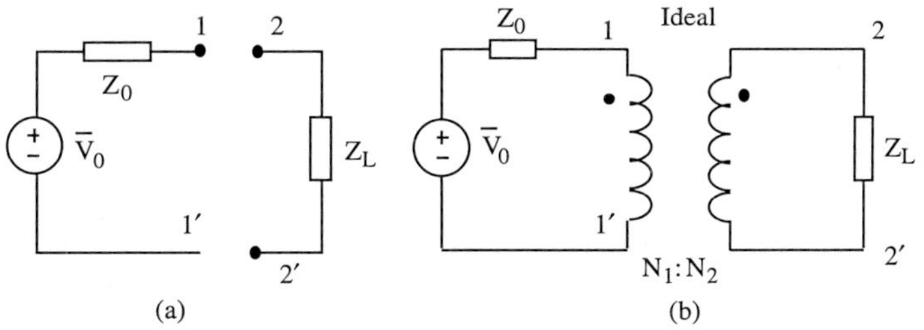

Figure 4.27: Impedance matching by ideal transformer

load Z_L; let the rest of the network be represented by means of voltage $\overline{V}_o$ in series with impedance Z_o. We know that maximum power can be transferred to the load if either (1) $Z_L = Z_o^*$ or (2) $|Z_L| = |Z_o|$ under the constraint of R_L/X_L being constant. Whereas it is difficult to achieve the transference of maximum power as per method (1), the magnitudes of Z_L and Z_o can be made almost equal by interposing an ideal transformer between the source and the load, as shown in Figure 4.27(b).

Turns ratio N_1/N_2 is so selected that $\mid Z_o \mid \approx (N_1/N_2)^2 \mid Z_L \mid$, which is equal to the impedance looking into terminals 1 1′.

Example 4.7 In Figure 4.28, the input resistance is 100Ω and there are

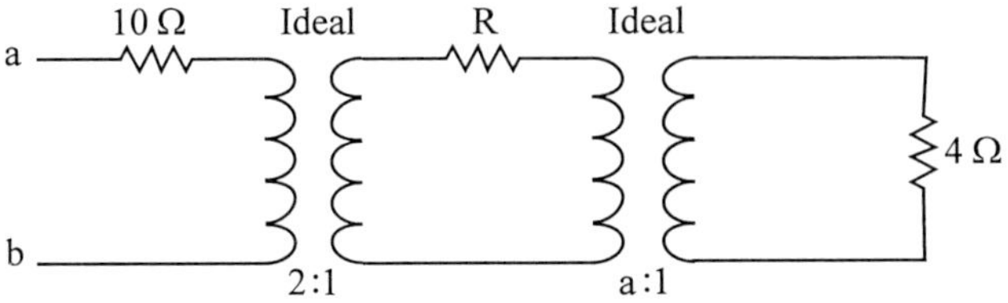

Figure 4.28: Impedance-matching example

two ideal transformers. The value of R is 9Ω. For maximum power transfer to the 4Ω resistance, find the turns ratio 'a' of the transformer connected to the load.

Solution: The resistance 4Ω is retained and the rest of the network is equivalized at its terminals to compute Rth. Thus we have Rth=((100/4)+9/a*a) Ω. For maximum power transfer to the 4 ohm resistor we set Rth=4. Simplifying, we solve for a=2.915.

4.4.3 Power transformer

The power transformer is an essential component of the power system, whether it is at the generating, transmission, or distribution side. It steps up the voltage at the generating point, transmits it economically over long distances, and then steps it down at the sub-transmission and distribution levels for use by both commercial units and individual homes. A physical transformer consists of two windings mounted on a core, generally of a ferromagnetic material of high permeability, as shown in Figure 4.29. The

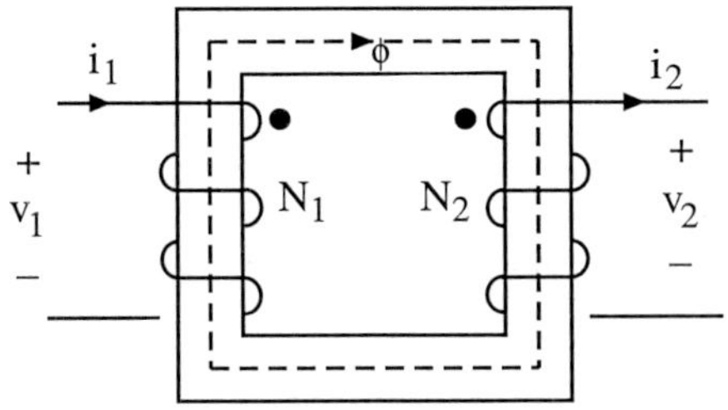

Figure 4.29: Two-winding transformer

construction of the core is such as to keep the leakage flux to a minimum. The winding to which power is supplied is called the "primary" winding, and the other one is called the "secondary" winding. In the case of power transformers, the terms high-voltage (HV) and low-voltage (LV) windings are also used, depending on the voltage levels.

We assume it to be an ideal transformer. This implies

a) No leakage flux.

b) Winding resistances are neglected.

c) Core has infinite permeability. Hence, reluctance of the core is zero and negligible current is required to set up the magnetic field.

d) Magnetic core is lossless.

Let the voltage impressed across the primary winding be given by

$$v_1(t) = V_{m1} \cos \omega t \tag{4.41}$$

Since $v_1(t) = N_1 \frac{d\phi}{dt}$, we have

$$\begin{aligned} \phi(t) &= \frac{1}{N_1} \int V_{m1} \cos \omega t dt \\ &= \frac{V_{m1}}{N_1} \frac{\sin \omega t}{\omega} \end{aligned} \tag{4.42}$$

Referring to (4.42), the maximum value of the flux $\phi_{\max}$ is given by

$$\phi_{\max} = \frac{V_{m1}}{N_1 \omega} = \frac{V_{m1}}{2\pi f N_1}$$

Hence,

$$V_{m1} = 2\pi f N_1 \phi_{\max} \tag{4.43}$$

The rms value of the voltage is given by

$$\begin{aligned} V_1 &= \frac{V_{m1}}{\sqrt{2}} \\ &= \sqrt{2}\pi f N_1 \phi_{\max} \\ &= 4.44 f N_1 \phi_{\max} \end{aligned} \tag{4.44}$$

where $\phi_{\max}$ is in webers. This is a basic relation in the design of power transformers with an iron core.

Example 4.8 A certain transformer having 200 turns on the primary winding and 400 turns on the secondary winding is supplied at its rated voltage of 230 V, 60 Hz on the primary side. (a) Find the maximum value of the flux density if the cross-sectional area is 0.005 m^2, and mean core length of the flux path is $\ell = 0.5$ m. (b) Assume that the B-H curve is linear and that it requires a magnetic field intensity of 300 At/m to establish a flux density of 1 weber/m^2 in the material. On this basis, calculate the rms value of the magnetizing current required with the secondary winding open.

(a)

$$\begin{aligned} V_1 &= 4.44 f N_1 \phi_{\max} \text{ V} \\ V_1 &= 230V, f = 60 \text{ Hz}, N_1 = 200 \end{aligned}$$

Hence,

$$\phi_{\max} = \frac{230}{4.44 \times 60 \times 200} = 4.32 \times 10^{-3} \text{ webers}$$

Therefore,

$$B_m = \frac{4.32 \times 10^{-3}}{0.005} = 0.864 \text{ webers/m}^2$$

(b) To establish 1 Wb/m^2, we need $H = 300At/\text{m}$. Hence, for $B_m = 0.864Wb/\text{m}^2$ the value of H_m is given by $0.864 \times 300 = 259$ At/m. Since $N = 200$ and $H_m \ell = N i_m$ the peak value of the current i_m is $(259)(0.5)/200 = 0.6475$ A. Hence, $I_{\text{rms}} = 0.46$ A is the magnetizing current on the primary side.

4.4.4 Equivalent circuit of transformer with linear core

We relax the assumption of the previous section with respect to (a) and (b), namely, the leakage flux and resistance. Consider the two-winding transformer as shown in Figure 4.30(a). Let v_2 be the voltage across R_L. Let the resistances of the primary and secondary windings be R_1 and R_2, respectively. Viewing the two windings as coupled coils, the differential equations are written as

$$\begin{aligned} v_1 &= i_1 R_1 + L_1 \frac{di_1}{dt} - M \frac{di_2}{dt} \\ 0 &= i_2 R_2 + L_2 \frac{di_2}{dt} - M \frac{di_1}{dt} + i_2 R_L \\ v_2 &= i_2 R_L \end{aligned} \tag{4.45}$$

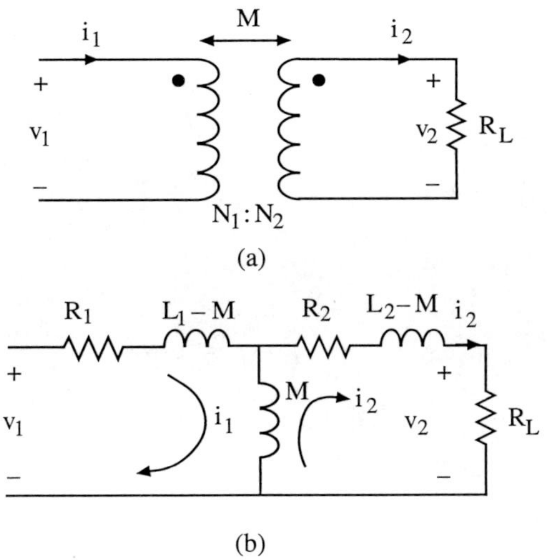

Figure 4.30: Physical transformer and a possible equivalent circuit

L_1 and L_2 are the self inductances and M is the mutual inductance.

An equivalent circuit can be developed from (4.45) by identifying a common voltage term. (4.45) is rearranged as follows

$$\begin{aligned} v_1 &= i_1 R_1 + (L_1 - M)\frac{di_1}{dt} + M\frac{d}{dt}(i_1 - i_2) \\ 0 &= i_2 R_2 + (L_2 - M)\frac{di_2}{dt} + M\frac{d}{dt}(i_2 - i_1) + i_2 R_L \\ v_2 &= i_2 R_L \end{aligned} \tag{4.46}$$

$(i_1 - i_2)$ is the current through M from top down. The equivalent circuit is shown in Figure 4.30(b). In this circuit, $L_1 - M$ or $L_2 - M$ can be less than zero, if $\sqrt{(L_1/L_2)} < k$ or $\sqrt{L_2/L_1)} < k$, respectively, and, hence the usefulness of the circuit is limited. To overcome this limitation, (4.45) is rearranged by multiplying the equations on the secondary side by a and with i_2 replaced by i_2/a, where $a = N_1/N_2 =$ turns ratio. This substitution does not mathematically change (4.45). They become

$$\begin{aligned} v_1 &= i_1 R_1 + L_1 \frac{di_1}{dt} - aM \frac{d}{dt}\left(\frac{i_2}{a}\right) \\ 0 &= \left(\frac{i_2}{a}\right) a^2 R_2 + a^2 L_2 \frac{d}{dt}\left(\frac{i_2}{a}\right) - aM \frac{di_1}{dt} + \left(\frac{i_2}{a}\right) a^2 R_L \\ av_2 &= \left(\frac{i_2}{a}\right) a^2 R_L \end{aligned} \tag{4.47}$$

(4.47) results in the equivalent circuit in Figure 4.31 where the secondary

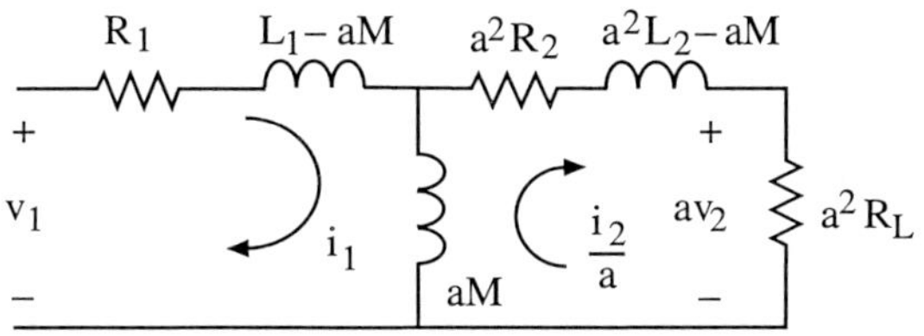

Figure 4.31: Another equivalent circuit of the two-winding transformer

winding variables are i_2/a and av_2. It is possible to give a physical interpretation to the various inductances in terms of the fluxes linking the two coils. Consider first the inductance

$$L_1 - aM = L_1 - \frac{N_1}{N_2} M$$

Since $L_1 i_1 = N_1 \phi_{11}$ and $N_2 \phi_{21} = M i_1$ from (4.12) and (4.13), we have

$$\begin{aligned} L_1 - aM &= N_1 \left(\frac{\phi_{11}}{i_1} - \frac{\phi_{21}}{i_1}\right) & (4.48) \\ &= N_1(\phi_{11} - \phi_{21})/i_1 \\ &= N_1(\phi_{\ell 1}/i_1) & (4.49) \end{aligned}$$

where $\phi_{\ell 1}$ is the flux linking only coil 1 and is the leakage flux of coil 1. ϕ_{21} is the flux linking coils 1 and 2 due to i_1. Hence, the expression $N_1(\phi_{\ell 1}/i_1)$ is termed the *leakage inductance of winding 1*. Through similar considerations, we can show that

$$a^2 L_2 - aM = \left(\frac{N_1}{N_2}\right)^2 \frac{N_2 \phi_{\ell 2}}{i_2}$$

where $N_2(\phi_{\ell 2}/i_2)$ is the *leakage inductance of winding 2*. Multiplication by $(N_1/N_2)^2$ makes the inductance "referred" to side 1. Since $\phi_{\ell 1}$ and $\phi_{\ell 2}$ are always greater than or equal to zero, we conclude that inductances $L_1 - aM$ and $a^2L_2 - aM$ are always positive, thus eliminating the possibility of negative elements in the equivalent circuit of Figure 4.31. The quantity aM appearing across the circuit is interpreted as the magnetizing inductance because it controls the amount of current necessary to establish the flux. In transformer terminology, the current through aM is called the *magnetizing current.* Physically, the net current through aM in the downward direction is given by

$$i_1 - \frac{i_2}{a} = i_1 - \frac{N_2 i_2}{N_1} = \frac{N_1 i_1 - N_2 i_2}{N_1} \tag{4.50}$$

Considering our dot conventions as shown in Figure 4.30(a), the numerator $N_1 i_1 - N_2 i_2$ is the net ampere-turns for the core, and it is proportional to the reluctance times the mutual flux.

In the equivalent circuit, the voltage and current variables of winding 2 are associated with the turns ratio a. To retain R_L with the variables v_2 and i_2 associated with it, we can invoke the concept of the ideal transformer and introduce it as shown in Figure 4.32.

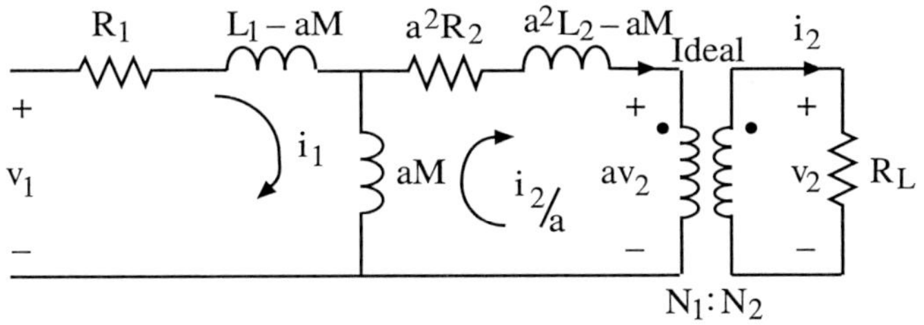

Figure 4.32: Complete equivalent circuit of the physical transformer

This equivalent circuit was arrived at on the basis of treating the two windings as coupled coils. The linear $\phi - i$ relationship is assumed for the magnetic core. No assumptions have been made regarding the degree of coupling between the coils. Hence, the equivalent circuit reflects all the features consistent with the linear $\phi - i$ relationship for the core.

Now, we inject some of the losses (core losses) that are inherent in the operation of the physical transformer to see how they can be accounted for in the linear model. There are two kinds of losses occurring in the physical transformer, besides those occurring in the resistances of the windings. These are

1. Hysteresis losses due to the nonlinear multi-valued nature of the $\phi - i$ relationship of the actual core.

2. Eddy current losses occurring in the laminations, resulting in an I^2R-type loss.

It is very difficult to compute both these losses analytically. Empirically, the expressions for these losses are given by

$$\text{Hysteresis loss } P_h = K_h f B_m^n \text{ W}$$

where B_m is the maximum flux density, and n (often assumed to be 1.6) and K_h are constants depending on the material of the core; and

$$\text{Eddy current loss } P_e = K_e f^2 B_m^2 \text{ W}$$

where K_e is a constant that depends on the resistivity of the material of the laminations. The sum of these two losses represents the constant losses for the transformer and depends only on the value of B_m. They are called *core* or *iron losses*. For a given constant frequency of operation, the value of B_m depends on the applied voltage V_m. To account for these losses in the linear circuit model, it is therefore appropriate to represent them by means of a resistance R_c in parallel with the magnetizing inductance aM. Since these are on side 1, we denote them as R_{c1} and $(aM)_1$. The value of R_{c1} is generally very high. The model then becomes as shown in Figure 4.33.

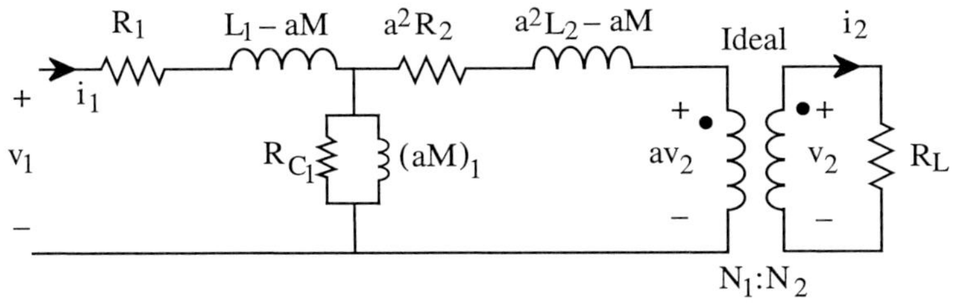

Figure 4.33: Complete equivalent circuit of transformer including effects of core losses

4.4.5 Transformer under sinusoidal steady-state conditions

If the applied voltage is $v_1 = V_{m1} \sin \omega t$ and the interest is focussed only on the steady-state operation, we can replace all elements by their impedances, and the voltage and current variables by their phasors. Define the various reactances as follows:

$$\begin{aligned}
\omega(L_1 - aM) = X_{\ell 1} &= \text{leakage reactance of winding 1} \\
\omega(aM) = X_{m1} &= \text{magnetizing reactance referred} \\
& \text{to winding 1} \\
\omega(L_2 - M/a) = X_{\ell 2} &= \text{leakage reactance of winding 2} \\
\omega(a^2 L_2 - aM) = a^2 X_{\ell 2} &= \text{leakage reactance of winding 2} \\
& \text{referred to side 1.}
\end{aligned}$$

The steady-state equivalent circuit is shown in Figure 4.34. The shunt ele-

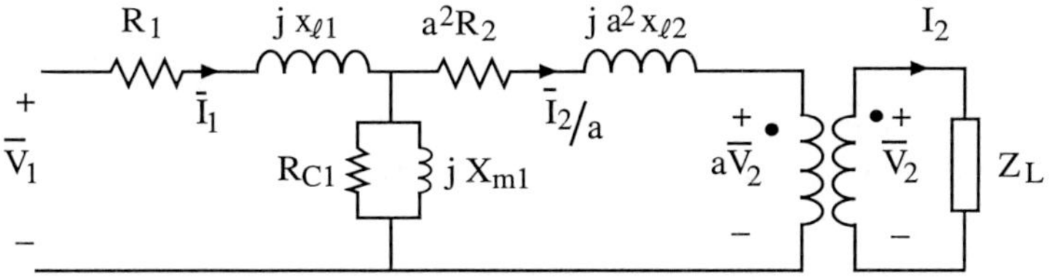

Figure 4.34: Equivalent circuit of transformer for sinusoidal steady-state

ments corresponding to the core losses and the magnetizing inductance are R_{c1} and X_{m1}, respectively. Also the resistance load R_L is replaced by a more general representation of Z_L. The ideal transformer can be omitted if the impedance Z_L is referred to winding 1, as shown in Figure 4.35. All the

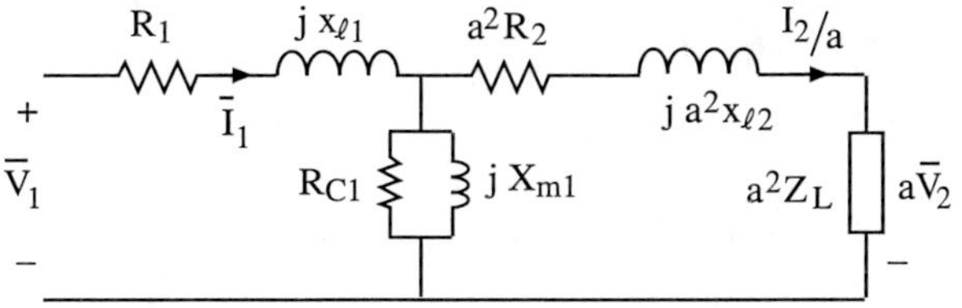

Figure 4.35: Equivalent circuit with all quantities referred to winding 1

quantities in this diagram are referred to winding 1.

Sometimes it is necessary to have the circuit with all the elements referred to winding 2. This can be done as in Figure 4.36.

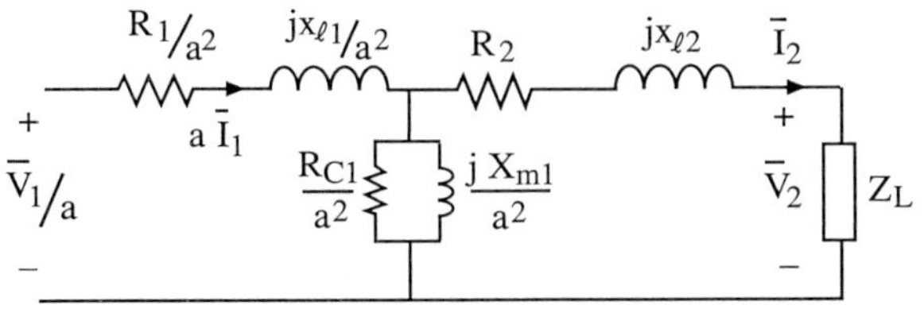

Figure 4.36: Equivalent circuit with all quantities referred to winding 2

4.4.6 Approximate equivalent circuit

Generally, problems involving transformers fall into two basic categories: 1) Given all impedances, including Z_L and desired $\overline{V}_2$, find $\overline{V}_1$; and 2) given $\overline{V}_1$ and all impedances, including Z_L, compute $\overline{V}_2$. Instead of the impedance Z_L, the power across the load as well as the PF can be specified. The computation of other performance indices, such as the voltage regulation and efficiency, is straightforward, once the quantities in the equivalent circuit are determined. However, in Figure 4.35 the parallel branch of R_{c1} and jX_{m1} makes the computation somewhat difficult because "topologically" there are three loops. It is recognized that in most practical transformers, the constant losses and hence, the current through R_{c1} is very small. Also, the magnetizing current through X_{m1} is also small. Hence, no serious numerical error is introduced if the parallel branch of R_{c1} and jX_{m1} is moved to the terminals of winding 1. This is illustrated in Figure 4.37 and is called the

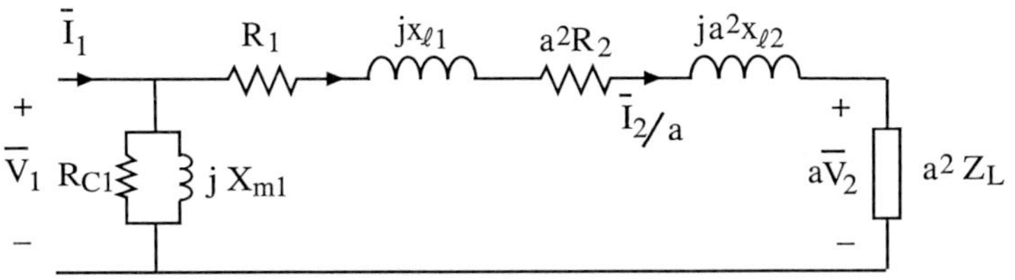

Figure 4.37: Approximate equivalent circuit

approximate equivalent circuit. Solving the circuit of Figure 4.37 is much easier than solving that of either Figure 4.35 or Figure 4.36. In Figure 4.37, we can define

$$R_{1\text{eq}} = R_1 + a^2 R_2 \tag{4.51}$$
$$X_{1\text{eq}} = X_{\ell 1} + a^2 X_{\ell 2} \tag{4.52}$$

as the equivalent resistance and reactance, respectively, referred to winding 1. This will further reduce the circuit to that which is shown in Figure 4.38.

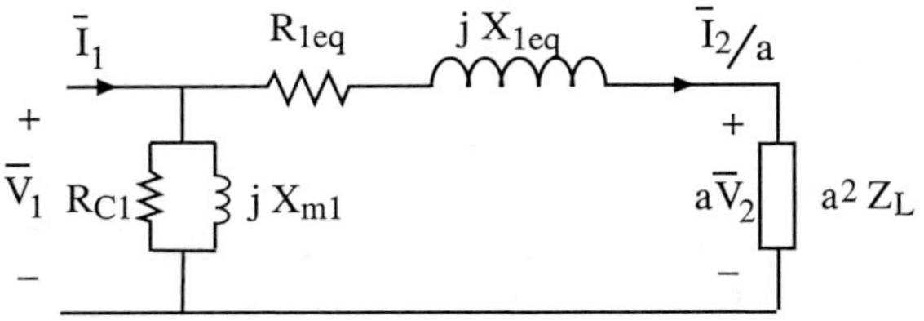

Figure 4.38: Approximate equivalent circuit with all quantities referred to winding 1

If we desire a similar equivalent circuit with all quantities referred to winding 2, it can be easily deduced from Figure 4.38. The impedances are divided by a^2, voltage is divided by a and the current is multiplied by a. The resulting circuit is shown in Figure 4.39,

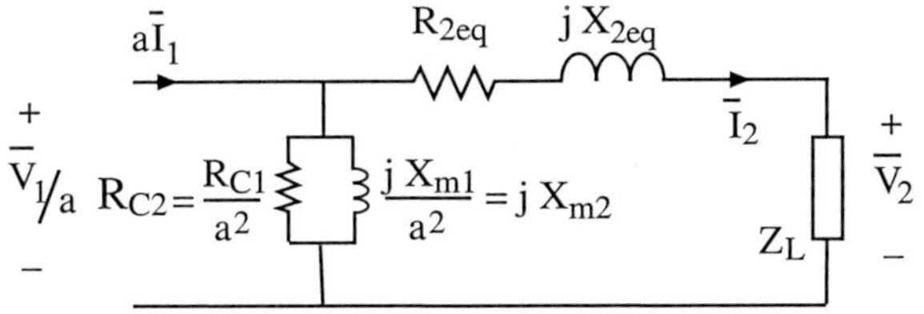

Figure 4.39: Approximate equivalent circuit with quantities referred to winding 2

where

$$
\begin{aligned}
R_{2\text{eq}} &= \frac{R_1}{a^2} + R_2 = R_{1\text{eq}}/a^2 \\
X_{2\text{eq}} &= \frac{X_{\ell 1}}{a^2} + X_{\ell 2} = X_{1\text{eq}}/a^2 \\
R_{c2} &= \frac{R_{c1}}{a^2} \\
X_{m2} &= \frac{X_{m1}}{a^2}
\end{aligned} \tag{4.53}
$$

4.4.7 Open- and short-circuit tests for a transformer

The parameters in the linear circuit model of a transformer can be determined experimentally by two simple tests called the *open-circuit test* and the *short-circuit test.* These tests are invariably made in the case of power transformers to supply relevant information to the user for modeling purposes and also to compute the performance.

In the power transformers, there are two windings called high-voltage (HV) and low-voltage (LV) windings. As a first approximation assuming negligible losses, the HV winding carries a lesser current, and the LV winding carries a higher current; the current ratios are inversely proportional to the voltage ratios. Discussion can also be carried out on the basis of calling the windings *primary* and *secondary.* This distinction is based on the actual usage of the transformer. Once the circuit parameters are determined, performance and efficiency of the transformer can be computed.

Open-circuit test (OC test)

The open circuit test measures the magnetizing reactance X_m and R_c. Because of the ready availability of a low-voltage source, this test is performed with all instrumentation on the LV side with the HV side being open-circuited. Figure 4.40 shows the test configuration. *Rated voltage* is applied

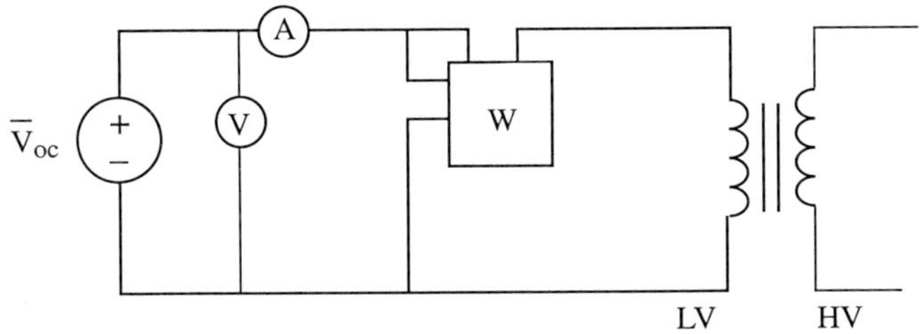

Figure 4.40: Open-circuit test

to the LV side. The wattmeter W has a current coil in series and a voltage coil in parallel to the source. From the approximate equivalent circuit as shown in Figure 4.38, the equivalent circuit for the open circuit test reduces to that shown in Figure 4.41. The meters in Figure 4.40 measure three quantities V_{oc}, I_{oc}, and P_{oc}, which are applied voltage, no load current, and watt meter reading, respectively. V_{oc} = rated voltage of the LV side. $\frac{V_{oc}^2}{R_c} =$

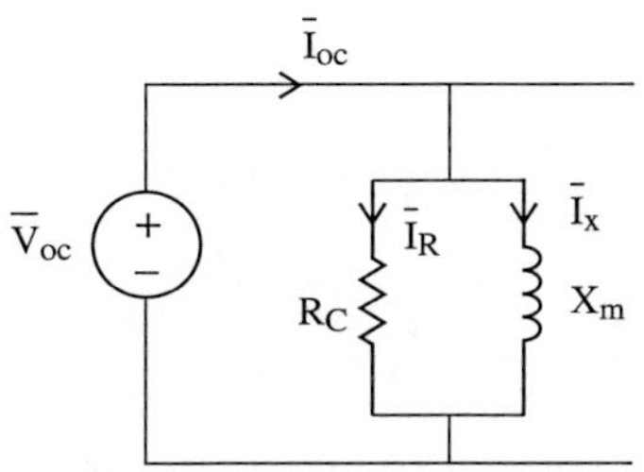

Figure 4.41: Equivalent circuit for open-circuit test

P_{oc} where P_{oc} is the reading of the wattmeter. Hence,

$$\begin{aligned} R_c &= \frac{V_{\text{oc}}^2}{P_{\text{oc}}} \\ I_R &= \frac{V_{\text{oc}}}{R_c} \\ \overline{I}_{\text{oc}} &= \overline{I}_R + \overline{I}_x \end{aligned}$$

Hence,

$$I_x = \sqrt{I_{\text{oc}}^2 - I_R^2}$$

since $\overline{I}_R$ and $\overline{I}_x$ are 90° out of phase.

$$X_m = \frac{V_{\text{oc}}}{I_x}$$

R_c and X_m are the values referred to the LV side. To refer them to the HV side, we need to use the turns ratio.

Short-circuit test (SC test)

In the short-circuit test, one of the windings is short-circuited and *rated current* is made to flow in the windings by applying an appropriate voltage. Since current on the HV side is smaller, all the instrumentation is on the HV side. The test configuration is shown in Figure 4.42. The equivalent circuit is shown in Figure 4.43. From the meters of Figure 4.42, three quantities are measured, namely, V_{sc}, I_{sc}, and P_{sc}, which are the applied voltage, short-circuit current, and watt meter reading, respectively. I_{sc} = rated current on the HV side. We have

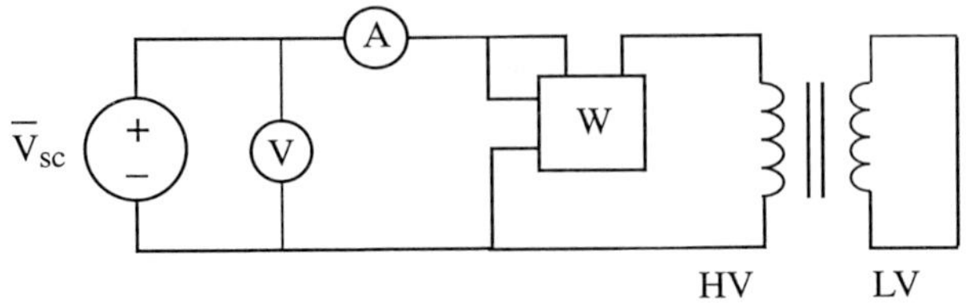

Figure 4.42: Short-circuit test

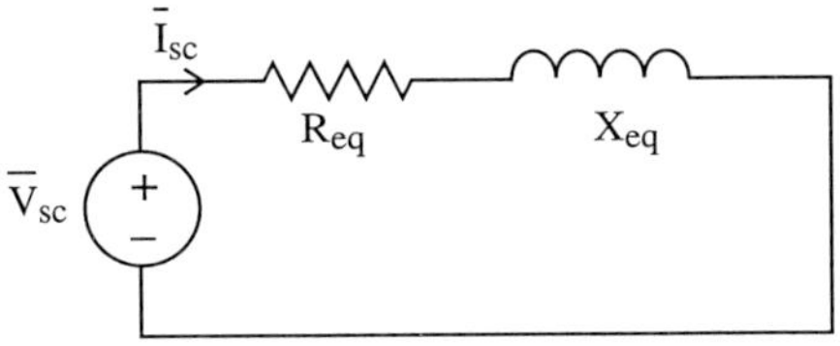

Figure 4.43: Equivalent circuit for short-circuit test

$$P_{sc} = I_{sc}^2 R_{eq}, R_{eq} = \frac{P_{sc}}{I_{sc}^2}$$

$$Z_{eq} = \frac{V_{sc}}{I_{sc}}, X_{eq} = \sqrt{Z_{eq}^2 - R_{eq}^2}$$

These are referred to the HV side.

To get the complete equivalent circuit, all impedances have to be referred to the same side.

Example 4.9 A 7.5 kVA, 440/220 V transformer has the following readings from OC and SC tests.

OC test (LV side)	$V_{oc} = 220, I_{oc} = 1A, P_{oc} = 50$ W
SC test (HV side)	$V_{sc} = 15V, I_{sc} = 17A, P_{sc} = 60$ W

Find the equivalent circuit parameters referred to the HV side.

OC test

$$\begin{aligned} R_c &= \frac{(220)^2}{50} = 968\Omega \\ I_R &= \frac{220}{968} = 0.227 \text{ A} \\ I_x &= \sqrt{1^2 - (0.227)^2} = 0.974 \text{ A} \\ X_m &= \frac{220}{0.974} = 225.9\Omega \end{aligned}$$

These are referred to the LV side.

SC test

$$\begin{aligned} R_{\text{eq}} &= \frac{60}{(17)^2} = 0.2076\Omega \\ Z_{\text{eq}} &= \frac{15}{17} = 0.882 \\ X_{\text{eq}} &= 0.8576 \end{aligned}$$

These quantities are referred to the HV side.

Since the turns ratio $a = 2$, then R_c' and X_m' (R_c and X_m, referred to the HV side) are $(968)2^2 = 3872\Omega$ and $(225.9)2^2 = 903.6\Omega$, respectively. The equivalent circuit, with all quantities referred to the HV side, is given in Figure 4.44.

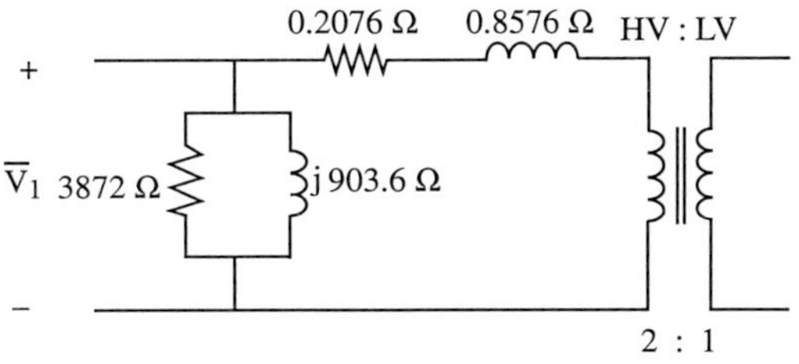

Figure 4.44: Equivalent circuit referred to HV side

4.5 Efficiency and Voltage Regulation

The transformer equivalent circuit is used in computing efficiency as well as voltage regulation under a wide variety of load conditions.

4.5.1 Efficiency

The efficiency of a transformer is defined as the ratio of output power to input power. It is usually expressed as a percentage. Thus,

$$\text{Efficiency } \eta = \frac{P_{\text{out}}}{P_{\text{in}}} = \frac{P_{\text{out}}}{P_{\text{out}} + \text{losses}} \times 100\%$$

Losses are the transformer copper loss P_c (also called I^2R loss) and the core losses P_i. Thus,

$$\eta = \frac{P_{\text{out}}}{P_{\text{out}} + P_c + P_i} \times 100\%$$

Alternatively, if power input is known, then

$$\eta = \frac{P_{\text{in}} - P_c - P_i}{P_{\text{in}}} \times 100\%$$

Example 4.10 A 150 kVA, 2400/240 V, 60 Hz transformer has $R_{\text{eq}} = 0.425\Omega$ and $X_{\text{eq}} = 0.914\Omega$, $R_c = 9931\Omega$ referred to the HV side. Find the efficiency when it supplies the load at rated voltage and rated current at 0.8 lag power factor at the low-voltage side.

Solution: Let the HV side be side 1 and the LV side be side 2. The equivalent circuit is shown in Figure 4.45.

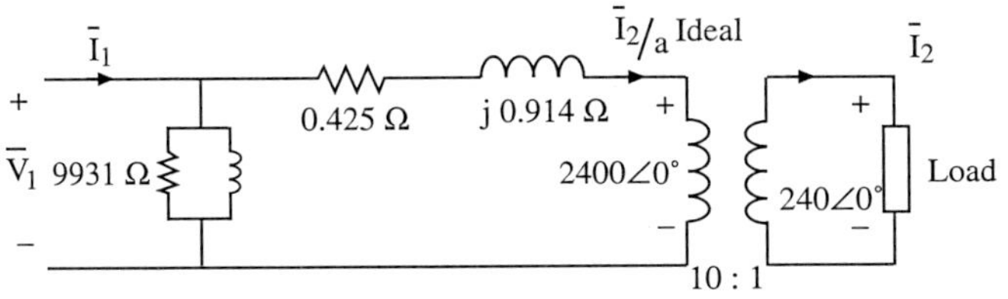

Figure 4.45: Equivalent circuit

$$\begin{aligned}
\text{Rated current on LV side} &= \frac{\text{Rated VA}}{\text{Rated V}} \\
&= \frac{150 \times 1000}{240} = 625 \text{ A} \\
\text{PF angle} &= \cos^{-1} 0.8 = 36.87^\circ
\end{aligned}$$

Hence, $\overline{I}_2 = 625\angle - 36.87°$. Turns ratio $a = 10$

$$\begin{aligned}
a\overline{V}_2 &= 2400\angle 0° \text{ V}, \frac{\overline{I}_2}{a} = 62.5\angle - 36.87° \text{ A} \\
\overline{V}_1 &= 2400\angle 0° + (0.425 + j0.914)62.5\angle - 36.87° \text{ V} \\
&= 2455.7\angle 0.69° \text{ V} \\
P_c &= (62.5)^2(0.425) = 1600 \text{ W} \\
P_i &= \frac{(2455.7)^2}{9931} = 607 \text{ W} \\
\eta &= \frac{150 \times 10^3 \times 0.8}{150 \times 10^3 \times 0.8 + 1660 + 607} = 0.9814 = 98.14\%
\end{aligned}$$

4.5.2 Voltage regulation

The percent regulation is defined as

$$\% \text{ voltage regulation} = \frac{V_{\text{no load}} - V_{\text{load}}}{V_{\text{load}}} \times 100$$

The voltage is measured at the terminals where the load is applied.

Example 4.11 Calculate the voltage regulation in Example 4.10. Under no load, there will be no current in the load, and hence, $\overline{I}_1 = \frac{\overline{I}_2}{a} = 0$. The voltage on the low-voltage side will therefore be $\frac{2455.7}{10} = 245.57$ V. Hence,

$$\begin{aligned}
\% \text{ voltage regulation} &= \frac{245.57 - 240}{240} \times 100 \\
&= 2.32\%
\end{aligned}$$

Example 4.12 Repeat Example 4.10 when the load is 100 kW at 0.8 PF lead. Also calculate the voltage regulation.

Solution:

$$
\begin{aligned}
a &= \frac{2400}{240} = 10 \\
\overline{V}_2 &= 240\angle 0^\circ \\
\overline{S}_2 &= \frac{100}{0.8}(0.8 - j0.6) = 100 - j75 \text{ kVA} \\
\overline{I}_2 &= \left(\frac{S_2^*}{V_2^*}\right) = \frac{(100 + j75)10^3}{240} = 520.83\angle 36.87^\circ \text{ A} \\
\overline{V}_1 &= a\overline{V}_2 + (0.425 + j0.914)(52.083\angle 36.87^\circ) \\
&= 2389.7\angle 1.23^\circ \text{ V} \\
P_c &= 0.425 \times (52.083)^2 = 1152.87 \text{ W} \\
P_i &= \frac{(2389.7)^2}{9931} = 575.03 \text{ W} \\
\eta &= \frac{P_{\text{out}}}{P_{\text{out}} + P_c + P_i} = \frac{100 \times 10^3}{100 \times 10^3 + 1152.87 + 575.03} \\
&\quad \times 100 \\
&= 98.3\% \\
\%\text{ voltage regulation} &= \frac{\frac{2389.7}{10} - 240}{240} = -0.43\%
\end{aligned}
$$

Note that with a leading PF, voltage regulation is negative.

4.6 Conclusion

In this section, we have discussed the principles underlying the phenomena of mutual inductance, magnetic circuits, and transformers. The convention regarding the sign of mutual inductance was carefully pointed out. The concept of reluctance was found useful in the analysis of magnetic circuits. In the analysis of physical transformers, we found it convenient to introduce the concept of the ideal transformer. An equivalent circuit of the physical transformer was developed. Both open- and short-circuit tests are explained to determine the parameters of the equivalent circuit.

4.7 Problems

4.1 For the circuit in Figure 4.46, find the inductance of the coil and the flux linkage when the current $i = 5$ A. The coil has $N = 200$ turns.

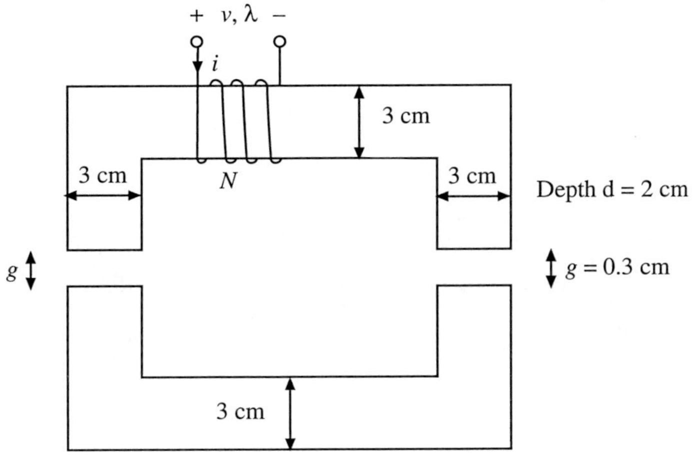

Figure 4.46: Magnetic circuit for Problem 4.1

Include the effect of fringing in the air gaps. The mean iron path length is 40 cm, and the relative permeability of the iron core is $\mu_r = 2000$.

4.2 In Figure 4.47, the reluctances are $\mathcal{R}_1 = 2 \times 10^6$ AT/Wb, $\mathcal{R}_2 = 2 \times$

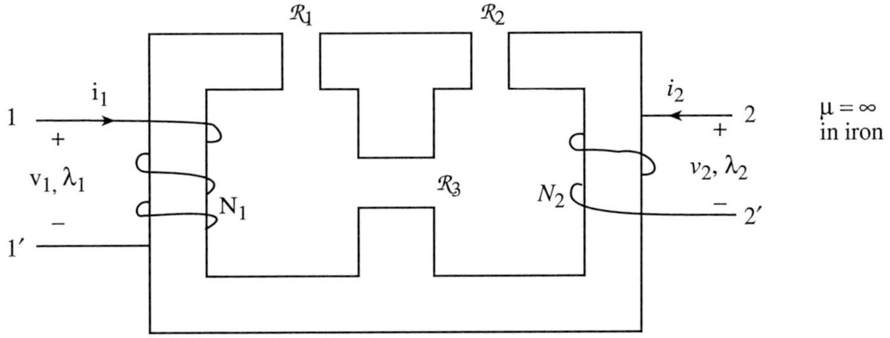

Figure 4.47: Magnetic circuit for Problem 4.2

10^6 AT/Wb, $\mathcal{R}_3 = 3 \times 10^6$ AT/Wb, $N_1 = 100$ turns, and $N_2 = 50$ turns.

a) Put dot markings on the two coils.

b) Find the flux linkages λ_1 and λ_2 in terms of i_1 and i_2.

c) Reverse the current direction in coil 2 and rewrite λ_1 and λ_2 with the new orientation. (Partial answers: $\lambda_1 = 3.125 \times 10^{-3} i_1 + 0.9375 \times 10^{-3} i_2$; $\lambda_2 = -0.9375 \times 10^{-3} i_1 - 0.7812 \times 10^{-3} i_2$)

4.3 In the magnetic circuit shown in Figure 4.48, for a current $i = 5$ A and

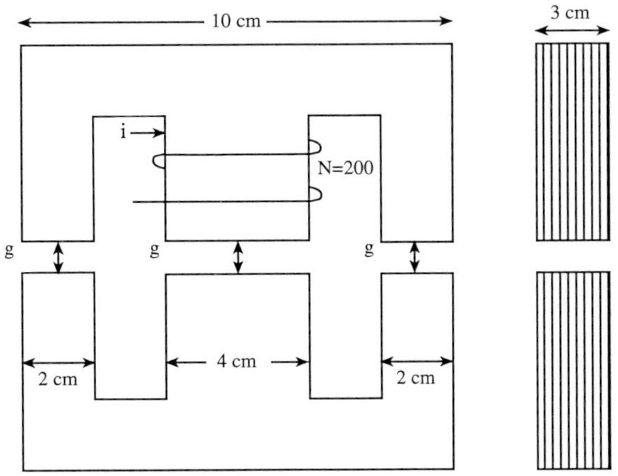

Figure 4.48: Magnetic circuit for Problem 4.3

$g = 0.25$ cm, find flux ϕ_c in the center leg and each of the outer legs. Account for fringing. $\mu = \infty$ in iron. (Answer: $\phi_c = 3.57 \times 10^{-4}$ Wb)

4.4 Figure 4.49 shows two mutually coupled coils. (a) Put the dot markings on the coils, (b) compute the flux linkages λ_1 and λ_2, and (c) determine the values of L_1, L_2, and M.

4.5 Two mutually coupled coils are shown in Figure 4.50

a) Put the dot markings in the two coils.

b) Find the self inductance of the two coils when the dotted terminals are connected together.

c) Repeat (b) when the dotted terminal of one coil is connected to the undotted terminal of the other coil.

(Hint: Write the loop equations in each case)

4.6 In the magnetic circuit of Figure 4.51, all legs have cross sections 1 cm by 1 cm.

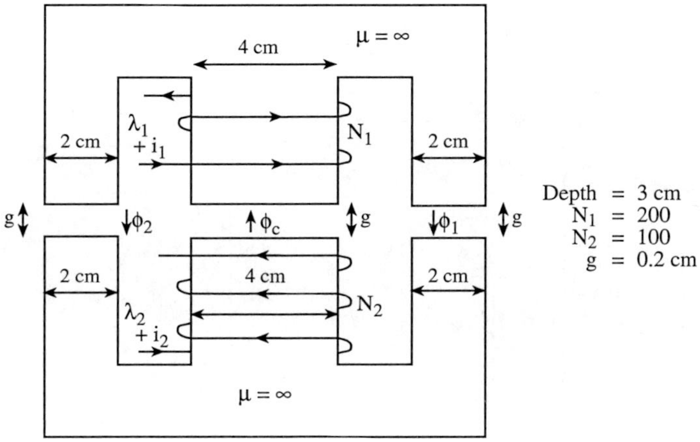

Figure 4.49: Magnetic circuit for Problem 4.4

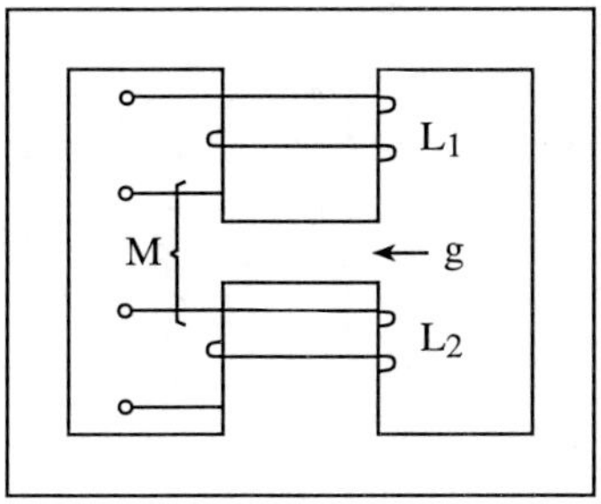

Figure 4.50: Magnetic circuit for Problem 4.5

a) Put polarity markings on the coils.

b) Determine the self inductances of coils 1 and 2.

c) Determine the mutual inductance.

4.7 The magnetic circuit shown in Figure 4.52 has a depth of 2 cm. The relative permeability of the iron core is $\mu_r = 1000$, each winding has $N = 400$ turns, and the two currents are $I_1 = 1$ A and $I_2 = 1.2$ A.

a) Find the total reluctance of the circuit and the flux linkage of each winding, including the effect of fringing.

b) Find the error made by neglecting the reluctance of the iron core.

c) Find the error made by neglecting the fringing in the air gap.

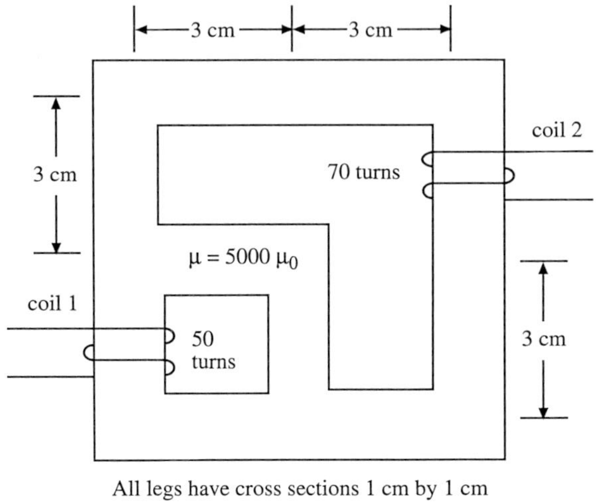

Figure 4.51: Magnetic circuit for Problem 4.6

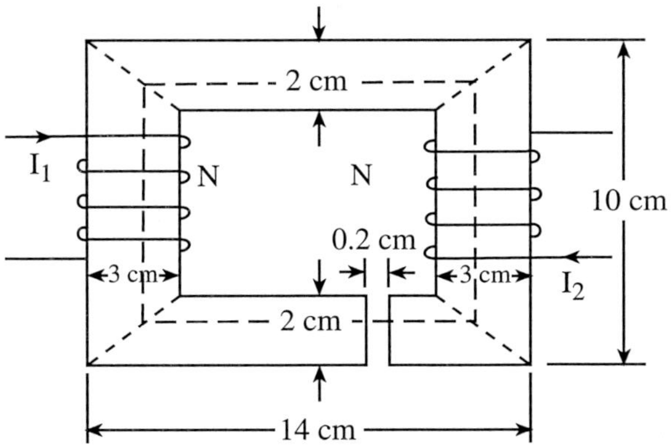

Figure 4.52: Magnetic circuit for Problem 4.7

d) Calculate the self and mutual inductance of the two windings.

Hint: Use the dashed center line to calculate the mean length of each leg.

4.8 Figure 4.53 shows the magnetic circuit of a relay. The relative permeability of the iron is $\mu_r = 1000$. Find the magneto-motive force (mmf) necessary to have a magnetic flux of 10^{-4} Wb in the core. (Neglect the fringing in the air gap).

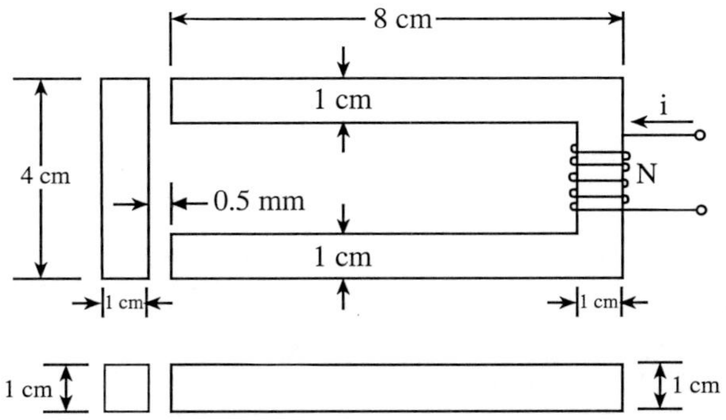

Figure 4.53: Magnetic circuit for Problem 4.8

4.9 Consider the two coupled coils in Figure 4.54.

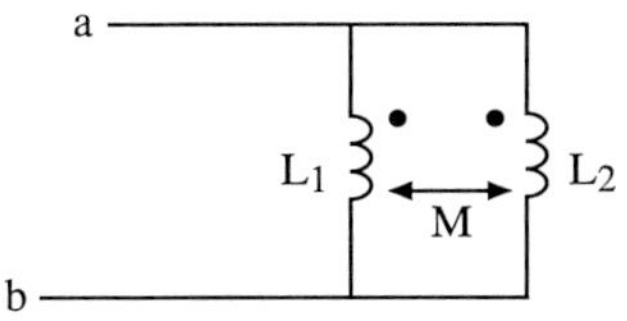

Figure 4.54: Coupled coils for Problem 4.9

a) Show that the inductance at the terminals ab is $\frac{L_1 L_2 - M^2}{L_1 + L_2 - 2M}$.

b) If the connections of coil 2 are reversed, show that $L_{ab} = \frac{L_1 L_2 - M^2}{L_1 + L_2 + 2M}$.

4.10 A 240/120 V, 4.8 kVA, 60 Hz, single-phase transformer is used to supply a 4.8 kVA resistive load (i.e., the load has a power factor of 1.0) at rated voltage of 120 V.

a) If you assume the transformer is ideal, what would be the magnitude of the expected primary side (240 V side) current?

b) Again, if you assume the transformer is ideal, what is the apparent impedance of the load viewed from the high-voltage side of the transformer?

Next, the transformer is tested to determine its parameters. Surprisingly, during the open circuit, $I_{\text{oc}} = 0$ and $P_{\text{oc}} = 0$. During the short-circuit test, in which a voltage is applied to the high-voltage side in order to get rated current to flow in the low-voltage side, $V_{\text{sc}} = 20$ V and $P_{\text{sc}} = 100$ W

c) Determine R_{eq} and X_{eq} referred to the HV side.

d) Determine X_m and R_c referred to the HV side.

e) Using the model values determined from the test data, what voltage must be applied to the high-voltage side in order to supply the 4.8 kVA resistive load at a rated voltage of 120 V?

4.11 A 2400/240 V transformer has $R_1 = 3\Omega$, $R_2 = 0.03\Omega$, $X_{\ell 1} = 6.5\Omega$, $X_{\ell 2} = 0.07\Omega$. (Side 1 is the HV side and side 2 is the LV side). $R_c = 20k\Omega$, $X_m = 200k\Omega$ both referred to side 1.

a) Find the percentage regulation if a load of 16 kW at 0.8 PF lag is supplied at the low-voltage end at *rated* voltage of 240 V.

b) Find efficiency at this load. Use the approximate equivalent circuit referred to the 2400 V side (side 1).

c) Repeat (a) and (b) for the same load and PF (lead).

(Answer: (a) 4.51%, (b) 95.6%, (c) −1%, 95.8%)

4.12 A transformer shown in Figure 4.55 with self inductances L_1 and L_2

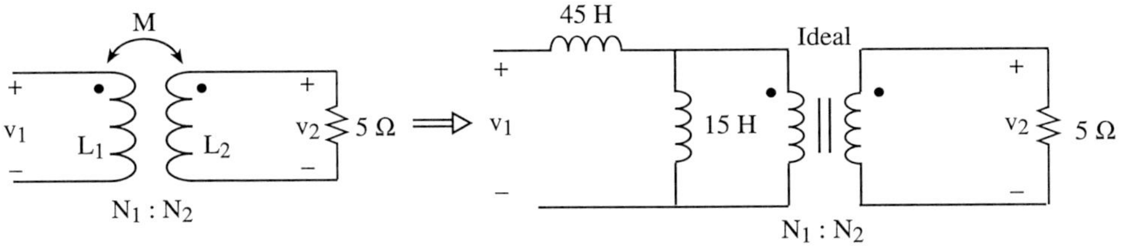

Figure 4.55: Transformer circuit for Problem 4.12

and mutual inductance M has a load of 5Ω connected across side 2. The turns ratio $N_1 : N_2 = 5$. The equivalent circuit is shown to the right.

a) Find L_1, L_2, M, and the coefficient of coupling k.

b) If the source is $v_1(t) = \sqrt{2}\ 100 \sin 10t$, write the loop equations in the phasor domain (you may write for either of the two figures).

(Partial answer: $L_1 = 60$ H, $k = 0.5$)

4.13 A 10 kVA, 2400/240 V transformer has $R_c = 340\Omega$ and $X_m = 250\Omega$ referred to the low-voltage side. R_{eq} and X_{eq} referred to the HV side are 9Ω and 15Ω, respectively. The transformer is supplying a load of 8 kVA on the low-voltage side at 0.94 PF lagging at 230 V. Find voltage regulation and efficiency. (Answer: Regulation = 2.07%, efficiency = 96.5%)

4.14 The circuit in Figure 4.56 shows a set of coupled coils.

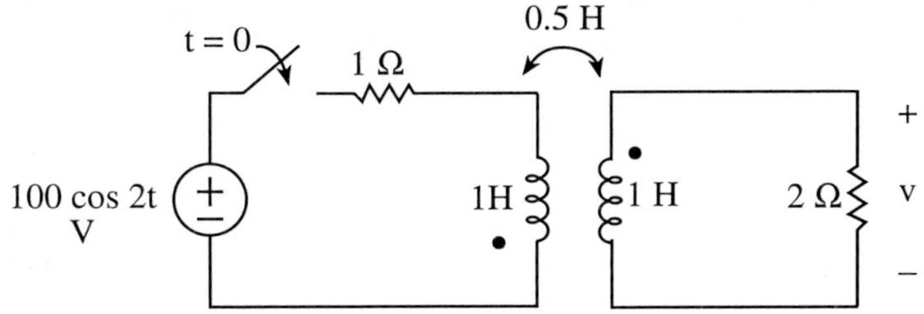

Figure 4.56: Coupled circuit for Problem 4.14

a) Write the loop equations.

b) Solve for $v(t)$ in the steady state using phasors.

4.15 For the circuit of Figure 4.57

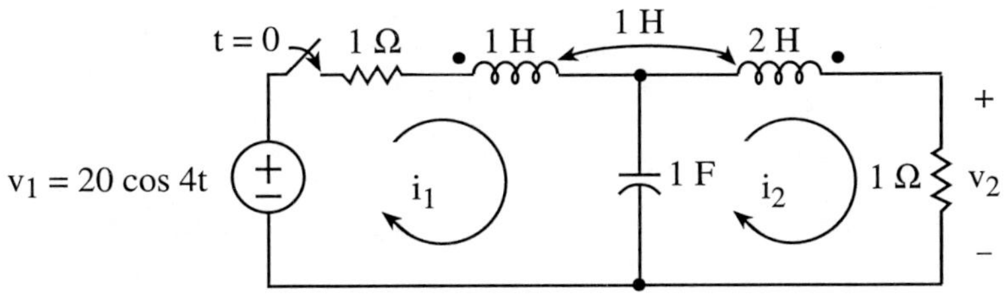

Figure 4.57: Coupled circuit for Problem 4.15

a) Write the loop equations for the two loops shown.

b) Solve for $v_2(t)$ in the steady state. Assume the initial conditions of the capacitor and inductors as zero.

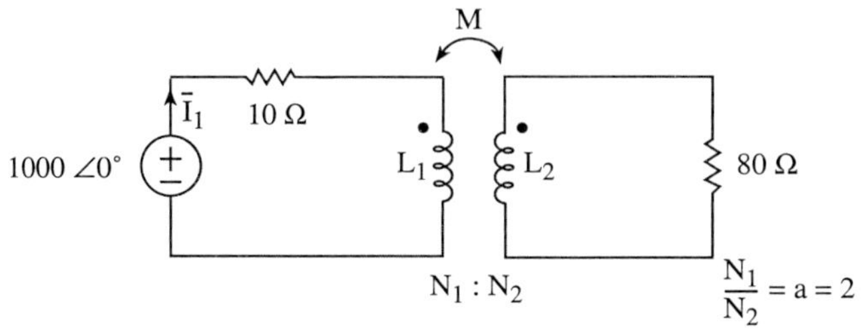

Figure 4.58: Coupled circuit for Problem 4.16

4.16 In the circuit of Figure 4.58, the parameters are $L_1 = 8$ H, $L_2 = 2$ H, $R_1 = 20\Omega$, $R_2 = 20\Omega$, and $k = 0.98$. The frequency $\omega = 1000$ rad/sec, $N_1/N_2 = 2$. Core losses are negligible. Draw the approximate equivalent circuit and compute the source current $\overline{I}_1$. (Answer: $\overline{I}_1 = 1.94\angle -39.5^\circ$ A)

4.17 Two identical but mutually coupled coils are connected in two ways, and inductance is measured in each case.

a) *In series* with the undotted terminal of one coil connected to the dotted terminal of the other coil. $L_{\text{eq}} = 0.08$ H.

b) *In series* with the undotted terminal of one coil connected to the undotted terminal of the other coil. $L_{\text{eq}} = 0.035$ H.

Find L, M, and k. (Answer: $L = 28.75$ mH; $k = 0.392$)

4.18 A single-phase 20 kVA, 2400/240 V, 60 Hz transformer has core loss $P_i = 120$ W and copper loss of 80 W at half-rated current. Find efficiency when the transformer delivers load at full-rated current at 0.8 PF.

4.19 A single-phase 100 kVA, 2000/20 V transformer has the following test results. *Open-circuit (OC) test* (instrumentation on LV side) 200 V, 8.0 A, 600 W. *Short-circuit (SC) test* (instrumentation on HV side) 100 V, 50 A, 2000 W.

a) Determine the approximate equivalent circuit referred to the HV side.

b) Find the voltage regulation at full-rated current delivering at rated voltage and PF of 0.8 lagging.

c) Repeat (b) for PF of 0.6 leading.

4.20 The output of a transformer is $V_2 I_2 \cos\theta$. The core losses is denoted as P_i and the copper loss is $I_2^2 R_{2eq}$. Show analytically that the maximum efficiency occurs when $P_i = I_2^2 R_{2eq}$. Assume V_2 and θ are constant.

4.21 A short-circuit test on a 50 kVA, 2400/240 V transfer is performed with instrumentation on the high-voltage side. The readings are $I = 20.8$ A, $V = 48$ V, $P = 617$ W.

a) Find R_{eq} and X_{eq} referred to the high-voltage side and the low-voltage side.

b) The low-voltage side now carries a full-rated load at 0.8 PF lagging and at rated voltage. Find the voltage at the high-voltage side needed to supply the load.

c) Find the voltage regulation with the load of part (b).

(Answer: (a) 1.43Ω, 1.81Ω; (b) 2446.5 V; (c) 1.9366%)

4.22 The results of the OC and SC test on a 150 kVA, 2400/240 V, 60 Hz transformer are as follows

OC test (LV side)	$V_{\text{oc}} = 240$ V, $I_{\text{oc}} = 16.75A$, $P_{\text{oc}} = 580$ W
SC test (HV side)	$V_{\text{sc}} = 63V$, $I_{\text{sc}} = 62.5A$, $P_{\text{sc}} = 1660$ W

Find parameters of the approximate equivalent circuit, referred to (a) HV side, and (b) LV side. In each case, draw the equivalent circuit.

(Partial answer: $R_{eq_1} = 0.425\Omega$, $X_{eq_1} = 0.915\Omega$, Side 1 = HV side)

4.23 In Problem 4.22, find the regulation and efficiency if the transformer supplies load at rated voltage and rated current on the LV side at a PF of 0.8 lag.

Chapter 5

ANALYSIS OF ELECTROMECHANICAL SYSTEMS USING ENERGY METHODS

5.1 Introduction

In this chapter, we will study magnetic circuits in which there is a moving member. It can be either a translational or rotational system. The purpose is to present techniques of systematic derivation of mathematical models (differential equations) for lumped-parameter electromechanical systems. The theory is outlined for N electrical coils and M mechanical variables. For application purposes, we assume only one mechanical variable, but there may be more than one set of electrical coils that interact to produce the force or torque on the mechanical system. This force or torque of electric origin will be derived from fundamental principles of stored energy in the magnetic field. In general, both the currents in the electrical coils and the force or torque are time varying. We get a set of coupled electromechanical differential equations. These equations are then put in the form of first-order differential equations which is called the state space form. Such a formulation is convenient for computer simulation, analysis, and design. The numerical solution of these equations is briefly discussed. Thus, a system viewpoint valid for a wide class of electromechanical systems is presented.

5.2 Translational Systems

5.2.1 Application of laws of electromagnetism

As discussed in Section 4.1, there are four basic laws derived from Maxwell's equations that are directly applicable. The variables involved are $\underline{H}$ (magnetic field intensity), $\underline{J}_f$ (current density), $\underline{B}$ (magnetic flux density), and $\underline{E}$ (electric field intensity). The relationship between $\underline{B}$ and $\underline{H}$ is assumed to be linear.

We illustrate these concepts on the simple system of Figure 5.1. N turns

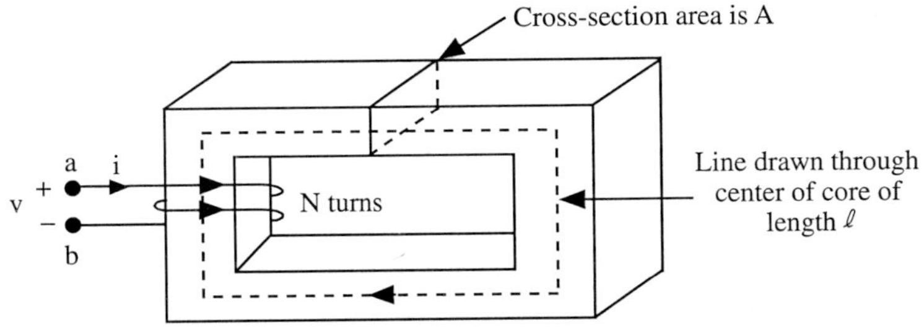

Figure 5.1: Illustrative example

of wire are wound on a magnetic core material uniformly. The current is flowing in the direction shown. There is flux inside the core, which according to the right hand rule, is in the direction shown. The flux is uniform across the cross section of the core.

Ampere's circuital law (ACL)

Take a vertical cross section. For simplicity, the direction $d\ell$ of the contour c is chosen to agree with the direction of $\underline{H}$ inside the core. Once c is chosen, we automatically have the surface s and the normal $\underline{n}$. Applying ACL along the contour in Figure 5.2,

$$\oint_c \underline{H} \bullet \underline{dl} = \int_s \underline{J}_f \bullet \underline{n} \ \ da \Rightarrow Hl = Ni \tag{5.1}$$

Faraday's law

The contour c is along the wire in the direction of the current in the coil and from b to a outside the coil, as shown in Figure 5.1. Imagine N spiral cross sections along the winding for the N turns. The surface s is sum of

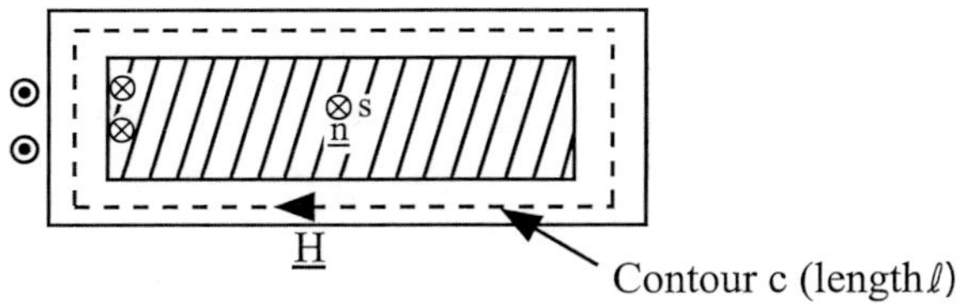

Figure 5.2: Contour c and surface s

the areas of the cross sections. Each cross section has the same area A and flux density B. For each area s_i,

$$\int_{s_i} \underline{B} \bullet \underline{n} \ \ da = B \times A \stackrel{\Delta}{=} \phi \tag{5.2}$$

where ϕ is the flux through the core. It is also the flux ϕ through one turn of the coil. Therefore, for N turns

$$\int_s \underline{B} \bullet \underline{n} \ \ da = \sum_{i=1}^{N} \int_{s_i} \underline{B} \cdot \underline{n} da = N \times B \times A = N\phi \tag{5.3}$$

$\oint_c \underline{E} \bullet \underline{dl}$ gives us $-v$ since E field along the conductor is zero. More specifically,

$$\begin{aligned} \oint_c \underline{E} \bullet \underline{dl} &= \int_b^a \underline{E} \bullet dl + \int_{\text{along cond.}} \underline{E} \bullet dl \\ &= \int_b^a E \bullet dl + 0 \\ &= -v \end{aligned} \tag{5.4}$$

Hence, by Faraday's law

$$\oint_c \underline{E} \bullet \underline{dl} = -\frac{d}{dt} \int_s \underline{B} \bullet \underline{n} \ \ da \tag{5.5}$$

becomes

$$-v = -\frac{d}{dt} N\phi \qquad v = \frac{d}{dt} N\phi \tag{5.6}$$

Define a quantity λ called flux linkage $\stackrel{\Delta}{=} N\phi$. Therefore, $v = \frac{d\lambda}{dt}$. Application of Gauss's law for a magnetic field depends on the geometry and *is needed* when different $\underline{H}$'s are present in an electromagnetic system. It is also used to establish continuity of fluxes. The conservation of charge law implies that the algebraic sum of currents at a node is zero, which is the same as Kirchhoff's current law (KCL). We are now in a position to apply these basic laws to electromechanical devices.

5.2.2 Structure of an electromechanical system

The structure of an electromechanical system is shown in Figure 5.3.

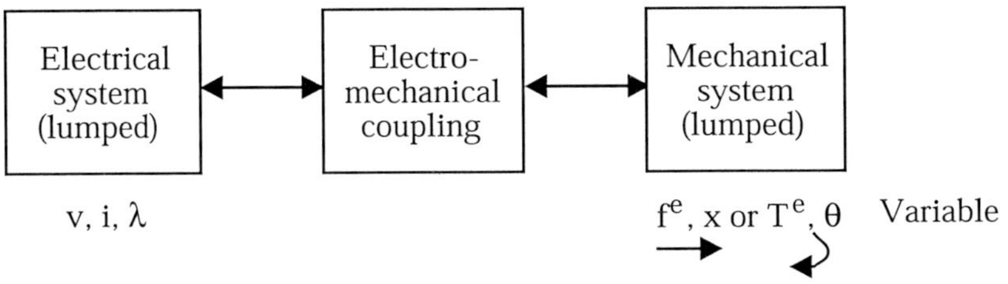

Figure 5.3: Electromechanical system

For analytical purposes, the electrical, mechanical, and coupling system systems are shown separately. Kirchhoff's laws are used to write the electrical equations, and Newton's law is used to write the mechanical equations. f^e and T^e are the force and torque of electric origin, respectively. λ depends explicitly on (i, x) or (i, θ). Considering the translational system, with one electrical and one mechanical variable, $\lambda = \lambda(i, x)$. We compute λ using ACL and Gauss' law. When the geometry is simple, the magnetic equivalent circuit can be used as in Chapter 4 to derive the flux linkages. Then by Faraday's law,

$$v = \frac{d\lambda}{dt} = \frac{\partial\lambda}{\partial i}\frac{di}{dt} + \frac{\partial\lambda}{\partial x}\frac{dx}{dt} \tag{5.7}$$

The first term proportional to $\frac{di}{dt}$ is termed the *transformer* voltage, and the second term proportional to $\frac{dx}{dt}$ is called the *speed voltage.*

Special case: Electrically linear system

In this case, we have

$$\lambda = L(x)i \tag{5.8}$$

Therefore,

$$\begin{aligned} v &= L(x)\frac{di}{dt} + i\frac{dL(x)}{dx}\frac{dx}{dt} \\ &= \text{transformer voltage} + \text{speed voltage} \end{aligned} \tag{5.9}$$

Further special case

For a given fixed geometry (no moving parts), then L is constant and $\frac{dL}{dx} = 0$. Hence,

$$\lambda = Li \tag{5.10}$$

$$v = L\frac{di}{dt} \tag{5.11}$$

This is the familiar characterization of an ideal inductor.

Multi-port system

Suppose we have an electromechanical system that has N electrical terminal pairs (N-ports) and M mechanical variables (M-ports) as shown in Figure 5.4. Let the variables be denoted by $i_1, i_2, \ldots, i_N$ and $x_1, x_2, \ldots, x_M$,

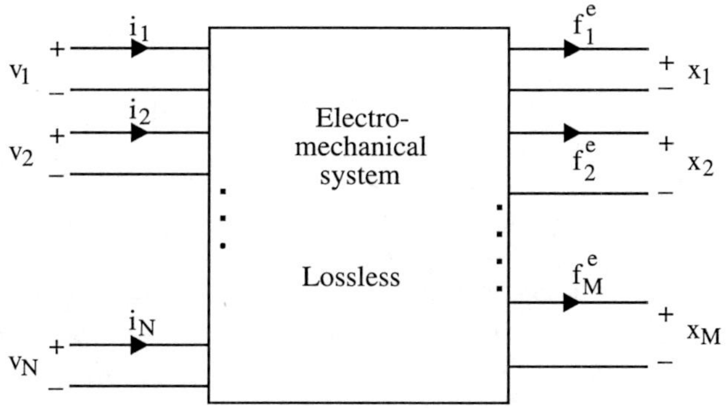

Figure 5.4: Multi-port system

respectively. Then, in principle, λ_k at any electrical port can be function of all the variables, i.e.,

$$\lambda_k = \lambda_k(i_1, i_2, \ldots, i_N, x_1, x_2, \ldots, x_M), \qquad k = 1, \ldots, N \tag{5.12}$$

Then,

$$v_k = \frac{d\lambda_k}{dt} = \sum_{j=1}^{N} \frac{\partial \lambda_k}{\partial i_j}\frac{di_j}{dt} + \sum_{j=1}^{M} \frac{\partial \lambda_k}{\partial x_j}\frac{dx_j}{dt} \qquad k = 1, 2, \ldots, N \tag{5.13}$$

Just like λ_k in (5.12), the force of electric origin at the mechanical ports can be functions of all the variables, i.e.,

$$f_i^e = f_i^e(i_1, i_2, \ldots, i_N, x_1, x_2, \ldots, x_M)$$

We will derive expressions for $f_i^e (i = 1, 2, \ldots, M)$ in Section 5.7.

Example 5.1 This example is adapted from [14].

Figure 5.5 shows the basic configuration used in tripping circuit breakers.

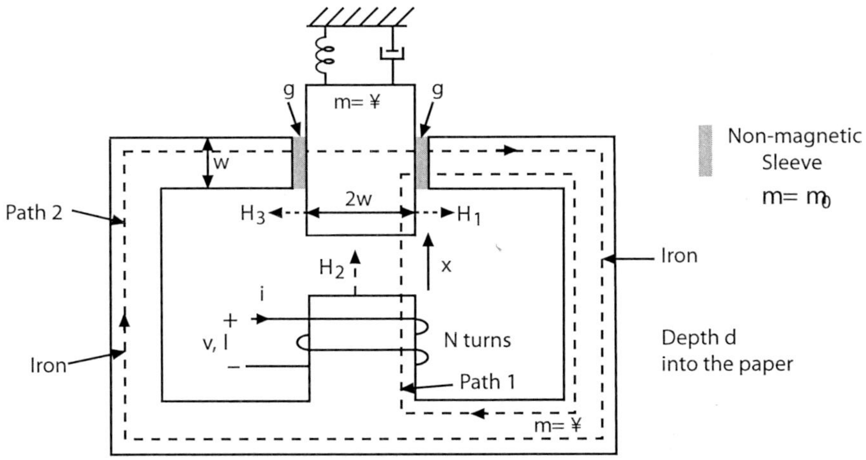

Figure 5.5: A magnetic field system

It consists of a fixed structure made of highly permeable material with an excitation winding of N turns. A movable plunger, also made of highly permeable material, is constrained by a non-magnetic sleeve to move in the x direction, as shown in the figure. This basic configuration is used for tripping circuit breakers, operating valves, etc., where a large force is required to move a member through a relatively small distance. To simplify analysis, 1) the permeability of iron $\mu = \infty$, 2) $g \gg w$, $x \gg 2w$, and 3) there is no leakage flux. In the iron, $B = \mu H$. Since $\mu = \infty$ and B is finite, we must have $H = 0$ in the iron. Hence, non-zero H exists only in the air gaps g and x. Find H_1, H_2, λ, and v.

Solution: Using ACL and Faraday's law Two paths are shown in Figure 5.5 for applying ACL.

Steps

1. Ampere's circuital law around path 2 gives $H_1 g - H_3 g = 0$, i.e., $H_1 = H_3$. Ampere's circuital law around path 1 gives

$$H_1 g + H_2 x = Ni \tag{5.14}$$

2. Gauss's law for the magnetic field for a closed surface surrounding the plunger gives

$$2(\mu_o H_1)(wd) - \mu_o H_2(2wd) = 0 \tag{5.15}$$

3. Solving (5.14) and (5.15) for H_1 and H_2, we get

$$H_1 = H_2 = \frac{Ni}{g + x} \tag{5.16}$$

4. Compute flux linkage of the coil. The flux ϕ through one turn of the coil is given by

$$\Phi = \underbrace{\mu_o H_2}_{B} \times \underbrace{(2wd)}_{A} = \frac{2wd\mu_o Ni}{(g + x)} \tag{5.17}$$

The flux linkage of the coil λ is therefore

$$\lambda = N\phi = \frac{2wd\mu_0 N^2 i}{(g + x)} \tag{5.18}$$

Note from (5.18) that this is an electrically linear system and hence, $\lambda = L(x)i$ where

$$L(x) = \frac{2wd\mu_0 N^2}{g + x} \tag{5.19}$$

Using Faraday's law, the voltage v is given by $v = d\lambda/dt$, which simplifies as $L(x)(di/dt) + i(dL(x)/dx)(dx/dt)$. This yields

$$v = \underbrace{\frac{2wd\mu_0 N^2}{(g + x)} \frac{di}{dt}}_{\text{Transformer voltage}} - \underbrace{\frac{2wd\mu_0 N^2 i}{(g + x)^2} \frac{dx}{dt}}_{\text{Speed voltage}} \tag{5.20}$$

Solution: Using magnetic equivalent circuit There are three air gaps (the "non-magnetic sleeves" of the plunger are also treated as air gaps) where there is finite H. Their reluctances are

$$\mathcal{R}_1 = \mathcal{R}_3 = \frac{g}{\mu_o w d}$$
$$\mathcal{R}_2 = \frac{x}{\mu_o 2wd}$$

The magnetic equivalent circuit is given in Figure. 5.6. Note the orientation

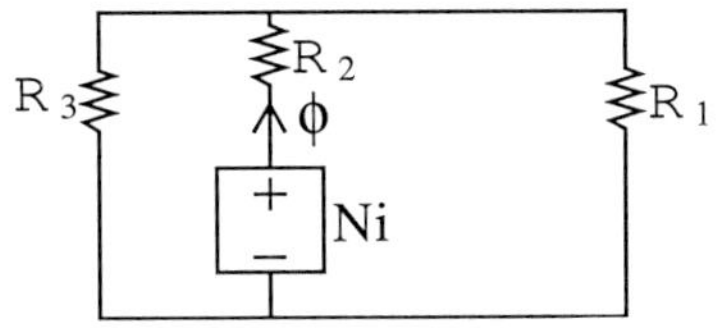

Figure 5.6: Magnetic equivalent circuit

of the mmf Ni and flux ϕ. Solve for ϕ, λ to agree with (5.17) and (5.18). This is left as an exercise. In translational systems, one can use either Maxwell's equations or the magnetic equivalent circuit approach. In rotational systems, the use of Maxwell's equations is easier.

5.3 Rotational Systems

Example 5.2 This example is also adapted from [14].

Consider the rotational electromechanical system in Figure 5.7. This is the basic configuration of many machines. Both rotor and stator have $\mu = \infty$. The radial air gap has length g. The cylindrical rotor is of length ℓ. At time t, θ is the angular displacement. Find λ_s, λ_r as a function of i_s, i_r, and θ. Also find v_s and v_r of the cylindrical rotor.

Solution: Using ACL and Faraday's law We will assume $g \ll R$ and ℓ. The magnetic field intensity in the air gap is radial. There are four regions with different values of H. We assume H to be constant in each region.

$$\begin{array}{ccccccc} 0 < \phi < \theta & , & \theta < \phi < \pi & , & \pi < \phi < \pi + \theta & , & \pi + \theta < \phi < 2\pi \\ H_{r1} & & H_{r2} & & H_{r3} & & H_{r4} \end{array}$$

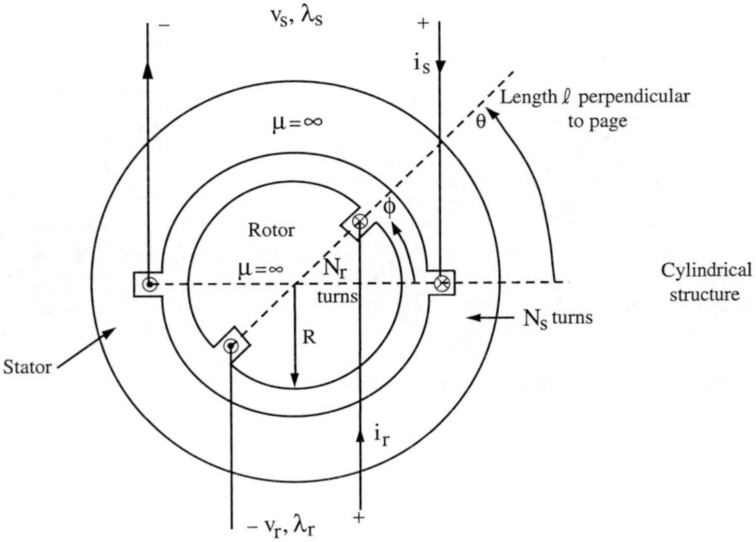

Figure 5.7: Doubly excited magnetic field system

1. From symmetry, we can conclude $H_{r1} = -H_{r3}$ and $H_{r2} = -H_{r4}$. We can also derive it and solve for the H_{ri}'s as follows. Apply ACL for the contours shown in Figure 5.8 and then apply Gauss's law.

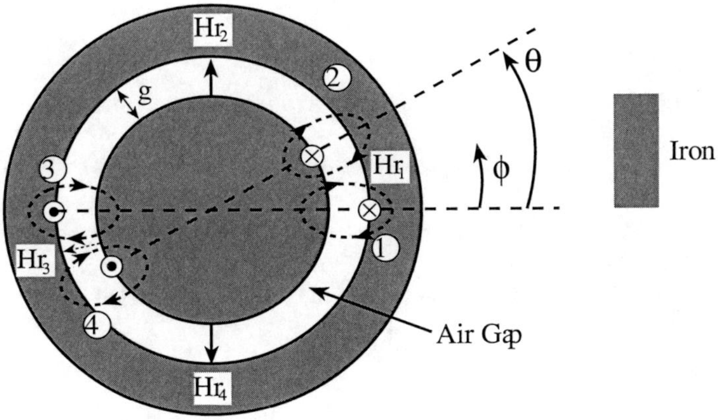

Figure 5.8: Contour paths

Ampere's circuital law

Contour 1
$$H_{r1} - H_{r4} = \frac{N_s i_s}{g} \quad (5.21)$$

Contour 2
$$H_{r2} - H_{r1} = \frac{N_r i_r}{g} \quad (5.22)$$

Contour 3
$$H_{r3} - H_{r2} = -\frac{N_s i_s}{g} \quad (5.23)$$

Contour 4
$$H_{r4} - H_{r3} = -\frac{N_r i_r}{g} \quad (5.24)$$

Gauss's law for magnetic fields

Enclose the rotor in a closed surface as shown in Figure 5.9. Applying

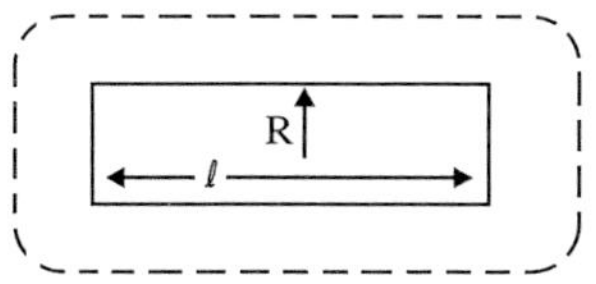

Figure 5.9: Closed surface around the rotor

Gauss's law to the closed surface, the total flux out of the rotor along the radial direction is zero, i.e.,

$$\oint_s \underline{B} \bullet \underline{n} da = 0$$

Multiplying by the circumferential area of each region, we get

$$\mu_0 l R(H_{r1}\theta + H_{r2}(\pi - \theta) + H_{r3}\theta + H_{r4}(\pi - \theta)) = 0$$

i.e.,

$$\mu_0 l R[\theta(H_{r1} - H_{r2} + H_{r3} - H_{r4}) + \pi(H_{r2} + H_{r4})] = 0 \quad (5.25)$$

Adding (5.21) and (5.23)

$$H_{r1} - H_{r2} + H_{r3} - H_{r4} = 0 \quad (5.26)$$

From (5.25) and using (5.26),

$$H_{r2} + H_{r4} = 0 \Rightarrow H_{r2} = -H_{r4} \quad (5.27)$$

Hence, from (5.26)

$$H_{r1} + H_{r3} = 0 \Rightarrow H_{r1} = -H_{r3} \tag{5.28}$$

2. To evaluate $H_{ri}(i = 1, 2, 3, 4)$, we proceed as follows. Subtracting (5.22) from (5.21) and using (5.27) gives

$$2H_{r1} = \frac{N_s i_s - N_r i_r}{g}$$

$$H_{r1} = \frac{N_s i_s - N_r i_r}{2g} = -H_{r3} \tag{5.29}$$

Subtracting (5.23) from (5.22) and using (5.28) gives

$$H_{r2} = \frac{N_s i_s + N_r i_r}{2g} = -H_{r4} \tag{5.30}$$

Thus (5.29) and (5.30) give $H_{ri}(i = 1, 2, 3, 4)$ in all the four regions.

3. *Computing flux linkages.* There are two coils, one on the stator and the other on the rotor.

Stator flux linkage λ_s

We first note that arc length of the coil with an angle θ is $R\theta$, where R is the radius. As illustrated in Figure 5.10, the flux ϕ_s through one turn

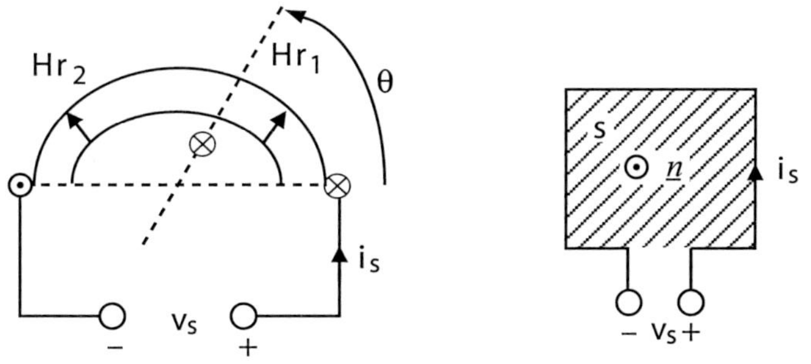

Figure 5.10: Flux through coil on the stator

of stator coil is given $\phi_s = \int_s \underline{B} \cdot \underline{n} da = \int_0^\theta \mu_o H_{r1} \ell R d\phi + \int_\theta^\pi \mu_o H_{r2} \ell R d\phi = (\mu_o H_{r1})(R\theta l) + (\mu_o H_{r2})(\pi - \theta)Rl$. The flux linkage is

$$\lambda_s = N_s \phi_s = N_s \mu_o H_{r1} R\theta l + N_s \mu_o H_{r2} R(\pi - \theta) l \tag{5.31}$$

Substitute for H_{r1} and H_{r2} from (5.29) and (5.30) in (5.31) and simplify:

$$\begin{aligned}\lambda_s &= \frac{N_s^2\mu_0 Rl\pi}{2g}i_s + N_sN_r\frac{\mu_0 Rl\pi}{2g}\left(1-\frac{2\theta}{\pi}\right)i_r \\ &= N_s^2L_0 i_s + N_sN_rL_0\left(1-\frac{2\theta}{\pi}\right)i_r \\ &= L_s i_s + L_m(\theta)i_r \qquad (5.32)\end{aligned}$$

where

$$\begin{aligned}L_s &= N_s^2L_o,\, L_o = \frac{\mu_0 Rl\pi}{2g},\, L_m(\theta) = N_sN_rL_o\left(1-\frac{2\theta}{\pi}\right) \\ \lambda_r &= N_r\int_s \underline{B}\bullet\underline{n}\; da \\ &= N_r\int_\theta^\pi \mu_0 H_{r2}\ell R d\phi + N_r\int_\pi^{\pi+\theta}\mu_0 H_{r3}\ell R d\phi. \qquad (5.33)\end{aligned}$$

Substituting for H_{r2} and H_{r3} from (5.29) and (5.30) and simplifying, we get

$$\begin{aligned}\lambda_r &= N_sN_rL_0\left(1-\frac{2\theta}{\pi}\right)i_s + N_r^2L_0 i_r \\ &= L_m(\theta)i_s + L_r i_r \qquad (5.34)\end{aligned}$$

Note that $L_r = N_r^2L_0$. The approach of using magnetic equivalent circuits is somewhat more complicated for rotational systems and hence, is not discussed.

Equations (5.32) and (5.34) represent an "electrically" linear system. L_s and L_r are constants, and L_m is a linear function of θ. The above relations are valid for $0 < \theta < \pi$. For $-\pi < \theta < 0$, simply replace θ by $-\theta$ so that $L_m = N_sN_rL_o\left(1+\frac{2\theta}{\pi}\right)$. The mutual inductance is sketched in Figure 5.11. Practical machines are designed so that $L_m = M\cos\theta$. This is accomplished by putting additional slots and windings around the periphery of the stator and the rotor. By a proper distribution of slots and the number of turns, L_m can be made the sinusoidal function shown in Figure 5.11. If $L_m = M\cos\theta$, then

$$v_s = \frac{d\lambda_s}{dt} = L_s\frac{di_s}{dt} + M\cos\theta\frac{di_r}{dt}\underbrace{-i_rM\sin\theta\frac{d\theta}{dt}}_{\text{Speed voltage}} \qquad (5.35)$$

$$v_r = \frac{d\lambda_r}{dt} = L_r\frac{di_r}{dt} + M\cos\theta\frac{di_s}{dt}\underbrace{-i_sM\sin\theta\frac{d\theta}{dt}}_{\text{Speed voltage}} \qquad (5.36)$$

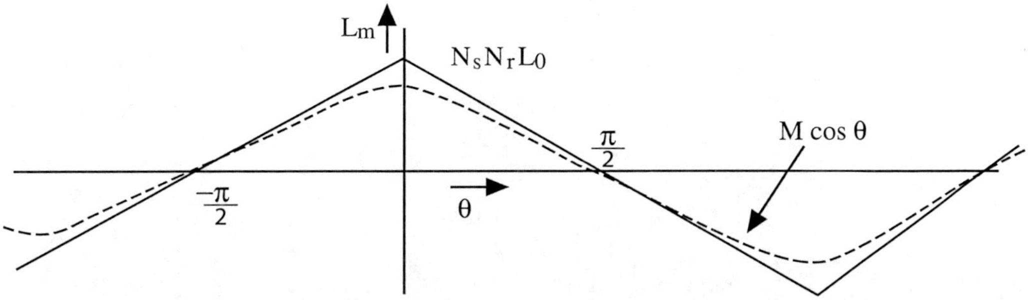

Figure 5.11: L_m as function of θ

The third term on the right sides of v_s and v_r are the "speed" voltages.

Electrically linear systems

For an electrically linear system, the $\lambda - i$ relations can be rewritten so that i's are expressed in terms of λ as follows.

(i) For one coil in a translational system,

$$\lambda = L(x)i \text{ and } i = \frac{\lambda}{L(x)} \tag{5.37}$$

(ii) For two coils in a translational system

$$\begin{aligned} \lambda_s &= L_s i_s + L_m(x) i_r \\ \lambda_r &= L_m(x) i_s + L_r i_r \end{aligned} \tag{5.38}$$

$$\begin{aligned} \begin{bmatrix} i_s \\ i_r \end{bmatrix} &= \begin{bmatrix} L_s & L_m(x) \\ L_m(x) & L_r \end{bmatrix}^{-1} \begin{bmatrix} \lambda_s \\ \lambda_r \end{bmatrix} \\ &= \frac{1}{L_s L_r - L_m^2(x)} \begin{bmatrix} L_r & -L_m(x) \\ -L_m(x) & L_s \end{bmatrix} \begin{bmatrix} \lambda_s \\ \lambda_r \end{bmatrix} \end{aligned} \tag{5.39}$$

$$i_s = \frac{1}{L_s L_r - L_m^2(x)} (L_r \lambda_s - L_m(x) \lambda_r) \tag{5.40}$$

$$i_r = \frac{1}{L_s L_r - L_m^2(x)} (-L_m(x) \lambda_s + L_s \lambda_r) \tag{5.41}$$

If it is a rotational system, x is replaced by θ.

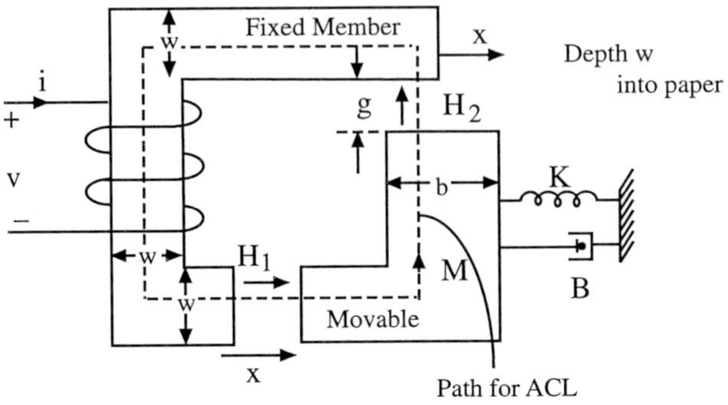

Figure 5.12: Electromechanical system

Example 5.3 Consider the magnetic circuit in Figure 5.12: $\mu = \infty$ in the iron path. Mass M moves only in the x direction.

1. Write the equations to solve for H_1 and H_2.

2. Compute $\lambda = \lambda(i, x)$ using both methods.

Solution: Using field theory For the path shown, the ACL gives

$$H_1 x + H_2 g = Ni \tag{5.42}$$

Gauss's law for the moving portion:

$$-(\mu_o H_1)w^2 + (\mu_o H_2)(b - x)w = 0 \tag{5.43}$$

Hence,

$$H_2 = \frac{H_1 w}{(b - x)}$$

Substitute H_2 in (5.42) to get

$$H_1 x + \frac{H_1 w g}{(b - x)} = Ni \tag{5.44}$$

Therefore, solving for H_1,

$$H_1 = \frac{Ni(b - x)}{x(b - x) + gw} \tag{5.45}$$

Hence,

$$H_2 = \left[\frac{Ni}{x(b-x)+gw}\right] w = \frac{Niw}{x(b-x)+gw} \tag{5.46}$$

The flux ϕ is given by

$$\phi = (\mu_o H_1) w^2 = \frac{\mu_o Ni(b-x)w^2}{x(b-x)+gw} \tag{5.47}$$

The flux linkages are

$$\begin{aligned} \lambda = N\phi &= \frac{\mu_o NNi(b-x)w^2}{x(b-x)+gw} \\ &= \mu_o \frac{N^2 iw^2(b-x)}{x(b-x)+gw} = L(x)i \end{aligned} \tag{5.48}$$

where

$$L(x) = \frac{\mu_o N^2 w^2 (b-x)}{x(b-x)+gw}$$

Solution: Using magnetic equivalent circuit The magnetic equivalent circuit is shown in Figure 5.13:

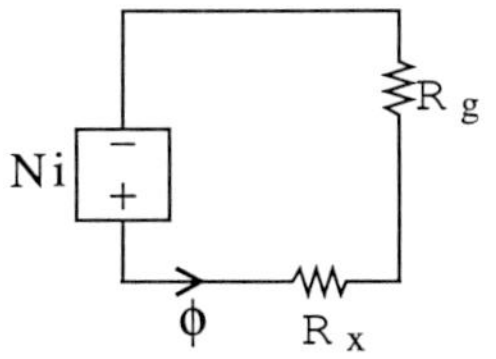

Figure 5.13: Magnetic equivalent circuit

$$\begin{aligned} \mathcal{R}_g &= \frac{g}{\mu_0 (b-x) w} \\ \mathcal{R}_x &= \frac{x}{\mu_0 w^2} \end{aligned}$$

$$\phi = \frac{Ni}{\mathcal{R}_g + \mathcal{R}_x} = \frac{N\,\mu_0 (b-x) w^2\, i}{gw + x(b-x)} \tag{5.49}$$

Then $\lambda = N\phi$ and agrees with (5.48).

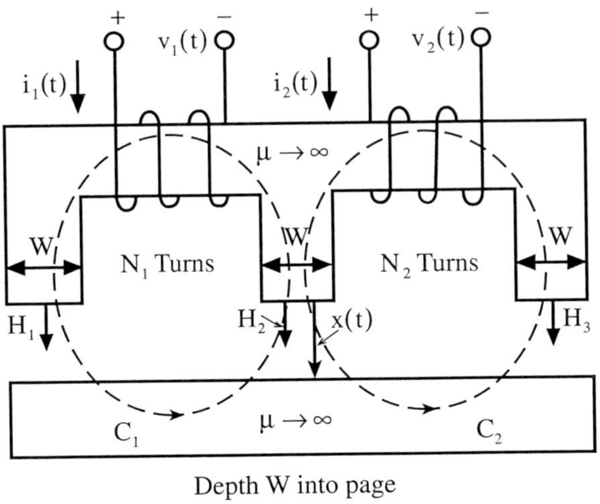

Figure 5.14: Electromechanical system

Example 5.4 The device shown in Figure 5.14 has an upper fixed piece and lower movable piece, both composed of *infinitely* permeable magnetic material, and both with depth W into the paper. Each of the three legs on the fixed upper piece has width W, and the distance between each of the legs and the lower movable piece is $x(t)$. Using both the magnetic equivalent circuit approach and the field theory approach,

1) calculate the flux linkages $\lambda_1(i_1, i_2, x)$ and $\lambda_2(i_1, i_2, x)$, and

2) identify the self and mutual inductances.

Solution: Using magnetic equivalent circuit The magnetic equivalent circuit is shown in Figure 5.15

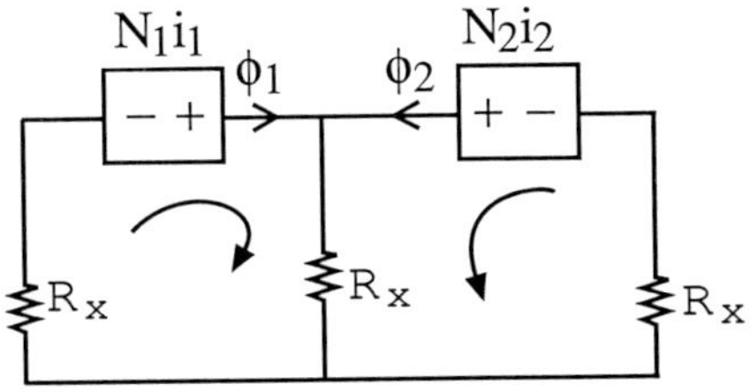

Figure 5.15: Magnetic equivalent circuit

$$\mathcal{R}_x = \frac{x}{\mu_0 A} = \frac{x}{\mu_0 W^2}$$

For the left loop:

$$N_1 i_1 = \mathcal{R}_x \phi_1 + \mathcal{R}_x(\phi_1 + \phi_2) = 2\mathcal{R}_x \phi_1 + \mathcal{R}_x \phi_2 \tag{5.50}$$

For the right loop:

$$N_2 i_2 = \mathcal{R}_x \phi_2 + \mathcal{R}_x(\phi_1 + \phi_2) = \mathcal{R}_x \phi_1 + 2\mathcal{R}_x \phi_2 \tag{5.51}$$

Solving (5.50) and (5.51)

$$\phi_1 = \frac{\mu_0 W^2}{3x}(2N_1 i_1 - N_2 i_2) \tag{5.52}$$

$$\phi_2 = \frac{\mu_0 W^2}{3x}(-N_1 i_1 + 2N_2 i_2) \tag{5.53}$$

The flux linkages are given by

$$\lambda_1 = N_1 \phi_1 = \frac{\mu_0 W^2}{3x}\left(2N_1^2 i_1 - N_1 N_2 i_2\right) \tag{5.54}$$

$$\lambda_2 = N_2 \phi_2 = \frac{\mu_0 W^2}{3x}\left(-N_1 N_2 i_1 + 2N_2^2 i_2\right) \tag{5.55}$$

We identify the self and mutual inductances as

$$L_{11} = \frac{\mu_0 W^2}{3x} 2N_1^2 \tag{5.56}$$

$$L_{12} = -\frac{\mu_0 W^2}{3x} N_1 N_2 = L_{21} \triangleq L_m \tag{5.57}$$

$$L_{22} = \frac{\mu_0 W^2 2N_2^2}{3x} \tag{5.58}$$

Conceptually, we can cast them as

$$\begin{aligned} \lambda_1 &= L_{11} i_1 + L_m i_2 \\ \lambda_2 &= L_m i_1 + L_{22} i_2 \end{aligned} \tag{5.59}$$

We have considered "lossless" systems so far, and any dissipative element, such as the resistance R on the electrical side, is taken outside of the coil. The coil can be excited by either a voltage or current source, as shown in Figure 5.16.

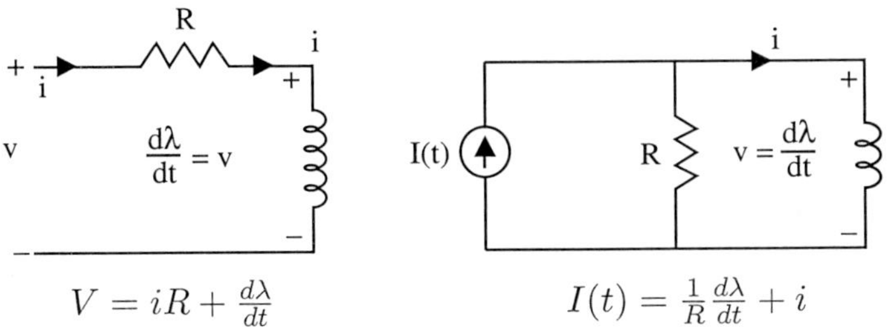

Figure 5.16: Electrical circuit

Solution: Using field theory Assume directions for H_1, H_2, and H_3 in the three air gaps as shown in Figure 5.14. The two contours c_1 and c_2 are both shown anti-clockwise. Using ACL along the two contours, we get

Contour c_1

$$-H_2 x + H_1 x = -N_1 i_1 \tag{5.60}$$

Contour c_2

$$H_2 x - H_3 x = N_2 i_2 \tag{5.61}$$

Using Gauss's law for a closed surface around the movable piece,

$$\mu_0 W^2 (H_1 + H_2 + H_3) = 0$$

i.e.,

$$H_1 + H_2 + H_3 = 0 \tag{5.62}$$

Solving (5.60), (5.61), and (5.62), we get (leaving the details out)

$$H_1 = \frac{-2N_1 i_1 + N_2 i_2}{3x}, \; H_2 = \frac{N_1 i_1 + N_2 i_2}{3x}$$

and

$$H_3 = \frac{N_1 i_1 - 2N_2 i_2}{3x} \tag{5.63}$$

Noting the current orientations in the two coils, the flux ϕ_1 from left to right in coil 1 is

$$\phi_1 = -\mu_0 W^2 H_1.$$

Hence,

$$\lambda_1 = N_1 \phi_1 \; = \; \frac{\mu_0 W^2}{3x}(2N_1^2 i_1 - N_1 N_2 i_2) \tag{5.64}$$

The flux ϕ_2 in coil is from right to left is

$$\phi_2 = -\mu_0 W^2 H_3$$

Hence,

$$\lambda_2 = N_2\phi_2 \quad = \quad \frac{\mu_0 W^2}{3x}(-N_1 N_2 i_1 + 2N_2^2 i_2) \tag{5.65}$$

We get the same results by both the methods.

5.4 Forces of Electric Origin Using Energy

So far we have seen how to compute flux linkages as a function of currents and the mechanical variables for systems with simple geometry. Specifically for a single electrical and single mechanical port (translational) system, the flux linkage is $\lambda = \lambda(i, x)$. The *force of electric origin* f^e is of the form $f^e = f^e(i, x)$ for the single electrical and single mechanical port system.

We will consider writing equations for the mechanical side later. We wish to compute f^e. Conceptually, we can think of an electrical subsystem coupled to the mechanical subsystem through a lossless magnetic field. f^e *always acts in the direction of positive* x. In the mechanical system, mass and spring are shown as two terminal elements. ℓ is the length of the spring when no force is acting on it as shown in Figure 5.17.

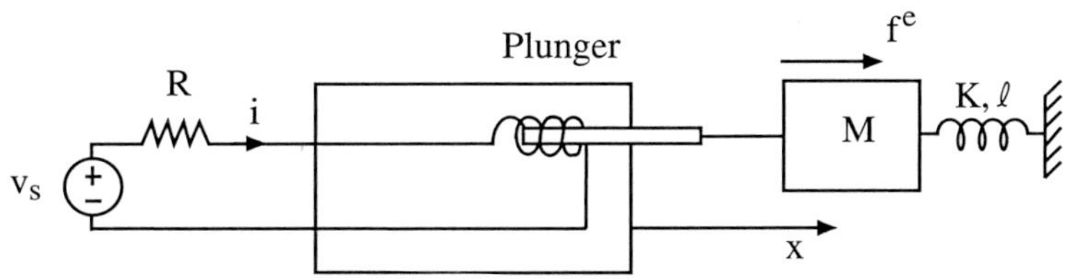

Figure 5.17: Electromechanical system

Energy considerations

The force of electric origin can also be expressed as $f^e(\lambda, x)$ since i can be solved for from $\lambda = \lambda(i, x)$. Theoretically, it is easier to get an expression for $f^e(\lambda, x)$. Figure 5.17 translates into the schematic diagram of Figure 5.18. The box in Figure 5.18 contains only the magnetic field whose value and therefore energy storage is affected by the electrical and mechanical variables. The coupling is assumed to be lossless, which means that energy can

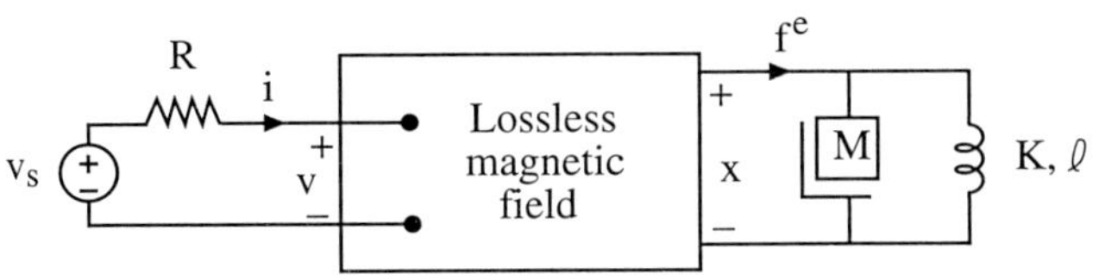

ℓ is the static equilibrium position of spring

Figure 5.18: Two-port representation

be stored through the terminals and also recovered through the terminals. If W_m is the total energy stored, then by the conservation of power principle, the rate of change of stored energy = electrical power in—mechanical power out, i.e.,

$$\begin{aligned} \frac{dW_m}{dt} &= vi - f^e \frac{dx}{dt} \\ &= i\frac{d\lambda}{dt} - f^e \frac{dx}{dt} \end{aligned} \tag{5.66}$$

since

$$v = \frac{d\lambda}{dt}$$

Multiplying (5.66) by dt, we get

$$dW_m = id\lambda - f^e dx \tag{5.67}$$

In (5.66) one can also think of $-f^e \frac{dx}{dt}$ as power input at the mechanical terminals. Of the four variables (λ, i, x, f^e), we can choose two of them, one electrical and one mechanical, independently without violating the physics of the problem. From (5.67), let us choose (λ, x) as independent variables. Consider Figure 5.19, where we wish to compute the stored energy in going from point a to point b along the path C. Since the coupling medium is conservative, the change in stored energy in going from a to b in the $\lambda - x$ plane is independent of the path of integration. We can therefore compute it along the paths A or B. Of the two, we can choose the one that is easiest in terms of integration.

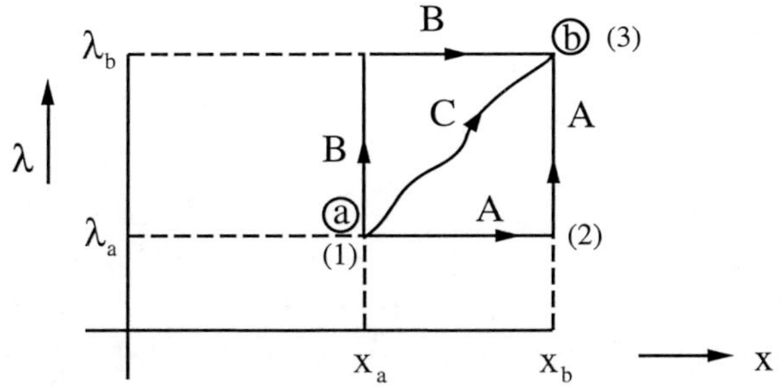

Figure 5.19: Path of integration in $\lambda - x$ plane

Path A

Along this path $d\lambda = 0$ along the segment 1–2 and $dx = 0$ along the segment 2–3. Hence,

$$W_m(\lambda_b, x_b) - W_m(\lambda_a, x_a) = -\int_{x_a}^{x_b} f^e(\lambda_a, x)dx + \int_{\lambda_a}^{\lambda_b} i(\lambda, x_b)d\lambda \quad (5.68)$$

Path B

Along this path we get

$$W_m(\lambda_b, x_b) - W_m(\lambda_a, x_a) = \int_{\lambda_a}^{\lambda_b} i(\lambda, x_a)d\lambda - \int_{x_a}^{x_b} f^e(\lambda_b, x)dx \quad (5.69)$$

Both (5.68) and (5.69) must give the same answer. We can also compute W_m along C if we know the equation for the path C as $\lambda = \lambda(x)$.

Physical significance of (5.68) and (5.69)

In (5.68), we first put energy into the system from the mechanical terminals while keeping λ as constant requiring doing work against f^e, and then put energy into the system from the electrical terminals keeping x constant. In (5.69), the order of integration is reversed.

If $\lambda_a = 0$, there is no force of electric origin to start with, i.e., $f^e(0, x) = 0$ in the first term in (5.68). Hence, path A is easier. Hence,

$$W_m(\lambda_b, x_b) - W_m(0, x_a) = \int_0^{\lambda_b} i(\lambda, x_b)d\lambda \quad (5.70)$$

Letting λ_b be any λ and assuming $W_m(0, x_a) = 0$, x_b becomes any x. Then

$$W_m(\lambda, x) = \int_0^{\lambda} i(\lambda, x) d\lambda \tag{5.71}$$

If path B is chosen, we need to know f^e to carry out the integration. Hence, path A is preferred. This means that if we put the system mechanically when no electrical force is present, then we can compute W_m through the electrical terminals alone.

Force energy relations

It is clear from (5.67) that magnetic field energy W_m can be computed as a function of λ and x. From (5.67), we have

$$dW_m = id\lambda - f^e dx \tag{5.72}$$

Since $W_m = W_m(\lambda, x)$, we can from elementary calculus compute the differential of $W_m(\lambda, x)$ as

$$dW_m = \frac{\partial W_m(\lambda, x)}{\partial \lambda} d\lambda + \frac{\partial W_m(\lambda, x)}{\partial x} dx \tag{5.73}$$

Comparing (5.72) and (5.73), we get

$$i = \frac{\partial W_m}{\partial \lambda}(\lambda, x) \tag{5.74}$$

and

$$f^e = -\frac{\partial W_m}{\partial x}(\lambda, x) \tag{5.75}$$

The expression for W_m is given by (5.71).

Example 5.5 This example is the same as Example 5.1 reproduced in Figure 5.20.

1. Compute the force of electric origin $f^e(\lambda, x)$.

2. Also express the force as $f^e(i, x)$.

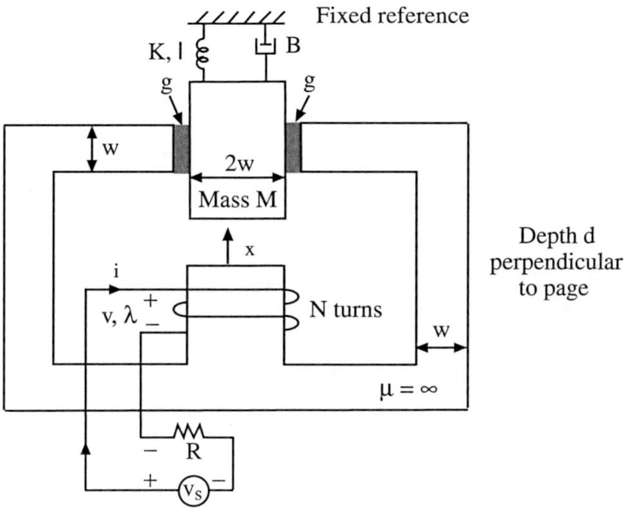

Figure 5.20: Electromechanical system

Solution: In Example 5.1, we have computed λ as

$$\begin{aligned}\lambda &= \frac{2wd\mu_0 N^2 i}{(g+x)} = \frac{2wd\mu_0 N^2}{g}\frac{i}{\left(1+\frac{x}{g}\right)} \\ &= L_0 \frac{i}{\left(1+\frac{x}{g}\right)}\end{aligned}$$

where

$$L_0 = \frac{2wd\mu_0 N^2}{g}$$

The expression for $\lambda(i, x)$ can easily be solved to express i as a function of (λ, x) as

$$i = \frac{\lambda}{L_0}\left(1+\frac{x}{g}\right)$$

Therefore, the stored energy W_m is

$$\begin{aligned}W_m &= \int_0^\lambda i(\lambda, x)d\lambda = \int_0^\lambda \frac{\lambda}{L_0}\left(1+\frac{x}{g}\right)d\lambda \\ &= \frac{\lambda^2}{2L_0}\left(1+\frac{x}{g}\right)\end{aligned}$$

The force of electric origin is given by

$$f^e = -\frac{\partial W_m(\lambda, x)}{\partial x} = -\frac{\lambda^2}{2L_0 g}$$

Using the expression for λ, we get

$$f^e(i, x) = -\frac{L_0^2 i^2}{2L_0 g\left(1 + \frac{x}{g}\right)^2} = -\frac{1}{2}\frac{L_0 i^2}{g\left(1 + \frac{x}{g}\right)^2}$$

Example 5.6 In Example 5.3, compute $f^e(\lambda, x)$.

Solution: λ has been computed as

$$\lambda = \frac{N^2 \mu_0 i(b - x)w^2}{[(b - x)x + gw]}$$

The current i is obtained from this expression as

$$i = \frac{\lambda[(b - x)x + gw]}{N^2 \mu_0 w^2 (b - x)}$$

Now,

$$W_m = \int_0^\lambda i(\lambda, x)d\lambda = \frac{\lambda^2[(b - x)x + gw]}{2N^2 \mu_0 w^2 (b - x)}$$

Hence,

$$\begin{aligned} f^e &= \frac{-\partial W_m}{\partial x} \\ &= \frac{-\lambda^2}{2N^2 \mu_0 w^2} \frac{[(gw + x(b - x)) + (b - 2x)(b - x)]}{(b - x)^2} \end{aligned}$$

To express f^e in terms of (i, x), substitute for λ.

$$\begin{aligned} f^e &= \frac{-(N^2 \mu_0 w^2)^2 (b - x)^2 i^2 ((b - x)x + gw) + (b - x)(b - 2x)}{2N^2 \mu_0 w^2 [(b - x)x + gw]^2 [b - x]^2} \\ &= -\frac{i^2}{2} N^2 \mu_0 w^2 \frac{[((b - x)x + gw) + (b - x)(b - 2x)]}{[(b - x)x + gw]^2} \\ &= -\frac{i^2}{2} N^2 \mu_0 w^2 \frac{(gw + b^2 - 2bx + x^2)}{(gw + bx - x^2)^2} \end{aligned}$$

5.5 Force of Electric Origin Using "Co-energy"

We have seen in the previous section that to compute $W_m(\lambda, x)$, we need to express $i = i(\lambda, x)$ from the flux linkage equation. This would require solving for i from $\lambda = \lambda(i, x)$. Particularly in multi-port systems, this could be quite complicated. We can avoid this problem by defining a quantity called *co-energy* directly computable from $\lambda = \lambda(i, x)$ and then using it to compute f^e. Consider

$$dW_m = id\lambda - f^e dx \tag{5.76}$$

Now

$$d(\lambda i) = id\lambda + \lambda di \tag{5.77}$$

From (5.77)

$$id\lambda = d(\lambda i) - \lambda di \tag{5.78}$$

Substitute (5.78) in (5.76) and rearrange the terms to get

$$dW_m \quad = \quad d(\lambda i) - \lambda di - f^e dx \tag{5.79}$$

Rearranging terms, we get

$$d\underbrace{(\lambda i - W_m)}_{W'_m} \quad = \quad \lambda di + f^e dx \tag{5.80}$$

Define $\lambda i - W_m \stackrel{\Delta}{=} W'_m$ as *co-energy.* From (5.80), it is clear that W'_m will be a function of (i, x). Now the state is (i, x). Hence, we can compute the co-energy W'_m as a function of (i, x). Figure 5.21 shows the (i, x) plane with

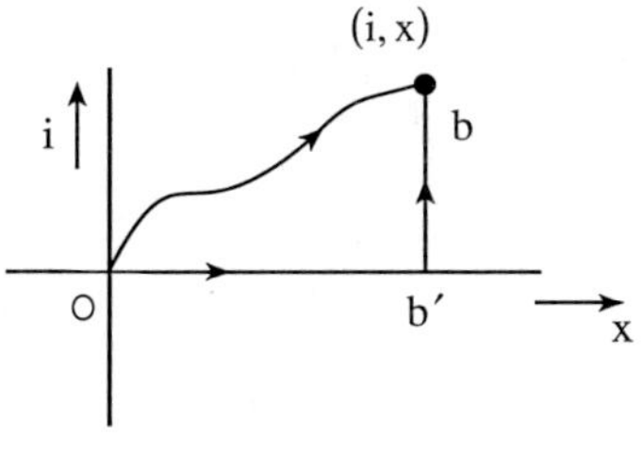

Figure 5.21: Integration in the $i - x$ plane

the origin being the state $(0, 0)$ for the sake of simplicity.

Starting with the origin, we assemble the mechanical system first along $0b'$ and then the electrical system along $b'b$ to reach (i, x). This means we integrate along $ob'b$. Since $f^e = 0$ along ob', there is no contribution from the second term in (5.80). Integration along $b'b$ gives

$$W'_m(i, x) = \int_0^i \lambda(i, x) di \tag{5.81}$$

Now mathematically the differential of $W'_m(i, x)$ is

$$dW'_m = \frac{\partial W'_m}{\partial i} di + \frac{\partial W'_m}{\partial x} dx \tag{5.82}$$

Compare (5.82) and (5.80), we get

$$\lambda = \frac{\partial W'_m}{\partial i}$$

and

$$f^e = \frac{\partial W'_m(i, x)}{\partial x} \tag{5.83}$$

Thus, to compute f^e, we compute W'_m and f^e directly from $\lambda(i, x)$ instead of computing $i = i(\lambda, x)$. In the final expression, if we substitute for i in terms of (λ, x), then we get f^e in terms of (λ, x).

Example 5.7 Find the force of electric origin f^e in Figure 5.22. The

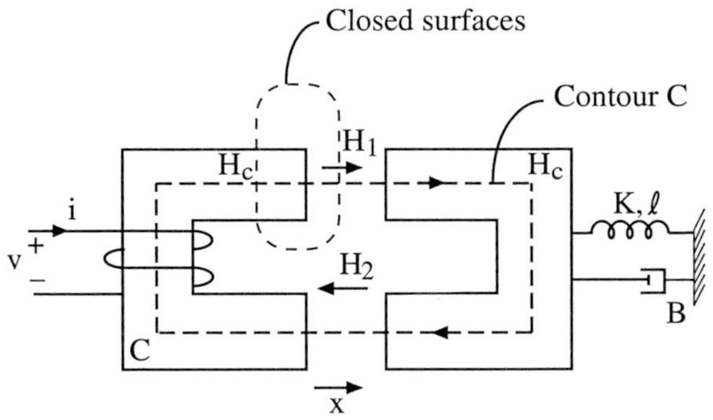

Figure 5.22: Electromechanical system

mean length of the magnetic path is ℓ_c, and the cross-sectional area is A. The magnetic circuit has finite μ.

Solution: Using field theory The contour C is as shown. Assume positive directions for H_1 and H_2 as shown. Ampere's circuital law gives

$$H_c \ell_c + H_1 x + H_2 x = Ni \tag{5.84}$$

Enclose the moving portion by a closed surface and apply Gauss's law. The result is

$$\mu_0 H_1 A = \mu_0 H_2 A$$

i.e.,

$$H_1 = H_2 \tag{5.85}$$

Applying Gauss's law to the closed surface around the upper fixed surface as shown gives

$$\begin{aligned} \mu H_c A &= \mu_0 H_1 A \\ H_c &= \frac{\mu_0}{\mu} H_1 \end{aligned} \tag{5.86}$$

From (5.84) and (5.85),

$$H_c \ell_c + 2 H_1 x = Ni \tag{5.87}$$

Substitute (5.86) in (5.87) and solve for H_1. The result is

$$H_1 = \frac{Ni}{\left(\frac{\mu_0}{\mu} \ell_c + 2x\right)}$$

The flux linkage of the coil is given by

$$\begin{aligned} \lambda &= N\phi \\ &= N \mu_0 H_1 \text{ A} \\ &= \frac{N \mu_0 N i}{\left(\frac{\mu_0}{\mu} \ell_c + 2x\right)} \text{ A} = \frac{N^2 i}{\left(\frac{\ell_c}{\mu A} + \frac{2x}{\mu_0 \text{ A}}\right)} \end{aligned} \tag{5.88}$$

The co-energy W'_m is

$$W'_m = \int_0^i \lambda(i, x) di = \frac{N^2 i^2}{2\left(\frac{\ell_c}{\mu A} + \frac{2x}{\mu_0 A}\right)} \tag{5.89}$$

The force of electric origin f^e is

$$f^e = \frac{\partial W'_m}{\partial x} \tag{5.90}$$

Thus,

$$f^e = \frac{-N^2 i^2}{2\left(\frac{\ell_c}{\mu \mathrm{A}} + \frac{2x}{\mu_0 A}\right)^2}\left(\frac{2}{\mu_0 \mathrm{~A}}\right) = \frac{-N^2}{\mu_0 A} \frac{i^2}{\left(\frac{\ell_c}{\mu A} + \frac{2x}{\mu_0 A}\right)^2} \tag{5.91}$$

Note that $\frac{\ell c}{\mu A}$ is the reluctance of the iron path, and $\frac{2x}{\mu_0 A}$ is the reluctance of the two air gaps in series.

Example 5.8 Repeat Example 5.7 using the concept of magnetic equivalent circuit as shown in Figure 5.23.

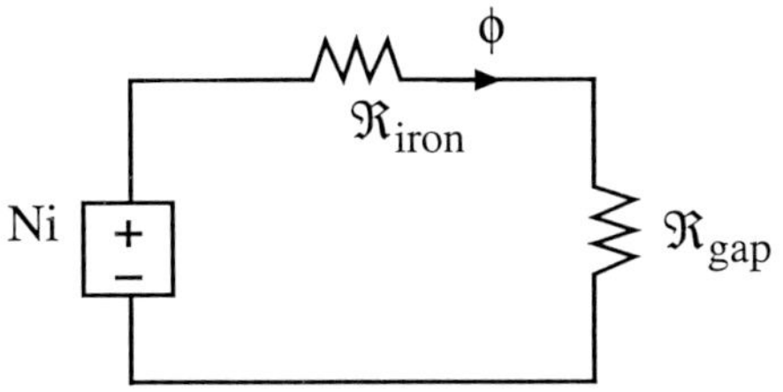

Figure 5.23: Magnetic equivalent circuit

Solution:

$$\mathcal{R}_{\text{iron}} = \frac{\ell_c}{\mu A}, \mathcal{R}_{\text{gap}} = \frac{2x}{\mu_0 \mathrm{~A}}$$

Since flux = mmf/reluctance,

$$\phi = \frac{Ni}{\mathcal{R}_{\text{iron}} + \mathcal{R}_{\text{gap}}} = \frac{Ni}{\frac{\ell_c}{\mu A} + \frac{2x}{\mu_0 A}}$$

Letting

$$\mathcal{R}_{\text{iron}} + \mathcal{R}_{\text{gap}} = \mathcal{R}(x),$$

we have

$$\begin{aligned}
\phi &= \frac{Ni}{\mathcal{R}(x)} \\
\lambda &= N\phi = \frac{N^2 i}{\mathcal{R}(x)} \\
W'_m &= \int_0^i \lambda(i,x)di = \frac{N^2 i^2}{2\mathcal{R}(x)}
\end{aligned}$$

The force of electric origin is

$$\begin{aligned} f^e &= \frac{\partial W'_m}{\partial x} = \frac{N^2 i^2}{2} \frac{d}{dx} \left(\frac{1}{\mathcal{R}(x)} \right) \\ \mathcal{R}(x) &= \frac{\ell_c}{\mu A} + \frac{2x}{\mu_0 A} \end{aligned}$$

Hence,

$$f^e = -\frac{N^2 i^2}{\mu_0 \text{ A} \left(\frac{\ell_c}{\mu A} + \frac{2x}{\mu_0 A} \right)^2} \tag{5.92}$$

This agrees with (5.91).

5.6 Energy and Co-energy in Multi-port Systems

5.6.1 Graphical interpretation of energy–co-energy

So far we have discussed energy and co-energy and how we can get the force of electric origin f^e from both of them. While we are used to the concept of energy, "co-energy" is a new concept. As shown in Sections 5.4 and 5.5, the force of electric origin f^e derived from either energy or co-energy expressions is identical. Depending on the system description, one or the other may be used. We now show that in "electrically" linear systems both *are numerically equal*, so that one need not distinguish between them in such systems.

Suppose $\lambda(i, x) = L(x)i$, which is electrically linear, then $i = \frac{\lambda(i,x)}{L(x)}$ and

$$W_m = \int_0^\lambda i d\lambda = \frac{\lambda^2(i, x)}{2L(x)} \tag{5.93}$$

$$W'_m = \int_0^i \lambda di = \frac{1}{2} L(x) i^2 \tag{5.94}$$

Since $\lambda(i, x) = L(x)i$, both (5.95) and (5.94) are equivalent. For a given value of x, the relationship between λ and i is linear, as shown in Figure 5.24. Interpretation of (5.82) and (5.83) leads to the following:

$$\begin{aligned} \text{Area A} &= \int_o^\lambda i(\lambda, x) d\lambda \text{ is the "energy"} \\ \text{Area B} &= \int_o^i \lambda(i, x) di \text{ is the "co-energy"} \end{aligned}$$

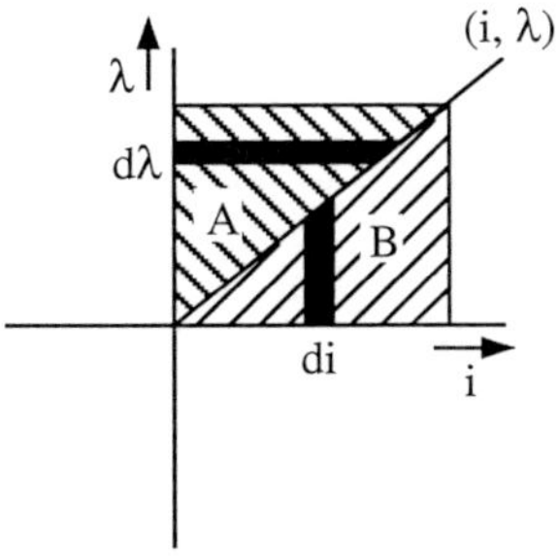

Figure 5.24: Electrically linear system

From geometrical considerations, area A = area B. Hence, numerically, energy and co-energy are equal.

If $\lambda(i, x)$ is a nonlinear function as shown in Figure 5.25, then area A

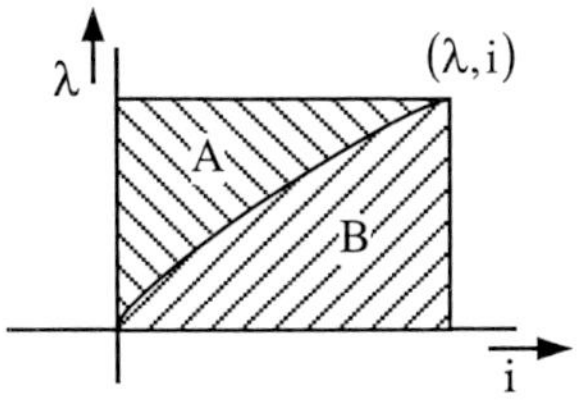

Figure 5.25: Electrically nonlinear system

$\neq$ area B. The force of electric origin is f^e. Derived using either energy or co-energy it will be the same. Note also that area A + area $B = \lambda i$, both for linear and nonlinear cases.

We will illustrate graphically that the force calculated from either energy or co-energy will be the same. Figure 5.26 shows the effect of an incremental change in Δx on the nominal $\lambda - i$ curve. Keeping λ fixed, if we change Δx incrementally, as shown in Figure 5.26 (a), there is a net decrease in energy W_m, shown by the shaded area as $-\Delta W_m$. The force of electric origin $f^e = -\frac{\partial W_m}{\partial x}$ can be interpreted as $-\frac{\Delta W_m}{\Delta x}$ with λ constant as $\Delta x \to 0$. Hence, the force f^e is equal to (shaded area)/Δx as $\Delta x \to 0$. On the other hand, the force $f^e = \frac{\partial W'_m}{\partial x}$ can be interpreted as the limit of $\frac{\Delta W'_m}{\Delta x}$ as $\Delta x \to 0$ with i equal constant (see Figure 5.26 (b). The shaded areas differ only by the small triangle abc of sides Δi and $\Delta\lambda$. In the limit, the shaded areas resulting from Δx at constant λ or constant i are equal. Thus, the computation of f^e is the same whether it is computed from energy or co-energy.

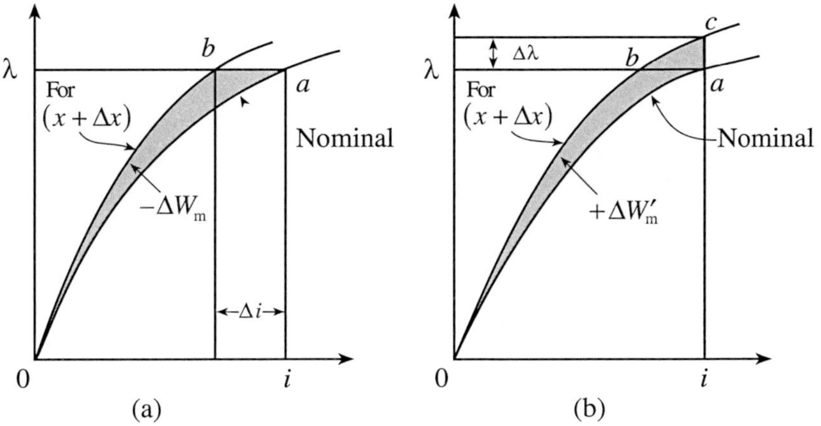

Figure 5.26: Effect of Δx on the energy and co-energy: (a) change of energy with λ held constant; (b) change of co-energy with i held constant [10]

5.6.2 Co-energy f^e for a two-electrical and one-mechanical port system

For the case of multi-electrical and multi-mechanical ports, we can extend the concepts of energy and co-energy in a straightforward manner. As a motivation, let us consider a two-electrical and one-mechanical port system with the flux linkages as $\lambda_1 = \lambda_1(i_1, i_2, x)$ and $\lambda_2 = \lambda_2(i_1, i_2, x)$. The state of the system is (i_1, i_2, x), and we consider the three-dimensional space as shown in Figure 5.27. Rate of change of stored energy is

$$\begin{aligned} \frac{dW_m}{dt} &= v_1 i_1 + v_2 i_2 - f^e \frac{dx}{dt} \\ &= i_1 \frac{d\lambda_1}{dt} + i_2 \frac{d\lambda_2}{dt} - f^e \frac{dx}{dt} \end{aligned} \tag{5.95}$$

Multiplying by dt

$$dW_m = i_1 d\lambda_1 + i_2 d\lambda_2 - f^e dx \tag{5.96}$$

Consider the differential of $d(\lambda_1 i_1 + \lambda_2 i_2)$ as

$$d(\lambda_1 i_1 + \lambda_2 i_2) = i_1 d\lambda_1 + \lambda_1 di_1 + i_2 d\lambda_2 + \lambda_2 di_2 \tag{5.97}$$

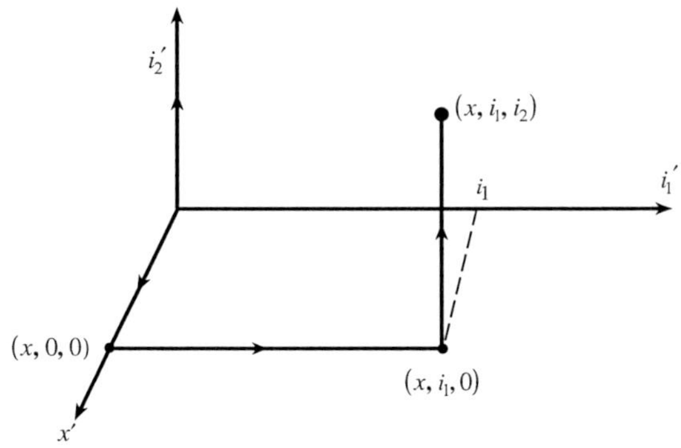

Figure 5.27: (i, x) plane for a two-electrical, one-mechanical system

From (5.97), we get

$$i_1 d\lambda_1 + i_2 d\lambda_2 = d(\lambda_1 i_1 + \lambda_2 i_2) - \lambda_1 di_1 - \lambda_2 di_2 \tag{5.98}$$

Substituting (5.98) in the right side of (5.96), and rearranging terms, we get

$$d(\lambda_1 i_1 + \lambda_2 i_2 - W_m) = \lambda_1 di_1 + \lambda_2 di_2 + f^e dx \tag{5.99}$$

Define W_m', the co-energy as

$$\lambda_1 i_1 + \lambda_2 i_2 - Wm \triangleq W_m' \tag{5.100}$$

Using (5.100) into (5.99), we get

$$dW_m' = \lambda_1 di_1 + \lambda_2 di_2 + f^e dx \tag{5.101}$$

As in the case of a singly excited system, (5.101) can be integrated by holding i_1 and i_2 fixed at zero and integrating over x. f^e is zero over this segment. One can then integrate over i_1 (while holding i_2 zero) and finally over i_2. The path of integration is shown in Figure 5.27. Thus,

$$W_m'(i_1, i_2, x) = \int_0^{i_1} \lambda_1(i'_1, 0, x) di'_1 + \int_0^{i_2} \lambda_2(i_1, i_2', x) di_2'. \tag{5.102}$$

5.6.3 Forces of electric origin in general multi-port systems

For the case of N electrical and M mechanical ports shown in Figure 5.4, we have the flux linkages as

$$\lambda_1(i_1, \ldots, i_N, x_1, \ldots, x_M), \ldots, \lambda_N(i_1, \ldots, i_N, x_1, \ldots, x_M)$$

Since it is cumbersome to invert these to find $i_1, \ldots, i_N$, it is easier to work with co-energy rather than energy.

Rate of change of stored energy is given by

$$\frac{dW_m}{dt} = v_1 i_1 +, \ldots, v_N i_N - f_1^e \frac{dx_1}{dt}, \ldots, -f_M^e \frac{dx_M}{dt} \tag{5.103}$$

Since $v_i = \frac{d\lambda_i}{dt}$, we have (cancelling dt on both sides)

$$dW_m = d\lambda_1 i_1 +, \ldots, +d\lambda_N i_N - f_1^e dx_1, \ldots, -f_M^e dx_M \tag{5.104}$$

Now

$$d(\lambda_1 i_1 +, \ldots, \lambda_N i_N) = (d\lambda_1 i_1 +, \ldots, d\lambda_N i_N) + \lambda_1 di_1 +, \ldots, \lambda_N di_N \tag{5.105}$$

Therefore,

$$d\lambda_1 i_1 +, \ldots, d\lambda_N i_N = d\left(\sum_{i=1}^{N} \lambda_i i_i\right) - \sum_{i=1}^{N} \lambda_i di_i \tag{5.106}$$

Substitute (5.106) in (5.104) and rearrange terms to get

$$\underbrace{d\left(\sum_{i=1}^{N} \lambda_i i_i - W_m\right)}_{W'_m} = \sum_{i=1}^{N} \lambda_i di + \sum_{i=1}^{M} f_i^e dx_i \tag{5.107}$$

Define

$$\sum_{i=1}^{N} \lambda_i i_i - W_m \triangleq W'_m = \text{ co-energy}$$

W'_m is the function of $(i, \ldots, i_N, x_1, \ldots, x_M)$. Mathematically,

$$dW'_m = \frac{\partial W'_m}{\partial i_1} di_1 +, \ldots, \frac{\partial W'_m}{\partial i_N} di_N + \frac{\partial W'_m}{\partial x_1} dx_1 +, \ldots, \frac{\partial W'_m}{\partial x_M} dx_M \tag{5.108}$$

The left sides of (5.107) and (5.108) are the same. Equating the right sides of these two equations and comparing terms, we obtain

$$\lambda_i = \frac{\partial W'_m}{\partial i_i} (i = 1, \ldots, N) \tag{5.109}$$

$$f_i^e = \frac{\partial W'_m}{\partial x_i} (i = 1, \ldots, M) \tag{5.110}$$

5.6.4 Computation of W'_m

We can use (5.107) and, as in Section 5.6.2, integrate along the x_i (i = 1,2, ..., M) axes first and then along each of the i_i (i = 1,2, ..., N) axes next. While integrating along x_i, $W'_m = 0$ since f_i^e is zero. We then get by integrating over $i_i (i = 1, 2, \ldots, N)$,

$$\begin{aligned} W'_m &= \int_0^{i_1} \lambda_1(i'_1, 0, \ldots, 0, x_1, x_2, \ldots, x_M) di'_1 \\ &\quad + \int_0^{i_2} \lambda_2(i_1, i'_2, 0, \ldots, x_1, x_2, \ldots, x_M) di'_2 + \ldots \\ &\quad + \int_0^{i_N} \lambda_N(i_1, i_2, \ldots, i_{N-1}, i'_N, x_1, \ldots, x_M) di'_N \end{aligned} \quad (5.111)$$

Note carefully that in each of the integrals, we have a different variable in the currents for integration. Hence, we put the ''prime sign" for the dummy variable of integration. For the special case of the two-electrical and two-mechanical port system discussed in Section 5.6.2, we can write W'_m and compute the forces of electric origin as

$$W'_m = \int_0^{i_1} \lambda_1(i'_1, 0, x_1, x_2) di'_1 + \int_0^{i_2} \lambda_2(i_1, i'_2, x_1, x_2) di'_2$$

$f_1^e = \frac{\partial W'_m}{\partial x_1}$ and $f_2^e = \frac{\partial W'_m}{\partial x_2}$.

Example 5.9 Given the flux linkages for a two-electrical two-mechanical port as

$$\begin{aligned} \lambda_1 &= a_1 x_1^2 i_1^3 + b x_2^2 x_1 i_2 \\ \lambda_2 &= b x_2^2 x_1 i_1 + c x_2^2 i_2^3 \end{aligned}$$

compute f_1^e and f_2^e

Solution:

$$
\begin{aligned}
W'_m &= \int_0^{i_1} (a_1 x_1^2 i_1'^3) di_1' + \int_0^{i_2} (b x_2^2 x_1 i_1 + c x_2^2 i_2'^3) di_2' \\
&= \frac{a_1 x_1^2 i_1^4}{4} + b x_2^2 x_1 i_1 i_2 + \frac{c x_2^2 i_2^4}{4} \\
f_1^e &= \frac{\partial W'_m}{\partial x_1} = \frac{a_1 x_1 i_1^4}{2} + b x_2^2 i_1 i_2 \\
f_2^e &= \frac{\partial W'_m}{\partial x_2} = 2 b x_2 x_1 i_1 i_2 + \frac{c x_2 i_2^4}{2}
\end{aligned}
$$

Example 5.10 A certain three-electrical port and two-mechanical port electromechanical system has the following flux-current relations where ϕ and ψ are rotational variables.

$$
\begin{aligned}
\lambda_1 &= L_{11} i_1 + M i_3 \cos(\phi - \psi) \\
\lambda_2 &= L_{22} i_2 + M i_3 \sin(\phi - \psi) \\
\lambda_3 &= L_{33} i_3 + M i_1 \cos(\phi - \psi) + M i_2 \sin(\phi - \psi)
\end{aligned}
$$

a) Compute the co-energy W'_m.

b) Compute the torques of electric origin at the two mechanical ports where the rotational variables are ϕ and ψ.

Solution:

$$
\begin{aligned}
W'_m &= \int_0^{i_1} \lambda_1(i_1', 0, 0, \phi, \psi) di_1' + \int_0^{i_2} \lambda_2(i_1, i_2', 0, \phi, \psi) di_2' \\
&\quad + \int_0^{i_3} \lambda_3(i_1, i_2, i_3', \phi, \psi) di_3' \\
&= \frac{1}{2} L_{11} i_1^2 + \frac{1}{2} L_{22} i_2^2 + \frac{1}{2} L_{33} i_3^2 + M i_1 i_3 \cos(\phi - \psi) + M i_2 i_3 \sin(\phi - \psi) \\
T_\phi^e &= \frac{\partial W'_m}{\partial \phi} = -M i_1 i_3 \sin(\phi - \psi) + M i_2 i_3 \cos(\phi - \psi) \\
T_\psi^e &= \frac{\partial W'_m}{\partial \psi} = M i_1 i_3 \sin(\phi - \psi) - M i_2 i_3 \cos(\phi - \psi)
\end{aligned}
$$

5.7 Energy Conservation

The fact that we are dealing with lossless coupled systems, in which the stored energy is a single-valued function of the independent variables, can be very useful in determining the energy conserving properties of electromechanical systems. In particular, systems that operate in a cyclic manner are good candidates for analysis.

5.7.1 Test for conservative nature of the system

The energy and power balance in an electromechanical system can be described through Figures 5.28 and 5.29. In the coupling medium, the field

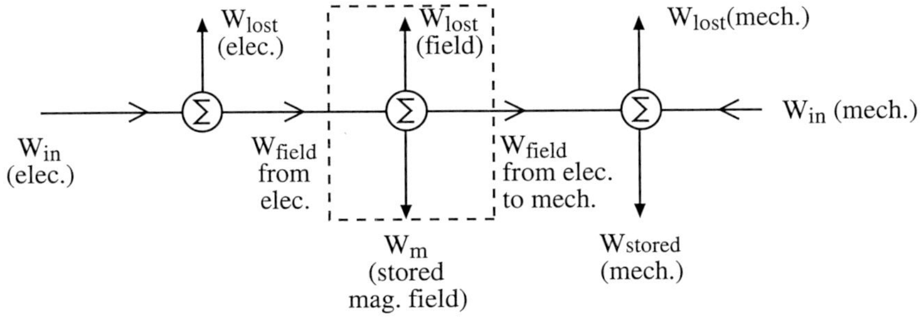

Figure 5.28: Energy balance

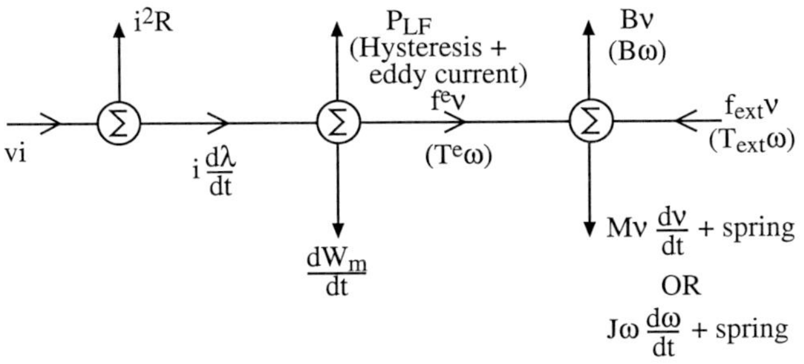

Figure 5.29: Power balance, $\dot{\theta} = \omega, \dot{x} = v$

losses would include both hysteresis and eddy current losses. The electrical and mechanical losses are external to the coupled system. Neglecting the field losses, we get the simple relation for the coupled system (shown dotted) as shown in Figure 5.30.

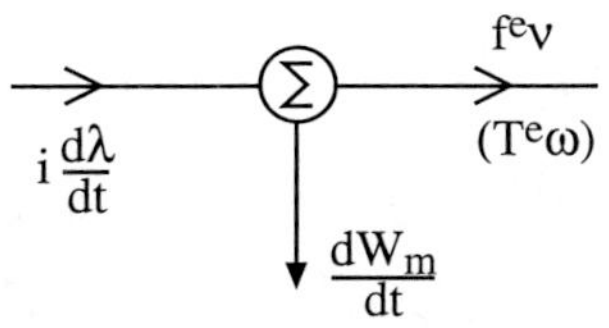

Figure 5.30: Energy balance for the conservative portion

From conservation of energy, we get

$$\frac{dW_m}{dt} = i\frac{d\lambda}{dt} - f^e\frac{dx}{dt} \tag{5.112}$$

$$dW_m = id\lambda - f^e dx \tag{5.113}$$

Only two of the four variables (i, λ, f^e, x) can be set independently. We can choose λ and x as the variables.

When λ and x are varied along a certain path, (5.113) can be used to evaluate the change in energy. A property of a conservative system is that its stored energy is a function of its *state* (i.e., particular values of λ and x) and does not depend on which path in the $\lambda - x$ plane was taken to reach that state.

It has been shown in Section 5.4 that letting $\lambda_a = 0$, $x_a = 0$, and $(\lambda_b x_b)$ be any (λ, x) in the $\lambda - x$ plane, the stored energy is given as $W_m = \int_0^\lambda i(\lambda', x)d\lambda'$. W_m is a function of (λ, x). Mathematically,

$$\frac{dW_m}{dt} = \frac{\partial W_m}{\partial \lambda}\frac{d\lambda}{dt} + \frac{\partial W_m}{\partial x}\frac{dx}{dt}$$

Comparing coefficients with (5.112),

$$f^e = -\frac{\partial W_m(\lambda, x)}{\partial x} \tag{5.114}$$

$$i = \frac{\partial W_m(\lambda, x)}{\partial \lambda} \tag{5.115}$$

We need a mathematical tool to verify the conservative nature of the system. It is given by the condition of "reciprocity" generally used in circuit theory. If we differentiate (5.114) with respect to λ and (5.115) with respect to x and note that $\frac{\partial^2 W_m}{\partial\lambda\partial x} = \frac{\partial^2 W_m}{\partial x\partial\lambda}$, then we get

$$\frac{\partial i(\lambda, x)}{\partial x} = -\frac{\partial f^e(\lambda, x)}{\partial \lambda} \tag{5.116}$$

This is a necessary and sufficient condition for the system to be conservative, from an energy viewpoint. Equation (5.113) can be rewritten as

$$dW_m = i(\lambda, x)d\lambda + (-f^e(\lambda, x)dx) \tag{5.117}$$

We can derive the test for the conservative nature of a system mathematically as follows. In general, given a function

$$I(x, y) = f(x, y)dx + g(x, y)dy \tag{5.118}$$

if the integral of I from (x_o, y_o) to (x, y) is to be path independent, then the condition is

$$\frac{\partial f}{\partial y} = \frac{\partial g}{\partial x} \tag{5.119}$$

This is also known as Green's theorem. Applying this test to (5.117), we get the condition (5.116) for conservativeness. The test for conservativeness in terms of co-energy is found from

$$dW'_m = \lambda(i, x)di + f^e(i, x)dx \tag{5.120}$$

and it is verified to be

$$\frac{\partial \lambda(i, x)}{\partial x} = \frac{\partial f^e(i, x)}{\partial i} \tag{5.121}$$

Thus, (5.116) and 5.121) are two ways of testing if the electromechanical system is conservative.

5.7.2 Two-electrical and one-mechanical port system

From (5.101),

$$dW'_m = \lambda_1 di_1 + \lambda_2 di_2 + f^e dx \tag{5.122}$$

From (5.109) and (5.110) for $N = 2$ and $M = 1$

$$\lambda_1 = \frac{\partial W'_m}{\partial i_1} \tag{5.123}$$

$$\lambda_2 = \frac{\partial W'_m}{\partial i_2} \tag{5.124}$$

$$f^e = \frac{\partial W'_m}{\partial x} \tag{5.125}$$

It is easily verified that the conditions for conservativeness are

$$\frac{\partial \lambda_1}{\partial x} = \frac{\partial f^e}{\partial i_1}, \frac{\partial \lambda_2}{\partial x} = \frac{\partial f^e}{\partial i_2}, \text{ and } \frac{\partial \lambda_1}{\partial i_2} = \frac{\partial \lambda_2}{\partial i_1} \tag{5.126}$$

This can be extended to multi-electrical and multi-mechanical port systems.

Example 5.11 [14] An electrically nonlinear magnetic field coupling network has the equations for i and f^e as

$$i = I_0 \left[\frac{\lambda/\lambda_0 + (\lambda/\lambda_0)^3}{1 + x/a}\right], f^e = \frac{I_0}{a}\left[\frac{\frac{1}{2}\lambda^2/\lambda_0 + \frac{1}{4}\lambda^4/\lambda_0^3}{(1 + x/a)^2}\right]$$

where I_0, λ_0, and a are positive constants.

a) Prove that this system is conservative.

b) Evaluate the stored energy at the point λ, x in variable space.

Solution: From (5.116), we have the condition for the system to be conservative as

$$\frac{\partial i}{\partial x} = -\frac{\partial f^e}{\partial \lambda}.$$

Now

$$\begin{aligned}
\frac{\partial i}{\partial x} &= -\frac{I_0}{a}\left[\frac{\lambda}{\lambda_0} + \left(\frac{\lambda}{\lambda_0}\right)^3\right] \Big/ \left(1 + \frac{x}{a}\right)^2 \\
-\frac{\partial f^e}{\partial \lambda} &= -\frac{I_0}{a}\left[\frac{\lambda}{\lambda_0} + \left(\frac{\lambda}{\lambda_0}\right)^3\right] \Big/ \left(1 + \frac{x}{a}\right)^2
\end{aligned}$$

Hence, the system is conservative.

$$\begin{aligned}
W_m(\lambda, x) &= \int_0^\lambda i(\lambda', x)d\lambda' \\
&= \frac{I_0}{\left[1 + \frac{x}{a}\right]}\left[\frac{1}{2}\frac{\lambda^2}{\lambda_0} + \frac{1}{4}\frac{\lambda^4}{\lambda_0^3}\right]
\end{aligned}$$

Example 5.12 Given $\lambda = 5x^2i^3$, find W_m, W_m', and $f^e(i, x)$ at $i = 10$ A, $x = 0.1$ m.

$$\begin{aligned}
W_m'(i, x) &= \int_o^i \lambda(i', x)di' = \frac{5x^2i^4}{4} \\
f^e &= \frac{\partial W_m'}{\partial x} = \frac{5}{2}xi^4 \\
W_m &= \lambda i - W_m' = 5x^2i^4 - \frac{5}{4}x^2i^4 = \frac{15}{4}x^2i^4 \\
W_m(10, 0.1) &= 375 \text{ Watt-sec}, W_m' = 125 \text{ Watt-sec} \\
f^e &= 2500 \text{ N-m}
\end{aligned}$$

Example 5.13 The relationship between λ and i is sketched in Figure 5.31 for two different values of x. Find W_m and W'_m at a and b.

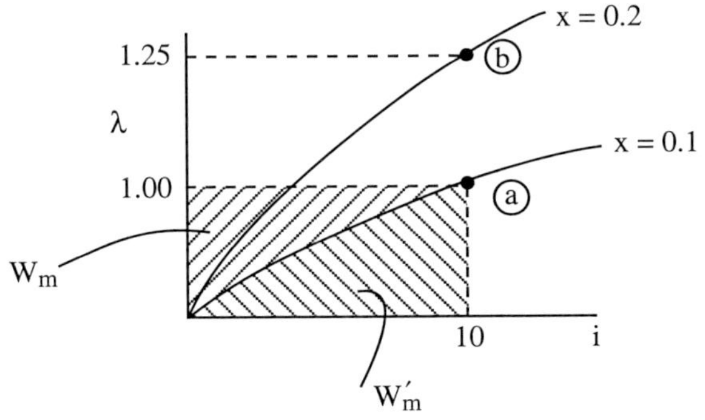

Figure 5.31: $\lambda - i$ curves for different values of x

Solution: At point a, W_m and W'_m are shown by the shaded areas. If we have an expression $\lambda(i, x)$, then $W'_m = \int_0^{10} \lambda(i', 0.1)di'$ and $W_m = \lambda i \mid_{i=10,\lambda=1}$ $-W'_m = 10 - W'_m$. We can compute W_m, W'_m at point b similarly.

5.7.3 Energy conversion between two points

In the $\lambda - i$ plane shown in Figure 5.31, to go from a to b, we get from the fundamental relation (5.117)

$$\begin{aligned} W_m(\lambda_b, x_b) - W_m(\lambda_a, x_a) &= \int_{\lambda_a}^{\lambda_b} i d\lambda - \int_{x_a}^{x_b} f^e dx \\ &= \int_{\lambda_a}^{\lambda_b} i d\lambda + \left[- \int_{x_a}^{x_b} f^e dx \right] \quad (5.127) \\ \Delta W_{m_{a \to b}} &= \text{EFE} \mid_{a \to b} + \text{EFM} \mid_{a \to b} \quad (5.128) \end{aligned}$$

where EFE stands for “energy from electrical” and EFM stands for “energy from mechanical.” To evaluate EFE and EFM, we need to specify a particular path. This concept of EFM is useful in studying energy conversion in devices that operate cyclically.

5.7.4 Energy conversion over a cycle

From (5.117),

$$dW_m = i d\lambda - f^e dx \quad (5.129)$$

Over a complete cycle, denoted as $\oint$, when the system state returns to the original state, $dW_m = 0$. Hence,

$$\begin{aligned} 0 &= \oint i d\lambda - \oint f^e dx \\ &= \oint i d\lambda + (-\oint f^e dx) \end{aligned} \tag{5.130}$$

From Figure 5.18 or Figure 5.30

$$\begin{aligned} i d\lambda &= \text{Energy from electrical (EFE)} \\ -f^e dx &= \text{Energy from mechanical (EFM)} \end{aligned}$$

Thus, over a complex cycle

$$\oint \text{EFE} + \oint \text{EFM} = 0 \tag{5.131}$$

An alternate notation is

$$\text{EFE}|_{\text{cycle}} + \text{EFM}|_{\text{cycle}} = 0 \tag{5.132}$$

Because there is no net change in stored energy, one can compute either $\text{EFE}|_{\text{cycle}}$ or $\text{EFM}|_{\text{cycle}}$. If $\text{EFE}|_{\text{cycle}} > 0$, then the system is operating as a *motor* and $\text{EFM}|_{\text{cycle}} < 0$. If $\text{EFM}|_{\text{cycle}} > 0$, then the system is operating as a *generator* and $\text{EFE}|_{\text{cycle}} < 0$. We will illustrate this through two examples.

Example 5.14 An electric machine has the $\theta - i$ relation as shown in Figure 5.32. The $\lambda - i$ relations are

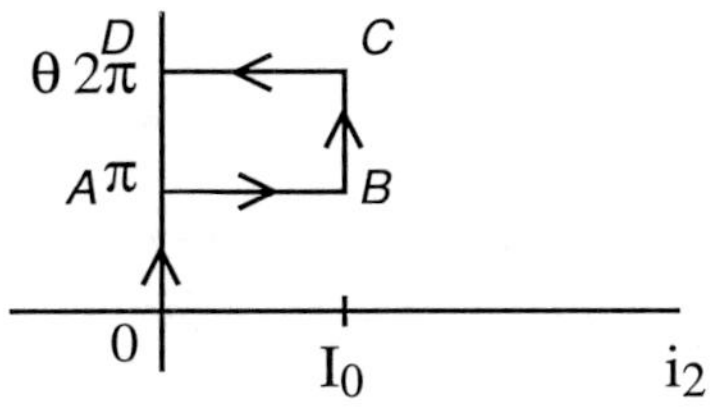

Figure 5.32: $\theta - i$ relation

$$\begin{aligned} \lambda_1 &= L_{11} i_1 + M \cos\theta \; i_2 \\ \lambda_2 &= M \cos\theta \; i_1 + L_{22} i_2 \end{aligned}$$

The machine is operated over the cycle shown above while $i_1 = I_o$. Find the energy converted from electrical to mechanical form for each cycle. Is this a motor or a generator?

Solution: We first find the torque of electric origin.

$$\begin{aligned} W'_m &= \frac{1}{2}L_{11}i_1^2 + M\cos\theta\ i_1 i_2 + \frac{1}{2}L_{22}i_2^2 \\ T^e &= \frac{\partial W'_m}{\partial \theta} = -M i_1 i_2 \sin\theta \end{aligned}$$

Since it is the $\theta - i$ plane, we compute $\int_{\theta_1}^{\theta_2} T^e d\theta$.

$$\begin{aligned} \text{EFM}\,|_{\text{cycle}} &= -\int_0^{2\pi} T^e d\theta = -\left[\int_0^{\pi} T^e d\theta + \int_{\pi}^{2\pi} T^e d\theta\right] \\ &= \int_0^{A} -T^e d\theta + \int_A^B -T^e d\theta + \int_B^C -T^e d\theta + \int_C^D -T^e d\theta \end{aligned}$$

It can be verified that all terms except the third one are zero. Hence,

$$\begin{aligned} \text{EFM}_{\text{cycle}} &= \int_B^C -T^e d\theta \quad = \quad \int_{\pi}^{2\pi} M I_0^2 \sin\theta d\theta \\ &= -2\,M I_0^2 \end{aligned}$$

Since $\text{EFM}|_{\text{cycle}} < 0$, it is motor action.

Example 5.15 Given an electromechanical system with $\lambda = \frac{L_o i}{(1+x/a)}$,

a) Find $f^e(\lambda, x)$.

b) Find EFM for the cycle $A \to B \to C \to D \to A$ shown in Figure 5.33.

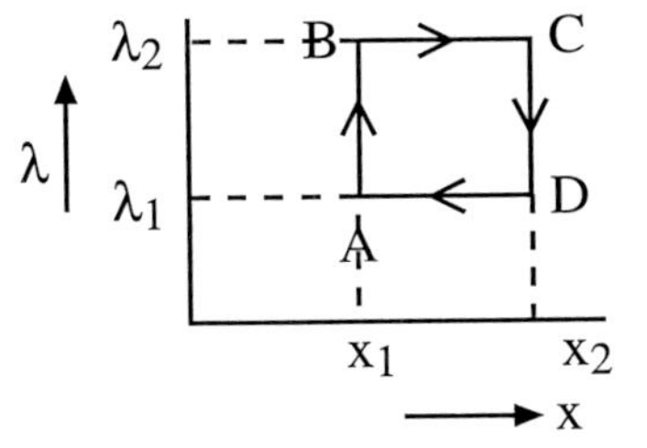

Figure 5.33: Flux displacement relationship [4]

c) Determine if the system is acting as a motor or a generator.

Solution:

$$
\begin{aligned}
W_m &= \int_0^\lambda i d\lambda' = \int_0^\lambda \frac{\lambda'(1+\frac{x}{a})}{L_0} d\lambda' = \frac{(1+\frac{x}{a})\lambda^2}{2L_0} \\
f^e &= -\frac{\partial W_m}{\partial x} = -\frac{\lambda^2}{2L_0 a}
\end{aligned}
$$

$$
\begin{aligned}
\text{EFM}\,|_{\text{cycle}} &= -\oint_{\text{cycle}} f^e dx = \int_A^B \frac{\lambda^2}{2L_0 a} dx + \int_B^C \frac{\lambda^2}{2L_0 a} dx \\
&\quad + \int_C^D \frac{\lambda^2}{2L_0 a} dx + \int_D^A \frac{\lambda^2}{2L_0 a} dx
\end{aligned}
$$

Of the four terms on the right side, the first term is zero since x is constant along AB and the third term is also zero since x is constant along CD. The only contribution is from the second and fourth terms. Therefore,

$$
\begin{aligned}
EFM\,|_{\text{cycle}} &= \frac{\lambda_2^2}{2L_0 a}(x_2 - x_1) + \frac{\lambda_1^2}{2L_0 a}(x_1 - x_2) \\
&= \frac{(x_2 - x_1)}{2L_0 a}(\lambda_2^2 - \lambda_1^2) > 0
\end{aligned}
$$

Since $x_2 > x_1$ and $\lambda_2 > \lambda_1$, $\text{EFM}|_{\text{cycle}} > 0$, and hence, the device acts like a generator.

Example 5.16 Compute $\text{EFE}|_{\text{cycle}}$ directly for Example 4.15.

Solution: For this, we need to redraw the cycle in the $\lambda - i$ plane. It is shown in Figure 5.34. The $\lambda - i$ curve is drawn for $x = x_1$ and $x = x_2$. i is

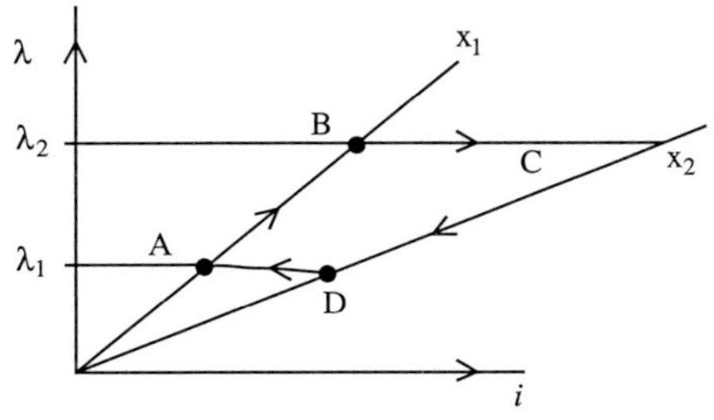

Figure 5.34: Flux current relationship

computed as $i = \frac{\lambda(1+x/a)}{L_0}$

$$\begin{aligned}\text{EFE}|_{\text{cycle}} &= \oint_{\text{cycle}} i d\lambda = \oint_{\text{cycle}} \frac{\lambda'(1+\frac{x}{a})}{L_o} d\lambda' \\ &= \int_A^B \frac{\lambda'(1+\frac{x_1}{a})}{L_0} d\lambda' + \int_C^D \frac{\lambda'(1+\frac{x_2}{a})}{L_0} d\lambda'\end{aligned}$$

The contribution from integrals along BC and DA are zero since λ is constant along these segments.

$$\begin{aligned}\text{EFE}|_{\text{cycle}} &= \frac{\lambda_2^2(1+\frac{x_1}{a})}{2L_0} - \frac{\lambda_1^2(1+\frac{x_1}{a})}{2L_0} + \frac{\lambda_1^2(1+\frac{x_2}{a})}{2L_0} - \frac{\lambda_2^2(1+\frac{x_2}{a})}{2L_0} \\ &= \frac{(\lambda_2^2-\lambda_1^2)}{2L_0 a}(x_1 - x_2) < 0\end{aligned}$$

Hence, it acts as a generator. This agrees with the result of Example 4.15.

5.8 Dynamics of Lumped Mechanical Systems

So far we have studied the electrical subsystem and its interaction with the lossless magnetic field system, which resulted in the force of electric origin acting on the mechanical system. We next study the dynamics of the mechanical system and finally get the overall dynamic model.

Lumped elements of the mechanical system are masses, springs, and dashpots or dampers. The masses store kinetic energy (KE), the springs store potential energy (PE), and the dashpots represent the dissipative elements. The elements in the electrical system are capacitance (C), inductance (L), and resistance (R). They store energies in C and L and dissipate energy in R. Instead of KCL and KVL, we have Newton's law to write the equations of motion in a mechanical system.

5.8.1 Mass-spring system

Consider a mass $M = \frac{W}{g}$ suspended from a spring of stiffness K that has units of force per unit length. In the static equilibrium position Figure 5.35(a) (also called equilibrium position, neutral position, or central position), the gravitational force $W = Mg$ is balanced by the spring force $K\ell$ where ℓ is the stretching of the spring due to the weight W. If we choose the equilibrium position as reference, then only forces due to displacement

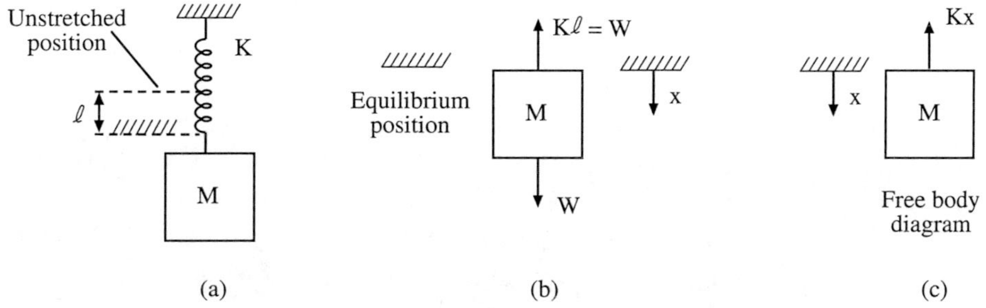

Figure 5.35: Translational system

from this position need be considered for the equation of motion, as shown in Figure 5.35(b). The free body diagram for this system is shown in Figure 5.35(c). Positive displacement x in the downward direction causes a force in the spring to pull the mass upward.

Newton's law states that *acceleration force in the positive x direction is equal to the algebraic sum of all the forces acting on the mass in the positive x direction.* In Figure 5.35(c),

$$M\ddot{x} = -Kx \text{ or } M\ddot{x} + Kx = 0 \tag{5.133}$$

We can also measure the distance of the mass from a fixed position such as the unstretched position of the spring as shown in Figure 5.36. Let this be

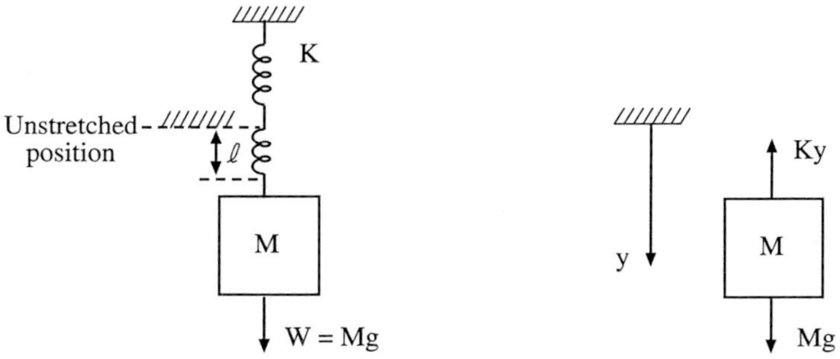

Figure 5.36: Free body diagram from a fixed position

denoted by y. Newton's law then becomes

$$M\ddot{y} = -Ky + Mg \tag{5.134}$$
$$M\ddot{y} + Ky = Mg \tag{5.135}$$

In equilibrium position, $y = \ell$, so that $Mg = K\ell$ and (5.135) becomes

$$M\ddot{y} + K(y - \ell) = 0 \tag{5.136}$$

This is equivalent to (5.133) by recognizing that $x = y - \ell$. The unstretched position is a *fixed reference* as opposed to the *equilibrium position reference.* When a fixed reference is used, both K and ℓ have to be specified for the spring, or one must compute ℓ, knowing the external forces.

5.8.2 Mass-spring system with dissipative element

Consider the mass M supported on a spring in Figure 5.37, and a spring-

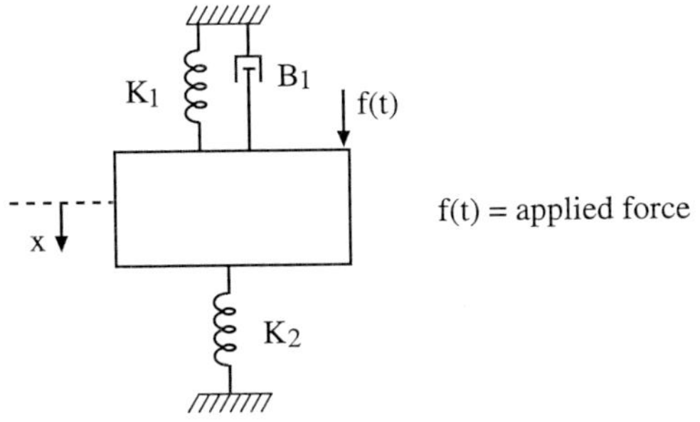

Figure 5.37: A single degree of freedom mechanical system

dashpot arrangement. $f(t)$ is the applied force. The positive x direction is as shown. x is measured from the *static equilibrium position.* The symbol for the dashpot is given in Figure 5.38. In an ideal dashpot, there is a

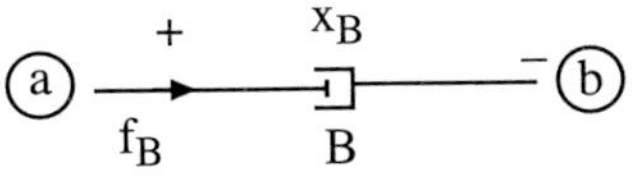

Figure 5.38: Two-terminal representation of a dashpot

force proportional to the relative velocity between the two nodes. This implies that if (a) is x_B units positive from (b), there is a force $f_B = B\frac{dx_B}{dt}$ in the direction shown. The free body diagram is given in Figure 5.39. We now apply Newton's law, which states that acceleration force in the positive x direction is equal to the sum of all the forces acting in the positive x direction.

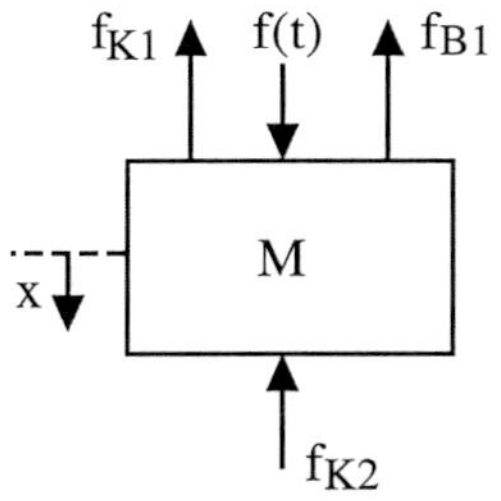

Figure 5.39: Free body diagram

Application of Newton's law in Figure 5.39 gives

$$\begin{aligned} M\ddot{x} &= f(t) - f_{K1} - f_{K2} - f_B \\ &= f(t) - K_1 x - K_2 x - B\frac{dx}{dt} \end{aligned}$$

Example 5.17 For the two-degree-of-freedom system in Figure 5.40, write

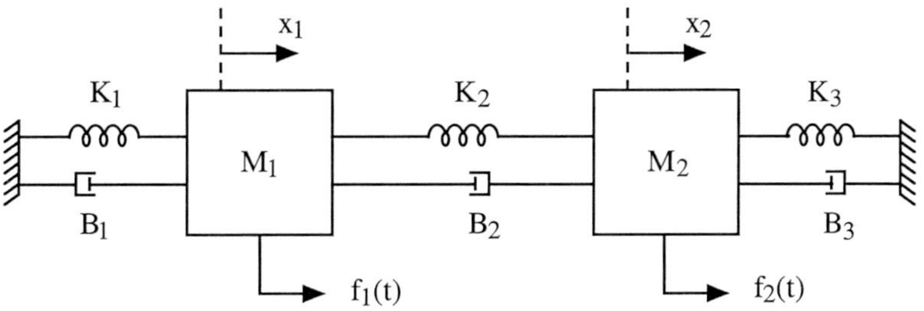

Figure 5.40: Two-degrees-of-freedom mechanical system

the mechanical equations. x_1 and x_2 are measured from the equilibrium position.

Solution: Define

$$x_2 - x_1 = x$$

The free body diagram is given in Figure 5.41. Applying Newton's law to each of the two masses, we get

$$\begin{aligned} M_1\ddot{x}_1 &= f_1(t) + K_2(x_2 - x_1) + B_2(\dot{x}_2 - \dot{x}_1) - B_1\dot{x}_1 - K_1 x_1 \\ M_2\ddot{x}_2 &= f_2(t) - B_2(\dot{x}_2 - \dot{x}_1) - K_2(x_2 - x_1) - B_3\dot{x}_2 - K_3 x_2 \end{aligned}$$

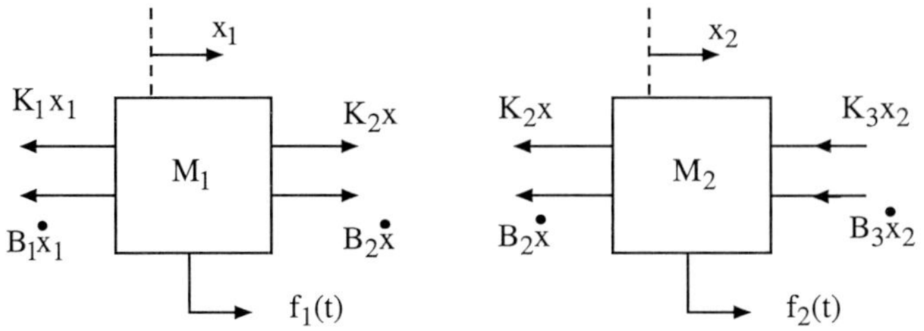

Figure 5.41: Free body diagram

Suppose $f_1(t)$ and $f_2(t)$ are unit step functions of magnitudes $\hat{F}_1$ and $\hat{F}_2$, respectively, then in steady state the derivatives of x_1 and x_2 are zero. We have as $t \to \infty$

$$\begin{bmatrix} K_1 + K_2 & -K_2 \\ -K_2 & K_2 + K_3 \end{bmatrix} \begin{bmatrix} x_1^{ss} \\ x_2^{ss} \end{bmatrix} = \begin{bmatrix} \hat{F}_1 \\ \hat{F}_2 \end{bmatrix}$$

We can solve for x_1^{ss} and x_2^{ss}, which will be the new equilibrium positions.

Example 5.18 This is an example of an automobile suspension system shown in Figure 5.42(a). K_1 is the spring, B the shock absorber, and K_2

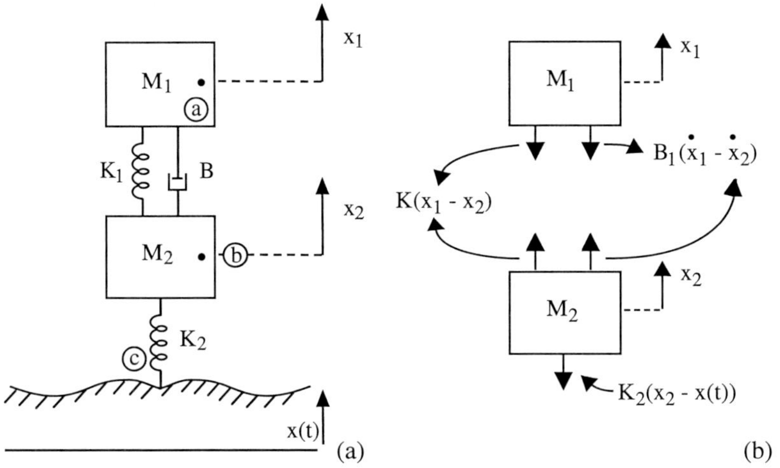

Figure 5.42: Automobile suspension system

the tire stiffness. M_1 is the mass of the auto and M_2 the wheel mass. x_1 and x_2 are measured from an equilibrium position. Write the dynamic equations.

Solution: The free body diagram is shown in Figure 5.42(b). The equations are

$$\begin{aligned} M_1 \frac{d^2 x_1}{dt^2} &= -K_1(x_1 - x_2) - B(\dot{x}_1 - \dot{x}_2) \\ M_2 \frac{d^2 x_2}{dt^2} &= K_1(x_1 - x_2) + B(\dot{x}_1 - \dot{x}_2) - K_2(x_2 - x(t)) \end{aligned}$$

Note that x_1 and x_2 are the dynamic variables and $x(t)$ acts as an input.

5.9 State Space Models

When we write the equations on the electrical and mechanical side, we have the complete dynamic description of the system. These equations are coupled, i.e., the electrical variables enter into the mechanical equations via the force of electric origin, and the mechanical variables enter into the electrical equations when the voltage $= \frac{d\lambda(i,x)}{dt}$ is computed. These coupled equations are then put in the form of a set of first-order differential equations (d.e.'s) for analysis. This is known as the *state space model* for the system.

Example 5.19 For the system in Figure 5.43, write both the electrical and mechanical equations of motion and put them in the state space form.

Solution: We have derived the flux linkages in Example 4.7 as

$$\lambda = \frac{N^2 i}{\left(\frac{\ell_c}{\mu A} + \frac{2x}{\mu_0 A}\right)}$$

Note that $\frac{\ell_c}{\mu A} = \mathcal{R}_c$ and $\frac{2x}{\mu_0 A} = \mathcal{R}_g(x)$. Hence, $\lambda = \frac{N^2 i}{\mathcal{R}_c + \mathcal{R}_g(x)}$. The co-energy is given by

$$W'_m = \frac{N^2 i^2}{2(\mathcal{R}_c + \mathcal{R}_g(x))} = \frac{N^2 i^2}{2\mathcal{R}(x)}$$

where

$$\mathcal{R}(x) = \mathcal{R}_c + \mathcal{R}_g(x)$$

The force f^e of electric origin is obtained by taking the derivative of W'_m with respect to x.

$$f^e = \frac{-N^2 i^2}{\mu_0 A \left(\mathcal{R}_c + \frac{2x}{\mu_0 A}\right)^2}$$

We now write the equations on the *electrical side*

$$\begin{aligned} v_s &= iR + \frac{d\lambda}{dt} \\ &= iR + \frac{N^2}{\left(\mathcal{R}_c + \frac{2x}{\mu_0 A}\right)} \frac{di}{dt} - \frac{N^2 i}{\left(\mathcal{R}_c + \frac{2x}{\mu_0 A}\right)^2} \frac{2}{\mu_0 A} \frac{dx}{dt} \end{aligned} \quad (5.137)$$

Note how x appears in the electrical equation.

Mechanical equations

Let ℓ be the static equilibrium position of the moving member. This is the same as the unstretched length of the spring with $i = 0$. We assume $\ell > 0$, i.e., the two pieces do not touch each other. The electromechanical system in Figure 5.43 is redrawn to show the static equilibrium position

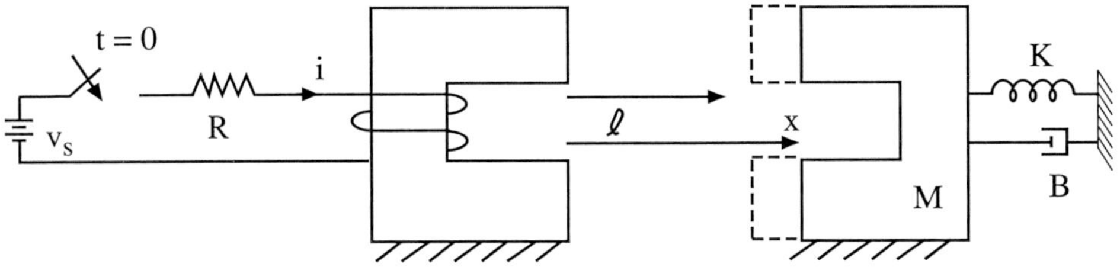

Figure 5.43: Electromechanical system redrawn to indicate static equilibrium

(shown dotted) as well as the dynamic position. If we measure the position of the moving member from the equilibrium position, then the mechanical equations have the variable $(x - \ell)$. Since $\frac{d^2(x-\ell)}{dt^2} = \frac{d^2x}{dt^2}$ and $\frac{d(x-\ell)}{dt} = \frac{dx}{dt}$, we have

$$M\frac{d^2x}{dt^2} + K(x - \ell) + B\frac{dx}{dt} = f^e = \frac{-N^2 i^2}{\mu_0 A \left(\mathcal{R}_c + \frac{2x}{\mu_0 A}\right)^2} \quad (5.138)$$

Equations (5.137) and (5.138) constitute the overall mathematical model of the electromechanical system. Note the coupling between the electrical and mechanical variables in the two sets of equations.

State space model

The state space model consists in writing (5.137) and (5.138) in the form of a set of three first-order differential equations. When put in this form, they can be easily integrated numerically using a computer. We define the *state variables* as x, $\frac{dx}{dt}$ (denoted by v), and i. We can write (5.137) and (5.138) in the form of three first-order differential equations by differentiating x, v, and i and expressing these derivatives in terms of x, v, i *only*, and any input to the system. Here the symbol v denotes velocity. Thus,

$$\frac{dx}{dt} = v \tag{5.139}$$

$$\frac{dv}{dt} = \frac{d^2x}{dt^2} \tag{5.140}$$

In (5.140), the right side is expressed in terms of x, v, and i using (5.138) and thus,

$$\frac{dv}{dt} = \frac{1}{M}\left[\left(\frac{-N^2i^2}{\mu_0 A\left(\mathcal{R}_c + \frac{2x}{\mu_0 A}\right)^2}\right) - K(x-\ell) - Bv\right] \tag{5.141}$$

From (5.137), the derivative of the current is

$$\frac{di}{dt} = \frac{1}{L(x)}\left[-iR + \frac{N^2 i}{\left(\mathcal{R}_c + \frac{2x}{\mu_0 A}\right)^2}\frac{2}{\mu_0 A}v + v_s\right] \tag{5.142}$$

where

$$L(x) = \frac{N^2}{\left(\mathcal{R}_c + \frac{2x}{\mu_0 A}\right)}$$

Equations (5.139), (5.141), and (5.142) are first-order differential equations and are called the *state space* equations of the electromechanical system. The initial conditions are $x(0) = \ell$, $v(0) = 0$, and $i(0) = 0$. $v_s(t)$ is a forcing function. More generally, we can also redefine the state variables mathematically as $x = x_1$, $v = x_2$, $i = x_3$, and $v_s = u$. The state space equations are then of the form

$$\begin{aligned} \dot{x}_1 &= f_1(x_1, x_2, x_3) \\ \dot{x}_2 &= f_2(x_1, x_2, x_3) \\ \dot{x}_3 &= f_3(x_1, x_2, x_3, u) \end{aligned} \tag{5.143}$$

In the matrix form, (5.143) is

$$\underline{\dot{x}} = \underline{f}(\underline{x}, \underline{u}), \underline{x}(0) = \begin{bmatrix} xx_1(0) \\ x_2(0) \\ x_3(0) \end{bmatrix}, \underline{u} = u, \text{a scalar} \tag{5.144}$$

where

$$\underline{x} = \begin{bmatrix} x_1 \\ x_2 \\ x_3 \end{bmatrix}, \underline{f} = \begin{bmatrix} f_1 \\ f_2 \\ f_3 \end{bmatrix}$$

$\underline{x}(0)$ are the initial conditions. In this case, $x_1(0) = \ell$, $x_2(0) = x_3(0) = 0$. This is the mathematical description of the electromechanical system. Many nonlinear physical systems are symbolically denoted in this manner. This representation is easier for computer simulation. It is a single input, three-state model. Given $u(t)$ and knowing $\underline{x}(0)$, we can integrate (5.144) to obtain the time domain response. This will be discussed in the next section.

Equilibrium points

Consider the equation $\underline{\dot{x}} = f(\underline{x}, \underline{u})$. If the input $\underline{u}$ is constant $= \underline{\hat{u}}$, then by setting $\underline{\dot{x}} = 0$, we get the algebraic equations $\underline{0} = \underline{f}(\underline{x}, \underline{\hat{u}})$. This may have several solutions for $\underline{x}$ denoted by $\underline{x}_1^e$, $\underline{x}_2^e$,.... These are called the *static equilibrium* solutions. To obtain them, we have to solve the nonlinear algebraic equations $\underline{0} = \underline{f}(\underline{x}, \hat{u})$. In small-dimensional systems, this can be done graphically.

For Example 4.19, the state space model is given by (5.139), (5.141), and (5.142). Setting derivatives equal to zero, we get $v^e = 0$ from (5.139) and hence, $i^e = \frac{v_s}{R}$ from (5.142). From (5.141), we get

$$-K(x - \ell) = \frac{N^2(i^e)^2}{\mu_0 A \left(\mathcal{R}_c + \frac{2x}{\mu_0 A}\right)^2} = -f^e(i^e, x) \tag{5.145}$$

Since $i^e = \frac{v_s}{R}$, the only unknown in (5.145) is x, and hence, we can plot $-K(x - \ell)$ and $-f^e(i^e, x)$ and get equilibrium values x^e graphically as the intersection of the two curves as shown in Figure 5.44. There are two values of equilibrium x^e, i.e., x_1^e and x_2^e. Suppose we change the spring constant K. For a higher value of K given by $- \cdot - \cdot -$, there may be only one intersection (x_3^e). For a weaker spring constant shown by $- \quad - \quad -$ there may be no intersection at all! ($x = 0$!). The technique of finding the equilibrium solutions for higher-order systems involves numerical techniques which are not discussed here.

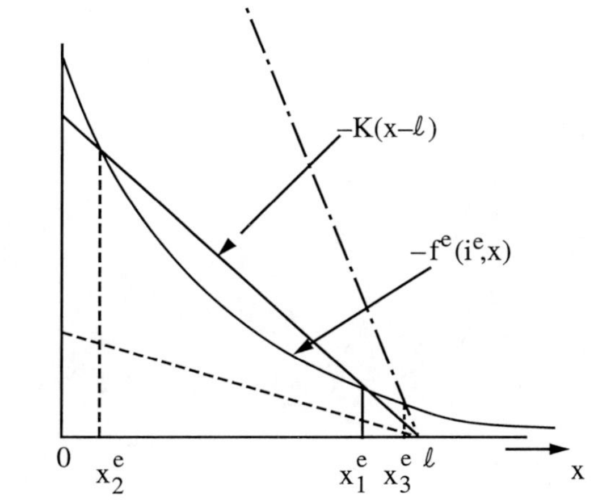

Figure 5.44: Finding equilibrium points graphically

5.10 Numerical Integration

There are two kinds of methods: a) explicit and b) implicit. Euler's method is an explicit one. Of the two, Euler's method is easier to implement for small systems. For large systems, the implicit method is better because it is numerically stable. We will explain only Euler's method.

5.10.1 Explicit method (Euler's method)

The equation is

$$\dot{\underline{x}} = \underline{f}(\underline{x}, \underline{u}), \quad \underline{x}(0) = \underline{x}_0 \tag{5.146}$$

where $\underline{x}$, $\underline{f}$, and $\underline{u}$ are vectors. We split the time interval into equally spaced time instants $t_0, t_1, t_2, \ldots, t_n, t_{n+1}, \ldots$, as shown in Figure 5.45. Suppose we

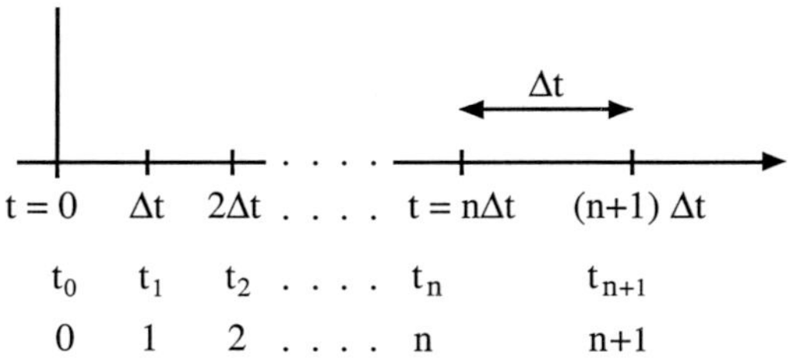

Figure 5.45: Notation for time steps

integrate from t_n to t_{n+1}. The integrand on the right side is kept constant at the value corresponding to the previous time instant. Thus,

$$\begin{aligned}
\int_{t_n}^{t_{n+1}} \underline{\dot{x}}(t)dt &= \int_{t_n}^{t_{n+1}} \underline{f}(\underline{x}, \underline{u})dt \\
\underline{x}(t_{n+1}) - \underline{x}(t_n) &= (t_{n+1} - t_n)\underline{f}(\underline{x}(t_n), \underline{u}(t_n)) \\
\underline{x}(t_{n+1}) &= \underline{x}(t_n) + \Delta t[\underline{f}(\underline{x}(t_n), \underline{u}(t_n))] \qquad (5.147)
\end{aligned}$$

where

$$\Delta t = t_{n+1} - t_n = \text{constant}, n = 0, 1, 2$$

Δt is called the integration step size. It is called explicit because all terms on the right side of (5.147) are known from the previous time step. We will denote the values of $\underline{x}$ at various time instants as $\underline{x}^{(1)}, \underline{x}^{(2)}, \ldots, \underline{x}^{(n)}, \underline{x}^{(n+1)}$. The various notations for time steps are shown in Figure 5.45.

There are other more complex algorithms in the numerical analysis literature. We will now give a few examples of numerical solution using Euler's method.

Example 5.21 Consider the scalar equation

$$\dot{x} = -(t + 2)x^2, x(0) = 1 \qquad (5.148)$$

Use Euler's method to compute $x(t)$ at $t = 0.1$, 0.2, and 0.3 seconds.

Solution: Choose

$$\Delta t = 0.10 \text{ second}$$

The algorithm is

$$x^{(n+1)} = x^{(n)} + \Delta t[f(x^{(n)}, t_n)], \quad n = 0, 1, 2, \ldots \qquad (5.149)$$

To obtain $x^{(1)}$, *let* $n = 0$, $t_0 = 0$ in (5.149)

$$x^{(0)} = 1, f(x^{(0)}, t_o) = -(0 + 2)1^2 = -2$$

Hence,

$$x^{(1)} = 1 + (0.1)(-2) = 1 - 0.2 = 0.8$$

To obtain $x^{(2)}$, *let* $n = 1$, $t_1 = 0.1$ From the previous time step,

$$f(x^{(1)}, t_1) = -(0.1 + 2)(0.8)^2 = -1.344$$

Therefore,

$$\begin{aligned} x^{(2)} &= x^{(1)} + \Delta t[f(x^{(1)}, t_1)] \\ &= 0.8 + (0.1)(-1.344) = 0.8 - 0.1344 = 0.6656 \end{aligned}$$

In this way, values for $n = 2, 3, \ldots$ can be calculated as

$$\begin{aligned} x^{(3)} &= 0.5681(t = 0.2 \text{ second}) \\ x^{(4)} &= 0.4939(t = 0.3 \text{ second}) \end{aligned}$$

Example 5.22 Find $i(t)$ by Euler's method in Figure 5.46.

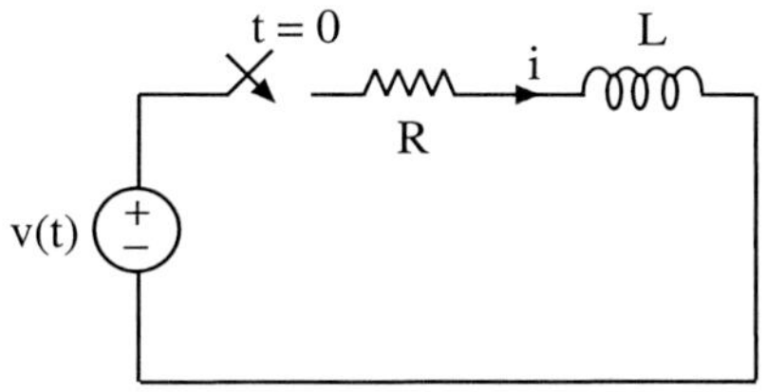

Figure 5.46: RL circuit

$R = (1 + 3i^2)\Omega$ $L = 1H$, $v(t) = 10t$ V.

Solution: Choose $\Delta t = 0.025$ second. The differential equation is

$$\begin{aligned} \frac{di}{dt} + iR &= v(t) \\ \frac{di}{dt} + (1 + 3i^2)i &= v(t), \quad i(0) = 0 \end{aligned}$$

Let $i = x, v(t) = u$

$$\begin{aligned} \frac{dx}{dt} &= -(1 + 3x^2)x + u(t) \\ &= f(x, t), x(0) = 0 = x^{(0)} \\ x^{(n+1)} &= x^{(n)} + \Delta t f(x^{(n)}, t_n) \quad n = 0, 1, 2, \ldots \end{aligned}$$

Choose $\Delta t = 0.025$ second.

To obtain $x^{(1)}$, $n = 0$, $t_0 = 0$

$$x^{(0)} = 0, \quad f(x^{(0)}, t_0) = 0, \quad u(0) = 0$$

Therefore,

$$x^{(1)} = 0$$

To obtain $x^{(2)}$, $n = 1$, $t_1 = 0.025$

From previous time step $x^{(1)} = 0$

$$u(0.025) = 0.25, \quad f(x^{(1)}, t_1) = -(1 + 0^2)0 + 0.25 = 0.25$$

Therefore,

$$x^{(2)} = x^{(1)} + (0.025)(0.25) = 0 + 0.00625 = 0.00625$$

Similarly one can calculate $x^{(3)}$, $x^{(4)}$, etc.

5.10.2 Application for higher-order systems

The same procedure is applicable when applied to a system of first-order differential equations. Let

$$\begin{aligned} \dot{x}_1 &= f_1(x_1, x_2, x_3, t) \\ \dot{x}_2 &= f_2(x_1, x_2, x_3, t) \\ \dot{x}_3 &= f_3(x_1, x_2, x_3, t) \end{aligned} \quad (5.150)$$

Application of Euler's method gives

$$\begin{aligned} x_1^{(n+1)} &= x_1^{(n)} + \Delta t(f_1(x_1^{(n)}, x_2^{(n)}, x_3^{(n)}, t_n)) \\ x_2^{(n+1)} &= x_2^{(n)} + \Delta t(f_2(x_1^{(n)}, x_2^{(n)}, x_3^{(n)}, t_n)) \\ x_3^{(n+1)} &= x_3^{(n)} + \Delta t(f_3(x_1^{(n)}, x_2^{(n)}, x_3^{(n)}, t_n)) \end{aligned} \quad (5.151)$$

Note that we have used the superscript notation to denote the time instant. Thus,

$$\begin{aligned} (n) \Rightarrow n\Delta t &= t_n \\ (n+1) \Rightarrow (n+1)\Delta t &= t_{n+1} \end{aligned}$$

Example 5.21 Integrate the following system of equations from $t = 0$ to $t = 0.02$ second with a step size of $\Delta t = 0.01$ second and compute them at $t = 0.01$ and 0.02 second.

$$\begin{aligned}\frac{dx_1}{dt} &= x_2\\ \frac{dx_2}{dt} &= x_2 - 10x_1^3 + 5\end{aligned}$$

The initial conditions are $x(0) = 0.2$ and $x_2(0) = 0$.

Solution: Using (5.151)

$$\begin{aligned}x_1(0.01) &= x_1(0) + \Delta t x_2(0)\\ &= 0.2 + (0.01)0 = 0.2\\ x_2(0.01) &= x_2(0) + \Delta t(x_2(0) - 10x_1^3(0) + 5)\\ &= 0 + .01(0 - 10(0.2)^3 + 5)\\ &= 0.0492\\ x_1(0.02) &= x_1(0.01) + \Delta t \;\; x_2(0.01)\\ &= 0.2 + .01(0.0492)\\ &= 0.2 + .00492\\ &= 0.200492\\ x_2(0.02) &= x_2(0.01) + 0.01(x_2(0.01) - 10x_1^3(0.01) + 5)\\ &= 0.0492 + .01(0.0492 - 110(0.2)^3 + 5)\\ &= 0.098892\end{aligned}$$

5.11 Conclusion

In this chapter, we have discussed the dynamics of electromechanical systems and its analysis using energy methods. Computation of forces and torques of electric origin using energy, and co-energy concepts was discussed. The concept of state space model, static equilibrium points, as well as an introduction to numerical integration of the state space model was also discussed. Euler's method is easily programmable on a hand calculator.

5.12 Problems

5.1 The device in Figure 5.47 has an upper fixed piece and lower movable

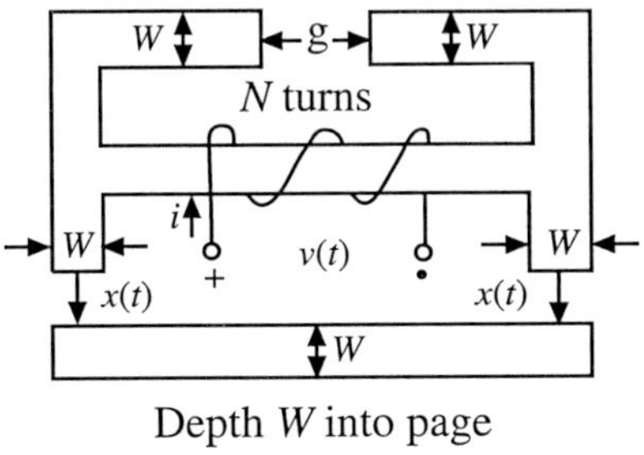

Figure 5.47: Electromechanical system

piece, both composed of infinitely permeable magnetic material, and both with depth W into the paper. The upper leg of the fixed piece has an air gap with spacing g, and the air gap between the fixed and movable piece is $x(t)$ on both sides. Each leg has width W. A variable current source $i(t)$ drives the terminals of an N-turn coil wrapped on the lower horizontal leg of the fixed piece.

(a) Find the flux $\lambda(i, x)$ linked by the electrical terminal pair,

$$\left(\lambda = \mu_0 W^2 \left(\frac{1}{2x} + \frac{1}{g}\right) N^2 i\right)$$

(b) Find the voltage $v(t)$ at the terminal pair.

(c) Find the force of electric origin $f^e(\lambda, x)$ using energy $W_m(\lambda, x)$.

(d) Find the force $f^e(i, x)$ using co-energy. Show that by substituting $\lambda = \lambda(i, x)$ in $f^e(\lambda, x)$ in part (c) you get the same result.

5.2 Consider the reluctance motor shown in Figure 5.48. The reluctance of the magnetic circuit and the inductance of the coil depend on the rotor position θ. It is assumed that this dependence is sinusoidal according to the formula

$$L(\theta) = L_1 + L_2 \cos(2\theta)$$

(a) Compute the inductances L_1 and L_2. The reluctance of the iron path can be neglected. Assume air gap cross-section area A_{g0} and A_{g1} for the air gaps g_0 and g_1, respectively.

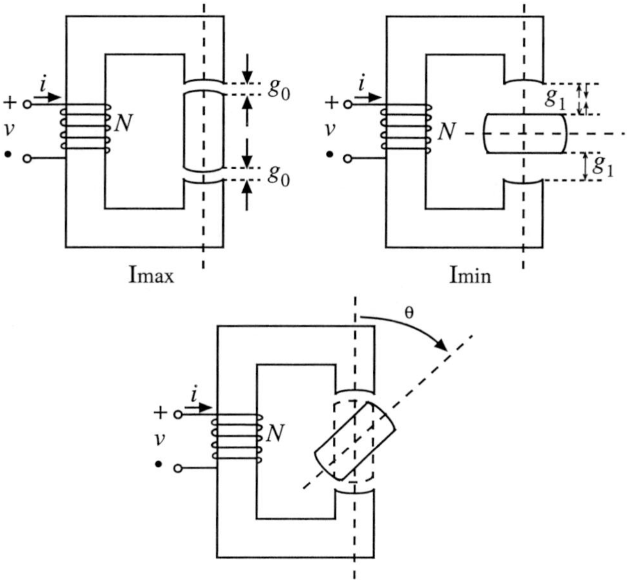

Figure 5.48: Reluctance motor

(b) Calculate the torque of electric origin as a function of θ. (Partial answer: $T^e = -i^2 L_2 \sin(2\theta)$)

5.3 The system shown in Figure 5.49 consists of a fixed left piece and a

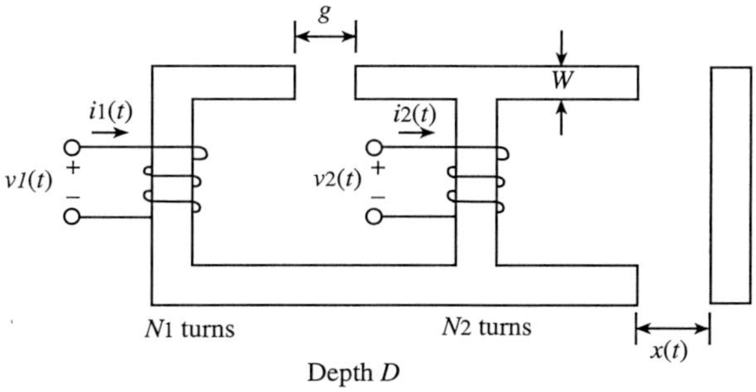

Figure 5.49: Electromechanical system

movable right piece, both pieces are constructed from infinitely permeable material. Each leg of the fixed piece has width W, and both pieces have depth D into the paper.

(a) Draw a static magnetic circuit that represents the magnetic field system.

(b) Find the fluxes ϕ_1 and ϕ_2 through the two coils and the flux linkages λ_1 and λ_2.

(c) Find the co-energy and the force of electric origin acting on the movable piece.

5.4 The magnetic field system shown in Figure 5.50 consists of a fixed piece

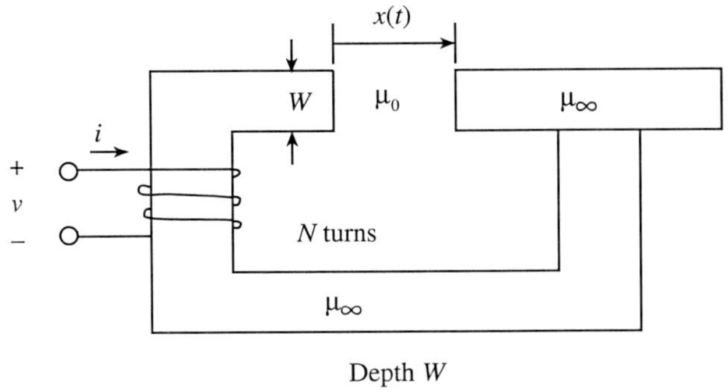

Figure 5.50: Electromechanical system

and a movable core. Both pieces are made from infinitely permeable material and have depth W into the paper. The air gap has permeability μ_0 and length $x(t)$. A variable current source $i(t)$ drives the terminals of an N-turn coil wound on the fixed piece.

Compute $W_m(\lambda, x)$, $W_m'(i, x)$, and $f^e(i, x)$ on the movable piece.

5.5 The device shown in Figure 5.51 has a fixed center core and two movable side pieces. The device is made from infinitely permeable material, and all three pieces have depth D into the paper. A variable current source $i(t)$ drives the terminals of an N-turn coil wound on the fixed core.

(a) Find the flux $\lambda(i, x, y)$ linked by the electrical terminal pair.

(b) Calculate the voltage $v(t)$ at the terminal pair.

(c) Compute the forces f_x^e and f_y^e acting on the two pieces. (Partial answer: $F_x^e = -\mu_o D N^2 i^2 a / 4x^2$)

5.6 In the electromechanical system in Figure 5.52, the iron portion has infinite permeability. Find $\lambda(i, x)$, $W_m'(i, x)$, and $f^e(i, x)$.

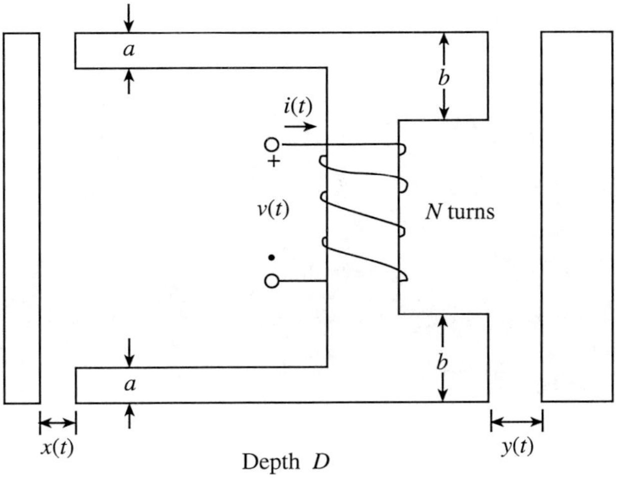

Figure 5.51: Electromechanical system

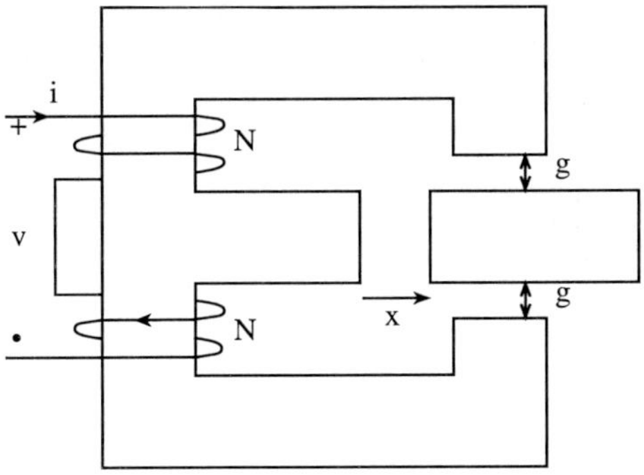

Figure 5.52: Electromechanical system

5.7 An electromechanical system with three electrical ports and one rotational mechanical port (θ) has the following equations for flux linkages

$$\begin{aligned}\lambda_1 &= L_{11}i_1 + (M\cos\theta)i_3\\ \lambda_2 &= L_{22}i_2 + (M\sin\theta)i_3\\ \lambda_3 &= (M\cos\theta)i_1 + (M\sin\theta)i_2 + L_{33}\,i_3\end{aligned}$$

where L_{11}, M, L_{22}, and L_{33} are constants.

(a) Compute the co-energy W'_m.

(b) Compute the torque of electric origin $T_\theta^e(i_1, i_2, i_3, \theta)$.

5.8 An electromechanical system has two electrical and one mechanical ports. The $\lambda - i$ relations are:

$$\begin{bmatrix} \lambda_1 \\ \lambda_2 \end{bmatrix} = \begin{bmatrix} (5 + \cos 2\theta) \times 10^{-3} & 0.1 \cos \theta \\ 0.1 \cos \theta & 50 + 10 \cos 2\theta \end{bmatrix} \begin{bmatrix} i_1 \\ i_2 \end{bmatrix}$$

(a) Find the torque of electrical origin.

(b) Evaluate the torque at $i_1 = 1$ A and $i_2 = 0.01$ A.

(c) Find the maximum torque possible and the angle at which it occurs.

(Answer: $\theta = \pm 49.6°$, $T^e = \mp 2.74 \times 10^{-3}$ N-m)

5.9 A machine has the following $\lambda - i$ relation

$$\begin{aligned} \lambda_{as} &= (L_0 + L_1 \cos 2\theta) i_{as} + (M \cos \theta) i_r \\ \lambda_{bs} &= (L_0 - L_1 \cos 2\theta) i_{bs} + (M \sin \theta) i_r \\ \lambda_r &= (M \cos \theta) i_{as} + (M \sin \theta) i_{bs} + (L_2 \cos 4\theta) i_r \end{aligned}$$

Find the co-energy and the torque of electrical origin. (Partial answer: $T^e = -L_1(\sin 2\theta) i_{as}^2 + \ldots$)

5.10 The co-energy of a device is $W'_m = \frac{i^3}{4x}$.

(a) Find λ as a function of i and x.

(b) Find the force of electrical origin.

(c) Find the energy stored in the coupling field W_m.

(Hint: $W_m + W'_m = \lambda i$)

5.11 A rotating machine has the following electrical terminal relations:

$$\begin{aligned} \lambda_s &= (L \cos 2\theta) i_s + (M \sin \theta) i_r \\ \lambda_r &= (M \sin \theta) i_s + (L \cos 2\theta) i_r \end{aligned}$$

where θ is the angle of the rotor.

(a) If the machine is operated with $i_s = 0$ (i.e., the stator is open circuited) and $i_r = I$ (i.e., a constant DC rotor current), and is spun with $d\theta/dt = \omega$, determine $v_s(t)$ in terms of L, M, I, and ω.

(b) Calculate the co-energy, W'_m, in terms of L, M, i_s, i_r, and θ.

(c) Find the torque of electric origin on the rotor.

5.12 Given $W_m = 5x\lambda^2$, find the energy transferred from the mechanical system to the coupling field as x moves from 0.1 to 0.2 m along a constant current $i = 8$ A path. (Answer: 16 joules)

5.13 The following $\lambda - i$ relations for a three electrical and two mechanical port are given:

$$\begin{aligned}
\lambda_1 &= L_{11}i_1 + Mi_3\cos(\theta - \phi) \\
\lambda_2 &= L_{22}i_2 + Mi_3\sin(\theta - \phi) \\
\lambda_3 &= L_{33}i_3 + Mi_1\cos(\theta - \phi) + Mi_2\sin(\theta - \phi)
\end{aligned}$$

Find the torques of electric origin T_θ^e and T_ϕ^e.

5.14 A translational electromechanical system has the following equations:

$$\begin{aligned}
\frac{d^2x}{dt^2} + 0.1\left(\frac{dx}{dt}\right)i + x^3 - x &= 0 \\
2\left(\frac{dx}{dt}\right)i + x\left(\frac{di}{dt}\right) + i - 5 &= 0
\end{aligned}$$

(a) Put them in state space form.

(b) Find the possible equilibrium points.

5.15 The magnetic system in Figure 5.53 has co-energy

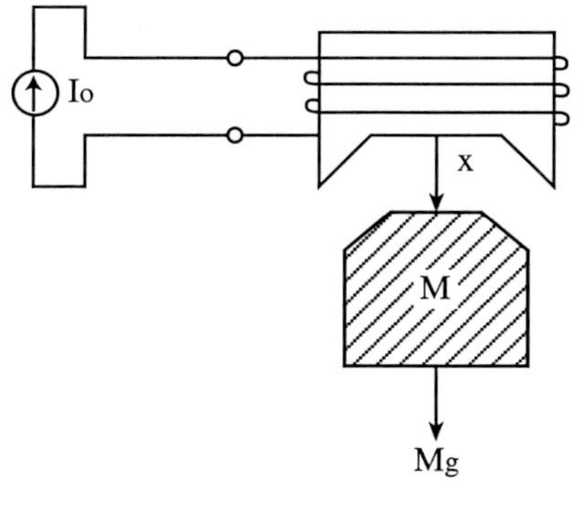

Figure 5.53: Magnetic system

$$W'_m = \frac{1}{2}\frac{L_0}{\left(1+\frac{x}{a}\right)}I_0^2, x > 0$$

The equation of motion is given by

$$M\frac{d^2x}{dt^2} = Mg + f^e$$

where f^e is the force of electric origin.

(a) Find the equilibrium point $x^e > 0$.

(b) Find the minimum value of I_0 for which the equilibrium exists.

5.16 The system in Figure 5.54 has a movable piece with mass M, restrained

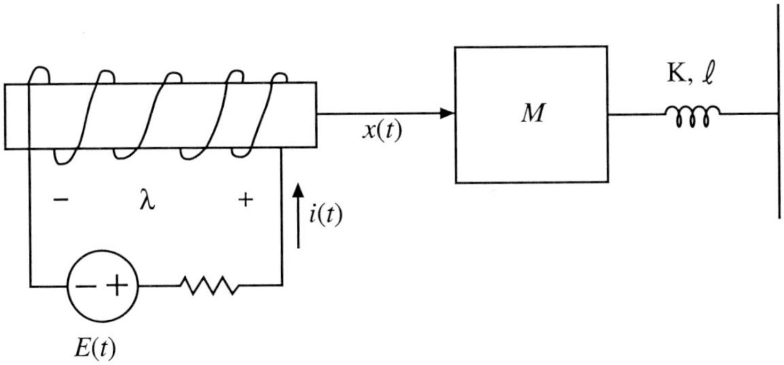

Figure 5.54: Electromechanical system

by a spring. Its position $x(t)$ with respect to the face of the fixed core determines the inductance $L(x)$ as measured at the terminals of a coil wound on the core. The inductance has been found to take the form

$$L(x) = L_0 e^{-x/a}$$

The movable piece is attached to a fixed wall through a spring with spring constant K, which exerts no force when $x = \ell$. The coil is driven by the voltage source $E(t)$, which connects to the coil through a resistance R.

(a) Calculate the force of electric origin acting on the mass.

(b) Write the electrical and mechanical equations of motion.

(c) Write a state space description of the system using the state vector

$$\underline{x}(t) = \begin{pmatrix} x(t) \\ v(t) \\ i(t) \end{pmatrix}$$

where $v(t) = \frac{dx}{dt}$.

(d) Determine the values of the equilibrium velocity v^e and the current i^e if $E(t) = E_0 = constant$

(e) Write an equation to compute the equilibrium displacement, x^e, of the movable piece (do not try to solve this equation for x^e).

5.17 In the system of Figure 5.55, the distance x of the mass M is measured from a fixed reference. $\lambda(i, x)$ is given as $\lambda(i, x) = \frac{Ci^2}{x}$. The gravitational force acting on M is given by Mg.

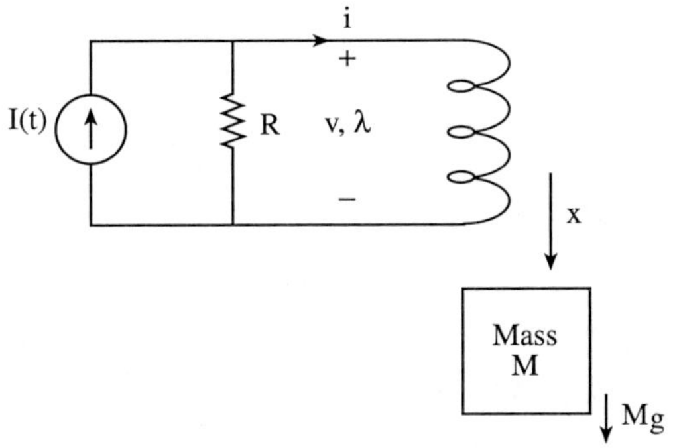

Figure 5.55: Electromechanical system

(a) Find the force of electric origin.

(b) Write the equations of the system in state space form. $I(t)$ is a specified function of time.

(c) Determine $W_m(\lambda, x)$.

5.18 In the following state space model

$$\begin{aligned} \dot{x}_1 &= x_2 \\ \dot{x}_2 &= -0.1x_2x_3 - x_1^3 + x_1 \\ \dot{x}_3 &= -2x_2x_3 - x_3 + 5 \end{aligned}$$

Initial conditions are $x_1(0) = 1$, $x_2(0) = 0.5$, and $x_3(0) = 10$ at $t = 0$.

(a) With a time step of 0.1 second, find using Euler's method x_1, x_2, and x_3 at $t = 0.1$ second and 0.2 second.

(b) Find all the possible static equilibrium points.

(Partial answer: x_1, x_2, and x_3 at 0.1 second are 1.05, 0.45 and 8.5, respectively)

5.19 A certain electromechanical system has the following dynamic equations:

$$\begin{aligned} \frac{d^2x}{dt^2} &= -\frac{0.0005i^2}{x^2} - 400(x-0.1) - 80\frac{dx}{dt} \\ V &= i + \frac{0.001}{x}\frac{di}{dt} - \frac{0.001}{x^2}i\frac{dx}{dt} \end{aligned}$$

where x is the displacement on the mechanical side, V is the applied voltage, and i is the current in the coil. The initial conditions at $t = 0$ are $x(0) = 0.1$, $\frac{dx}{dt}\mid_{t=0} = 0$, and $i(0) = 0$ and a voltage $V = 10$ volts is applied.

(a) Using a time step of $\Delta t = 0.001$ second find x, $\frac{dx}{dt}$ and i at $t = 0.001$ and 0.002 second.

(b) Determine the steady state value of the current i.

5.20 For the state space model of Problem 5.18, write a computer program using Euler's method to compute x_1, x_2, and x_3 up to 5 seconds using the time step of 0.1 second. Verify the steady state value with the equilibrium points obtained.

5.21 If $\lambda = \frac{\propto i^3}{(x+a)}$ where $\propto$ and a are constants, find both $W'_m(i, x)$ and $W_m(\lambda, x)$.

5.22 A two-electrical and one-mechanical rotational port system has the following $\lambda - i$ relations.

$$\begin{aligned} \lambda_1 &= (L_0 + M\cos 2\theta)i_1 + M\sin 2\theta i_2 \\ \lambda_2 &= M\sin 2\theta i_1 + (L_0 - M\cos 2\theta)i_2 \end{aligned}$$

(a) Determine the torque $T^e(i_1, i_2, \theta)$ of electric origin.

(b) If $i_1 = I\cos\omega_s t$, $i_2 = I\sin\omega_s t$, and the rotor has a constant angular velocity ω_m such that $\theta = \omega_m t + \delta$ (where δ = constant), evaluate the instantaneous torque. (Hint: Use the trigonometric identities $\cos^2 x - \sin^2 x = \cos 2x$, $\sin(x-y) = \sin x\cos y - \cos x \sin y$ and $\sin 2x = 2\sin x\cos x$). (Answer: $MI^2\sin(2\omega_s t - 2\theta)$)

5.23 Given the state space model below with $M = 0.001$ Kg, $K = 125$ N-m, $B = 0.1$ N sec/m, $\ell = 4.922 \times 10^{-3}$ m.

$$\begin{aligned}\frac{dx}{dt} &= v\\ \frac{dv}{dt} &= \frac{1}{M}[f^e(x,i) - K(x-\ell) - Bv]\\ \frac{di}{dt} &= \frac{2v}{x} + 1500 - i\end{aligned}$$

where

$$f^e(x,i) = -\frac{(8.854\times 10^{-12})(0.2)i^2}{2x^2}$$

Graphically sketch and estimate the two equilibrium points. (Answer: $x_1^e = 2.62$ mm, $x_2^e = 3.85$ mm)

5.24 Simulate by numerical integration the system

$$\begin{aligned}\dot{x}_1 &= x_2\\ \dot{x}_2 &= -5\sin x_1 - x_2\end{aligned}$$

with initial conditions $x_1 = 1$ radians, and $x_2 = 0$. Use Euler's method with a step of 0.1 second. Compute for $t = 0.1$, 0.2, and 0.3 second. Also write a computer program and simulate until steady state is reached.

5.25 A single-phase reluctance motor has $\lambda(i,\theta) = (L_0 - L_2\cos 2\theta)$. The motor has moment of inertia J and is restrained by a torsional spring with spring constant K, which is relaxed with $\theta = 0$.

(a) Write the mechanical equation of motion for $\theta(t)$.

(b) Assuming $i = I$ = constant, write the algebraic equation to determine equilibrium points.

(c) Under what condition does the system have more than one equilibrium point? (Hint: Sketch the equations in (b))

5.26 The state space model of a certain electromechanical system is given by

$$\begin{aligned}\frac{dx_1}{dt} &= x_2\\ \frac{dx_2}{dt} &= \frac{-x_3^2}{x_1^2} - 40(x_1 - 0.1) - 80x_2\\ \frac{dx_3}{dt} &= \frac{x_2 x_3}{x_1} + 1000x_1 - 1000x_1 x_3\end{aligned}$$

Initial conditions are $x_1(0) = 0.1$, $x_2(0) = 0$, and $x_3(0) = 0$. Find values of x_1, x_2, and x_3 using Euler's method at $t = 0.001$ and 0.002 seconds. Use $\Delta t = 0.001$ second. (Partial answer: At $t = 0.002$ seconds, $x_1 = 0.1$, $x_2 = -0.001$, and $x_3 = 0.19$)

5.27 (a) Formulate the state space model for the system by defining

$$\begin{aligned}x = x_1, \frac{dx}{dt} = x_2, i &= x_3\\ \frac{d^2x}{dt^2} + (0.1)\frac{dx}{dt}i + x^3 - x &= 0\\ 2\left(\frac{dx}{dt}\right)i + x\left(\frac{di}{dt}\right) + i - 5 &= 0\end{aligned}$$

(b) For $x(0) = 1$, $\left.\frac{dx}{dt}\right|_{t=0} = 0.5$, and $i(0) = 0$, numerically integrate the equations for two time steps with $\Delta t = 0.1$ second. (Partial answer: $x(0.2) = 1.1$, $\left.\frac{dx}{dt}\right|_{t=0.2} = 0.4867$, $i(0.2) = 0.881$)

5.28 This is a computer exercise.

(a) In the system shown in Figure 5.56, calculate the terminal relation $\lambda(i, x)$. The units of x should be meters. The fixed air gaps in the top portion are 1 mm in length each.

(b) You are advised to use a program such as MATHEMATICA or MATLAB to solve this part. The system has the plunger connected to a spring of spring constant $K = 1000$ N-m. The spring is relaxed when $x = 1$ cm, (i.e., $x(0) = 1$ cm). The mass of the plunger is 0.1 Kg, and the system has a friction corresponding to a damping $B = 1$ N-s/m. The coil is connected to a voltage source in series with a 20Ω resistor. Ignore gravity.

(i) Write the equations for this system in state space form.

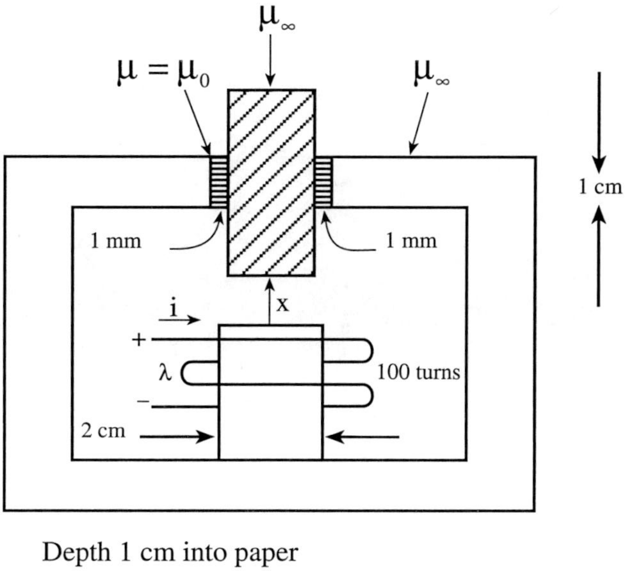

Figure 5.56: Electromechanical system

(ii) At $t = 0$, the system is at rest with $x = 0.01\text{m} = 1$ cm and zero current. The voltage is a step of height 160 V. Find the numerical solution for the position and current as a function of time. Continue the integration for 0.5 second. Have the computer plot your results for x and i.

(iii) Repeat ii) for a step of height 240 V. Continue the plots for 0.061 second.

(iv) If the voltage is equal to 160 V, find (numerically) any static equilibrium points for this system. Plot the spring and magnetic forces as a function of x with I constant. Are these equilibrium points stable or unstable?

(v) For each of the equilibrium points of part (iv), find the numerical solution if the plunger is released from the rest with x one mm greater than the equilibrium value.

5.29 For a two-electrical and two-mechanical port system, use co-energy to show that the test for conservativeness is

$$
\begin{aligned}
\frac{\partial f_1^e}{\partial i_1} &= \frac{\partial \lambda_1}{\partial x_1} \\
\frac{\partial f_2^e}{\partial i_2} &= \frac{\partial \lambda_2}{\partial x_2} \\
\frac{\partial \lambda_1}{\partial i_2} &= \frac{\partial \lambda_2}{\partial x_1} \\
\frac{\partial f_1^e}{\partial x_2} &= \frac{\partial f_2^e}{\partial x_1}
\end{aligned}
$$

Verify these relations for Example 5.9

Chapter 6

STABILITY OF ELECTROMECHANICAL SYSTEMS

6.1 Introduction

So far we have discussed techniques for developing the dynamic model of an electromechanical system. The models are described by nonlinear differential equations. We gave a brief introduction to the numerical differential equations for a given set of initial conditions and input. The stability of the dynamic nonlinear system is of great interest in terms of operation. This is true of an individual device as well as a large interconnected system such as the power system. In this chapter, we will introduce the necessary tools for stability analysis. While the emphasis is on electromechanical systems, these tools are equally valid for large power systems.

6.2 Linear and Nonlinear Stability

In Chapter 5, we discussed (a) the time-domain solution of the nonlinear dynamic system $\dot{\underline{x}} = \underline{f}(\underline{x}, u)$ by numerical integration and (b) computation of static equilibrium points of the dynamic system graphically. In the case of higher-order systems, numerical techniques such as Newton's method are used for computing equilibrium points.

In many cases, it will be helpful to know if the static equilibrium point(s) obtained are stable or not. For large disturbances in the state $\underline{x}$ or input $\underline{u}$, time-domain simulations are always necessary. For *small* disturbances around the equilibrium point, a linearized analysis is helpful to determine if the equilibrium point is stable or not by merely computing the eigenvalues of a matrix. For large disturbances, a direct method using energy functions can in some cases be used to assess stability without time-domain simulations.

6.3 Linearization

Given the state space model

$$\dot{\underline{x}} = \underline{f}(\underline{x}, \underline{u}) \tag{6.1}$$

and constant input $\hat{u}$, it is possible to compute the equilibrium point(s) by well-known numerical methods. Suppose $\underline{x} = \underline{x}^e$ is one such equilibrium point. The stability of the equilibrium point is important. Equation (6.1) may, for example, represent a large interconnected power grid. The equilibrium point will represent the current operating steady-state regime. The physical system representing (6.1) may be subjected to small or big disturbances. In the case of power grids, small disturbances may lead to oscillations of power in transmission lines at a low frequency of 1–3 Hz. If not properly damped, these may lead to cascading outages. The big disturbances may be in the form of a fault or lighting strike. For small disturbances around $\underline{x}^e$ and small changes in $\underline{x}$ and $\hat{\underline{u}}$, the stability can be judged by a linear model obtained by linearizing (6.1) around $\underline{x}^e$ and $\hat{\underline{u}}$.

6.3.1 Scalar case

$$\dot{x} = f(x, u) \tag{6.2}$$

Expand $f(x, u)$ in a Taylor series around x^e and $\hat{u}$ and retain only the linear terms. Define

$$\Delta x = x - x^e, \ \Delta u = u - \hat{u}$$

i.e., $x = x^e + \Delta x$ and $u = \hat{u} + \Delta u$. Expanding $f(x, u)$ in a Taylor series around x^e and $\hat{u}$

$$f(x, u) \quad = \quad f(x^e, \hat{u}) + \left.\frac{\partial f}{\partial x}\right|_{x^e} (x - x^e) + \left.\frac{\partial f}{\partial u}\right|_{x^e} (u - \hat{u}) + \text{h.o.t.} \tag{6.3}$$

where h.o.t. stands for "higher order terms". Since $f(x^e, \hat{u}) = 0$ and $\frac{d}{dt}(x^e + \Delta x) = \frac{d\Delta x}{dt}$, by substituting (6.3) in (6.2) and neglecting higher order terms, we get

$$\Delta \dot{x} = \left.\frac{\partial f}{\partial x}\right|_{x^e} \Delta x + \left.\frac{\partial f}{\partial u}\right|_{x^e} \Delta u \tag{6.4}$$

where $|_0$ denotes evaluation of the partial derivative at the equilibrium point, $(x^e, \hat{u})$.

6.3.2 Extension to second-order systems

$$\begin{aligned} \dot{x}_1 &= f_1(x_1, x_2, u) \\ \dot{x}_2 &= f_2(x_1, x_2, u) \end{aligned} \tag{6.5}$$

Let $\Delta x_1 = x_1 - x_1^e$, $\Delta x_2 = x_2 - x_2^e$, and $\Delta u = u - \hat{u}$. Linearizing (6.5) around the equilibrium point, we get

$$\begin{bmatrix} \Delta \dot{x}_1 \\ \Delta \dot{x}_2 \end{bmatrix} = \begin{bmatrix} \left.\frac{\partial f_1}{\partial x_1}\right|_{x^e} & \left.\frac{\partial f_1}{\partial x_2}\right|_{x^e} \\ \left.\frac{\partial f_2}{\partial x_1}\right|_{x^e} & \left.\frac{\partial f_2}{\partial x_2}\right|_{x^e} \end{bmatrix} \begin{bmatrix} \Delta x_1 \\ \Delta x_2 \end{bmatrix} + \begin{bmatrix} \left.\frac{\partial f_1}{\partial u}\right|_{x^e} \\ \left.\frac{\partial f_2}{\partial u}\right|_{x^e} \end{bmatrix} \Delta u \tag{6.6}$$

This can be extended to higher-order systems. For an electromechanical system, the state variables are x, v, and i and the input is a scalar u.

$$\begin{aligned} \frac{dx}{dt} &= f_1(x, v, i, u) \\ \frac{dv}{dt} &= f_2(x, v, i, u) \\ \frac{di}{dt} &= f_3(x, v, i, u) \end{aligned} \tag{6.7}$$

Let the equilibrium point be x^e, v^e, and i^e for a constant input $\hat{u}$. Linearizing (6.7)

$$\begin{aligned} \begin{bmatrix} \frac{d}{dt}\Delta x \\ \frac{d}{dt}\Delta v \\ \frac{d}{dt}\Delta i \end{bmatrix} &= \left.\begin{bmatrix} \frac{\partial f_1}{\partial x} & \frac{\partial f_1}{\partial v} & \frac{\partial f_1}{\partial i} \\ \frac{\partial f_2}{\partial x} & \frac{\partial f_2}{\partial v} & \frac{\partial f_2}{\partial i} \\ \frac{\partial f_3}{\partial x} & \frac{\partial f_3}{\partial v} & \frac{\partial f_3}{\partial i} \end{bmatrix}\right|_{x^e} \begin{bmatrix} \Delta x \\ \Delta v \\ \Delta i \end{bmatrix} + \left.\begin{bmatrix} \frac{\partial f_1}{\partial u} \\ \frac{\partial f_2}{\partial u} \\ \frac{\partial f_3}{\partial u} \end{bmatrix}\right|_{x^e} \Delta u \\ &= [A] \begin{bmatrix} \Delta x \\ \Delta v \\ \Delta i \end{bmatrix} + [b]\Delta u \end{aligned} \tag{6.8}$$

The eigenvalues of A determine the stability for small disturbances around the equilibrium point. Eigenvalues of A are obtained by solving for $det[A - \lambda I] = \mid A - \lambda I \mid = 0$.

If all the eigenvalues lie in the left-half plane (i.e., real parts < 0), we say the system is stable. If an eigenvalue lies in right-half plane, then the system is unstable. If a pair of complex eigenvalues lies on the imaginary axis, the system is *marginally stable.*

6.4 Stability of a Second-order System

Consider the nonlinear equation

$$M\frac{d^2x}{dt^2} + B\frac{dx}{dt} = f(x,u) \tag{6.9}$$

whose linearized form (after dividing by M) is (assuming constant u)

$$\begin{aligned}\frac{d^2\Delta x}{dt^2} + \frac{B}{M}\frac{d}{dt}\Delta x &= \frac{1}{M}\left.\frac{\partial f(x)}{\partial x}\right|_{x^e}\Delta x \\ &= -\omega_0^2\Delta x \end{aligned}\tag{6.10}$$

where

$$\frac{1}{M}\left.\frac{\partial f}{\partial x}\right|_{x^e} \triangleq -\omega_0^2$$

Define

$$\begin{aligned}\Delta x &= \Delta x_1 \\ \Delta \dot{x} &= \Delta x_2\end{aligned}$$

The state space form for (6.9) becomes

$$\begin{aligned}\Delta \dot{x}_1 &= \Delta x_2 \\ \Delta \dot{x}_2 &= -\frac{B}{M}\Delta x_2 - \omega_0^2\Delta x_1\end{aligned}\tag{6.11}$$

$$\begin{bmatrix}\Delta \dot{x}_1 \\ \Delta \dot{x}_2\end{bmatrix} = \begin{bmatrix}0 & 1 \\ -\omega_0^2 & -\frac{B}{M}\end{bmatrix}\begin{bmatrix}\Delta x_1 \\ \Delta x_2\end{bmatrix}\tag{6.12}$$

The characteristic equation is given by the determinant $\mid A - \lambda I \mid = 0$, i.e.,

$$\begin{aligned}
\begin{vmatrix} -\lambda & 1 \\ -\omega_0^2 & -\frac{B}{M} - \lambda \end{vmatrix} &= 0 \\
\lambda\left(\lambda + \frac{B}{M}\right) + \omega_0^2 &= 0 \\
\lambda^2 + \frac{B}{M}\lambda + \omega_0^2 &= 0
\end{aligned}$$

$$\lambda_1, \lambda_2 = -\frac{B}{2M} \pm \sqrt{\frac{B^2}{4M^2} - \omega_0^2}$$

Note that we can also get this by writing the characteristic equation of (6.9) as

$$s^2 + \frac{B}{M}s + \omega_0^2 = 0 \text{ and } s_{1,2} = \frac{-B}{2M} \pm \sqrt{\frac{B^2}{4M^2} - \omega_0^2}$$

and the roots of this equation are the same as the eigenvalues. The eigenvalue approach is useful for higher-order systems.

Case I $B > 0$, $M > 0$, $\omega_0^2 > 0$

Three separate possibilities are

1. $\frac{B^2}{4M^2} - \omega_0^2 > 0$, two unequal real roots, s_1, s_2.
2. $\frac{B^2}{4M^2} - \omega_0^2 = 0$, two equal real roots.
3. $\frac{B^2}{4M^2} - \omega_0^2 < 0$, two complex conjugate roots.

In case 1, both s_1 and s_2 will be negative. Then the solution for (6.9) is given by

$$\Delta x(t) = C_1 e^{s_1 t} + C_2 e^{s_2 t} \tag{6.13}$$

where C_1 and C_2 will depend on initial conditions. The equilibrium is *stable* since the real parts of s_1 and s_2 are < 0. The response $\Delta x(t)$ is called over damped. In case 2, $s_1 = s_2 = s$. Both roots are equal and negative.

$$\Delta x(t) = C_1 e^{st} + C_2 t e^{st} \tag{6.14}$$

As $t \to \infty$, $\Delta x(t) \to 0$. The response (6.14) is called critically damped. In case 3, $s_{1,2} = -\propto \pm j\beta$ ($\propto > 0$):

$$\begin{aligned}
\Delta x(t) &= C_1 e^{(-\propto + j\beta)t} + C_2^{(-\propto - j\beta)t} \\
&= e^{-\propto t}(A\cos\beta t + B\sin\beta t)
\end{aligned} \tag{6.15}$$

This is *stable* since real parts of $s_{1,2}$ are < 0. The response is called underdamped or oscillatory. Thus, if $\omega_0^2 > 0$, all three cases are *stable.*

Case II $B > 0$, $M > 0$, $\omega_0^2 < 0$

One of the roots s_1, s_2 will be real and positive. The equilibrium is *unstable.*

Special Case $B = 0, M > 0$

The characteristic equation is $s^2 + \omega_0^2 = 0$

$$s_1, s_2 = \pm\sqrt{-\omega_0^2}$$

Now

$$-\omega_0^2 = \frac{1}{M}\frac{\partial f}{\partial x}\bigg|_{x^e} \tag{6.16}$$

So $-\omega_0^2$ can be either > 0 or < 0 at $x = x^e$, depending on the sign of $\frac{\partial f}{\partial x}\big|_{x^e}$.

a) If $-\omega_0^2 > 0$, s_1, s_2 will be real and unequal in sign. Hence, the equilibrium point is unstable. Physically, this implies

$$\frac{\partial f}{\partial x}\bigg|_{x^e} > 0 \tag{6.17}$$

b) If $-\omega_0^2 < 0$, then s_1, s_2 will be complex conjugates with zero real part. Hence, the equilibrium point is *marginally stable.* Physically, this means

$$\frac{\partial f}{\partial x}\bigg|_{x^e} < 0 \tag{6.18}$$

We can give a physical significance to (6.17) and (6.18). Figure 6.1(a) shows a plot of $f(x)$ with positive slope of $f(x)$ at $x = x^e$. If the mass is moved a small distance to the right of the equilibrium point, the force f becomes positive and tends to increase the displacement still further. This makes x^e an unstable equilibrium point. If $\frac{\partial f}{\partial x}\big|_{x^e}$ is negative at $x = x^e$, then s_1 and s_2 are complex conjugates with zero real part. The mass will execute a sinusoidal motion around the equilibrium point. Figure 6.1(b) shows this case. If the mass moves to the right, the force becomes negative and tends to return the mass to $x = x^e$. In the absence of external disturbances, the

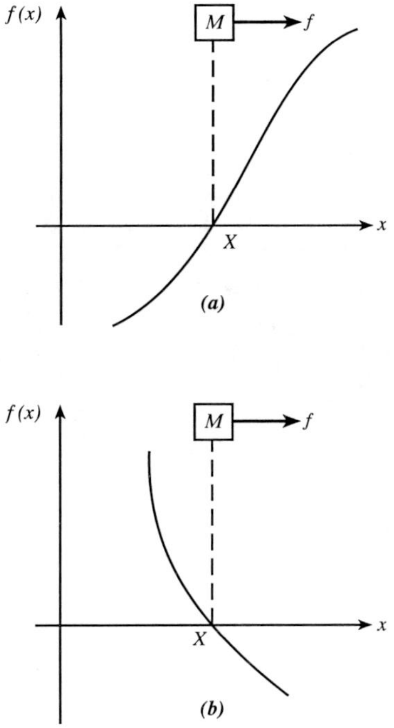

Figure 6.1: Graphical representations of a force $f(x)$ acting on a mass M with static equilibrium at $x = x^e$. [14]

mass will oscillate sinusoidally about the equilibrium point with constant amplitude; we call this motion stable since the response is bounded. If there was damping in the system, the amplitude of oscillation will decay until the mass came to rest at the equilibrium point.

Example 6.1 The magnetic system is the same as in Problem 6.15. It has co-energy

$$W'_m = \frac{1}{2}\frac{L_0}{(1+\frac{x}{a})}I_0^2, \; x > 0 \tag{6.19}$$

The equation of motion is given by

$$M\frac{d^2x}{dt^2} = Mg + f^e \tag{6.20}$$

where f^e is the force of electric origin.

a) Find the equilibrium point $x^e > 0$.

b) Find the minimum value of I_0 for which the equilibrium exists.

c) Determine if the equilibrium is stable.

Solution:

a) The force of electric origin f^e is

$$f^e = \frac{\partial W'_m}{\partial x} = -\frac{1}{2}\frac{L_0 I_0^2}{(1+\frac{x}{a})^2}\cdot\frac{1}{a} \tag{6.21}$$

The mechanical equation is

$$M\frac{d^2x}{dt^2} = Mg + f^e \tag{6.22}$$

For static equilibrium, set the derivative equal to zero. This yields

$$Mg = \frac{1}{2}L_0\frac{I_0^2}{\left(1+\frac{x}{a}\right)^2}\left(\frac{1}{a}\right)$$

Solving for x

$$1+\frac{x}{a} = \pm\sqrt{\frac{L_0 I_0^2}{2Mga}} \tag{6.23}$$

Take the positive sign on the right side of (6.23) for x^e to be > 0. Hence,

$$x^e = a\left(\sqrt{\frac{L_0 I_0^2}{2Mga}} - 1\right)$$

b) For $x^e > 0$, we must have $\frac{L_0 I_0^2}{2Mga} > 1$, i.e.,

$$I_0 > \sqrt{\frac{2Mga}{L_0}}$$

c) Linearizing the differential equation (6.22)

$$M\frac{d^2\Delta x}{dt^2} = \left.\frac{\partial f^e}{\partial x}\right|_{x=x^e}\Delta x = \frac{L_0 I_0^2}{(1+\frac{x^e}{a})^3}\cdot\frac{1}{a^2}\Delta x \tag{6.24}$$

Since

$$1+\frac{x^e}{a} > 0, \quad M\frac{d^2\Delta x}{dt^2} = K\Delta x$$

where $K > 0$. Therefore, the equilibrium is *unstable*.

Example 6.2 For the system of differential equations (6.25) and (6.26) describing an electromechanical system: (a) find equilibrium points (b) do linearization of the equations.

$$\frac{d^2\theta}{dt^2} = -4 + \theta i^2 \tag{6.25}$$

$$\frac{d}{dt}(2\theta i) + iR = v(t) \tag{6.26}$$

$$R = 1\Omega \text{ and } v(t) = 2 \text{ V}$$

Solution: Setting the derivatives equal to 0 and using the numerical values gives

$$\begin{aligned} 0 &= -4 + \theta^e (I^e)^2 \\ I^e(1) &= 2 \\ I^e &= 2, \theta^e = 1 \end{aligned}$$

Now linearize (6.25) and (6.26) around the equilibrium point

$$\begin{aligned} \frac{d^2\Delta\theta}{dt^2} &= i^2\big|_0 \Delta\theta + 2\theta i\big|_0 \Delta i \\ \frac{d^2\Delta\theta}{dt^2} &= 4\Delta\theta + 4\Delta i \end{aligned} \tag{6.27}$$

and

$$\Delta i + \frac{d}{dt}\left(2i\big|_0 \Delta\theta + 2\theta\big|_0 \Delta i\right) = 0 \tag{6.28}$$

$$\Delta i + 4\frac{d}{dt}\Delta\theta + 2\frac{d}{dt}\Delta i = 0 \tag{6.29}$$

The second-order system (6.27) in $\Delta\theta$ and (6.29) results in a third-order system.

Example 6.3 Express the linearized model of Example 6.2 in state space form and compute the eigenvalues.

Defining the state variables $\Delta\theta$, $\Delta\dot{\theta}$, and Δi as x_1, x_2, and x_3, respectively, the state space model is

$$\begin{bmatrix} \dot{x}_1 \\ \dot{x}_2 \\ \dot{x}_3 \end{bmatrix} = \begin{bmatrix} 0 & 1 & 0 \\ 4 & 0 & 4 \\ 0 & -2 & -1/2 \end{bmatrix} \begin{bmatrix} x_1 \\ x_2 \\ x_3 \end{bmatrix} \tag{6.30}$$

i.e., $\dot{x} = \underline{A}x$. The eigenvalues of $\underline{A}$ are $\lambda_1 = 0.4515$, and $\lambda_{1,2} = -0.4758 \pm j2.0502$.

6.5 Energy Function Methods for Nonlinear Systems

For large disturbances, the stability of a nonlinear dynamical system is analyzed best by a numerical simulation technique that may be computationally intensive. In many cases, useful information about the stability of the equilibrium point may be obtained by a direct method that avoids numerical integration. This technique is based on using the energy function approach. In its general form, it is formally known as Lyapunov's method. For conservative systems, the energy function approach gives satisfactory answers.

6.5.1 Conservative systems

In conservative systems, under disturbed conditions, the total energy remains constant, and this fact is used to monitor the evolution of the state variables, which in turn give information about the stability of the system. Consider the pendulum in Figure (6.2). It consists of a mass M connected

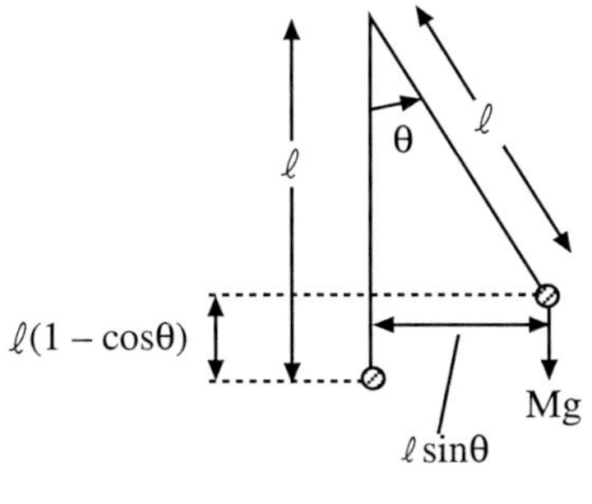

Figure 6.2: Simple pendulum

to a frictionless pivot through a rigid rod of length l. The deviation of the mass M from the vertical position is indicated by θ. The potential energy has to be defined with respect to a reference point.

Let $V(\theta) = 0$ at $\theta = 0$, i.e., $V(0) = 0$. Then, at any position θ, the potential energy is given by

$$V(\theta) = Mg\ell(1 - \cos\,\theta) \tag{6.31}$$

The moment of inertia $J = M\ell^2$. There are no forces other than gravity. Hence, summing the torques in a positive θ direction and equating it to the accelerating torque in the same direction gives

$$J\frac{d^2\theta}{dt^2} = -Mg(\ell \sin\theta) \tag{6.32}$$

Note that torque due to gravity acts opposite to positive θ. The system is *conservative*, i.e., there are no dissipative forces.

In such a case, the right side of (6.32) can be expressed as the negative derivative of a scalar potential function. In this case,

$$\begin{aligned} -Mg\ell\sin\theta &= -\frac{\partial}{\partial\theta}[Mg\ell(1-\cos\theta)] \\ &= \frac{-\partial V(\theta)}{\partial\theta} \end{aligned} \tag{6.33}$$

$MG\ell$ is added to $V(\theta)$ in (6.31) in order to make $V(\theta) = 0$ at $\theta = 0$. Equation (6.32) is rewritten as

$$J\frac{d^2\theta}{dt^2} = \frac{-\partial V(\theta)}{\partial\theta} \tag{6.34}$$

Equilibrium points are solutions of

$$-\frac{\partial V}{\partial\theta} = 0 \tag{6.35}$$

i.e.,

$$-Mg\ell \sin\theta = 0$$

Solving for θ in an interval of $-\pi$ to $+\pi$ gives

$$\theta^e = \pm\pi\ ,\ 0$$

Energy

$V(\theta)$ describes a surface in θ space. To explain this, consider (6.34):

$$J\frac{d^2\theta}{dt^2} + \frac{\partial V}{\partial\theta} = 0 \tag{6.36}$$

Multiply (6.36) by $\frac{d\theta}{dt}$ to yield

$$J\frac{d\theta}{dt}\frac{d^2\theta}{dt^2} + \frac{\partial V}{\partial\theta}\frac{d\theta}{dt} = 0 \tag{6.37}$$

Integrate (6.37) with respect to t to obtain

$$\underbrace{\frac{1}{2}J\left(\frac{d\theta}{dt}\right)^2}_{\text{Kinetic energy (KE)}} + \underbrace{V(\theta)}_{\text{Potential energy (PE)}} = E(\text{constant}) \tag{6.38}$$

Equation (6.38) essentially says that kinetic energy plus potential energy is equal to a constant $E\left(\theta, \frac{d\theta}{dt}\right)$.

This is another statement about the conservative nature of the system. $E\left(\theta, \frac{d\theta}{dt}\right)$ is called the *energy function*. Suppose at $t = 0$ you impart energy to the system, then $E(\theta, \dot{\theta})$ remains constant, and hence the energy is preserved. How do we make use of this property to get information about θ? This is done through the surface defined by $V(\theta)$: $\frac{-\partial V(\theta)}{\partial \theta} = 0$ defines equilibrium points, which are given by 0, $+\pi$ and $-\pi$.

Case I

Figure 6.3 shows the shape of the potential energy surface $V(\theta)$. Suppose

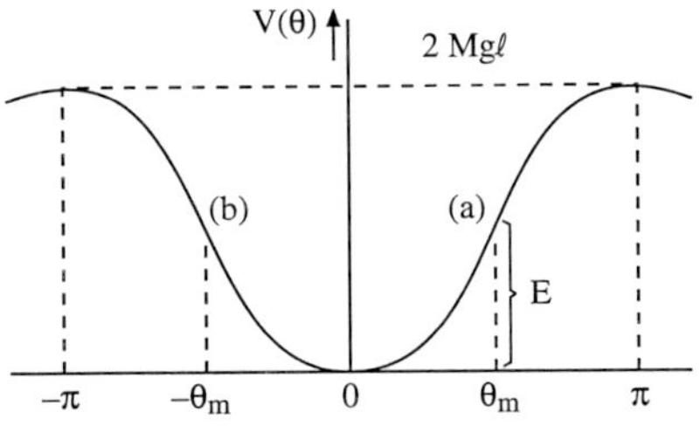

Figure 6.3: Potential energy *well*

the mass is released at $t = 0$, within $\theta = \theta_m$, $\frac{d\theta}{dt} = 0$. We then can visualize this as a ball rolling in the potential energy *well*. We can then visualize its movement from $t > 0$. Also in Figure 6.3, $V(\theta) = 2Mg\ell$ at $\theta \pm \pi$. This is directly obtained from (6.31). From (6.38)

$$\underbrace{\frac{1}{2}J\left(\frac{d\theta}{dt}\right)^2}_{\text{KE}} = E - V(\theta) \tag{6.39}$$

Hence, at $t = 0$, $E = V(\theta_m)$ since kinetic energy is zero at (a) in Figure 6.3, where $\frac{d\theta}{dt} = 0$ when the ball is released. The ball therefore oscillates in the potential energy *well* if $E < V(\pi)$ where $V(\pi) = 2Mg\ell$. At $\theta = 0$, the potential energy equals zero, but kinetic energy equals $E = V(\theta_m)$. Energy gets exchanged between the kinetic energy and potential energy portions for $t > 0$. As the ball moves from $\theta = \theta_m$ to $\theta = 0$, the potential energy gets converted into kinetic energy. After passing through $\theta = 0$, the ball moves up to (b) and then oscillates back. The same scenario is repeated if the ball is released at $\theta = -\theta_m$. $\theta = 0$ is therefore neither unstable or stable in the usual sense. In the control theory literature, it is *stable in the sense of Lyapunov* but not asymptotically stable.

Case II

At $t = 0$, $\theta = 0$, and $\frac{d\theta}{dt} \neq 0$. Let $\left.\frac{d\theta}{dt}\right|_{t=0} = \dot{\theta}_m$. Here the mass is at rest and an initial velocity is imparted to it. The total energy is

$$E = \frac{1}{2} J \left(\dot{\theta}_m\right)^2 + 0 \tag{6.40}$$

If the value of E in (6.40) is $> V(\pi)$, i.e., $> 2Mg\ell$ as shown in Figure 6.4, the ball will roll over the *hill* and will not return to $\theta = 0$. If $E < V(\pi)$, i.e.,

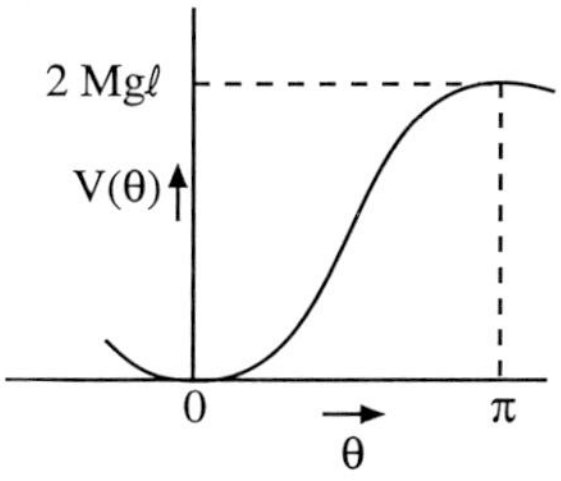

Figure 6.4: Potential energy *well*

$E < 2Mg\ell$, it will oscillate within the *well*.

Case III

At $t = 0$, $\theta = \theta_m$, $\frac{d\theta}{dt} = \theta_m$. This is a general case.

$$E = \frac{1}{2} J \left(\dot{\theta}_m\right)^2 + V(\theta_m) = V_{\mathrm{KE}}(\theta_m) + V(\theta_m) \tag{6.41}$$

If $E < 2Mg\ell$, then the ball *oscillates* within the potential *well*. If $E > 2Mg\ell$ the ball will go outside the well. E, V_{KE}, and V are shown in Figure (6.5) at a point $\theta = \theta_m$. This concept of a potential energy *well* is used in high-energy physics as well as power system stability.

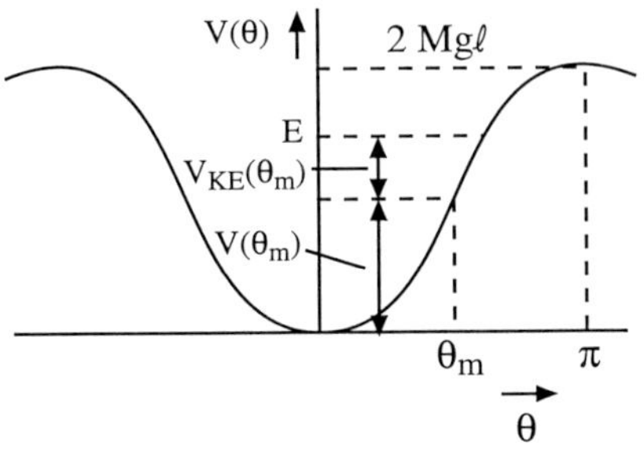

Figure 6.5: Potential energy *well*

6.6 Energy Functions in Electromechanical Systems

Consider the system in Figure 6.6. Let us assume that neither the electrical

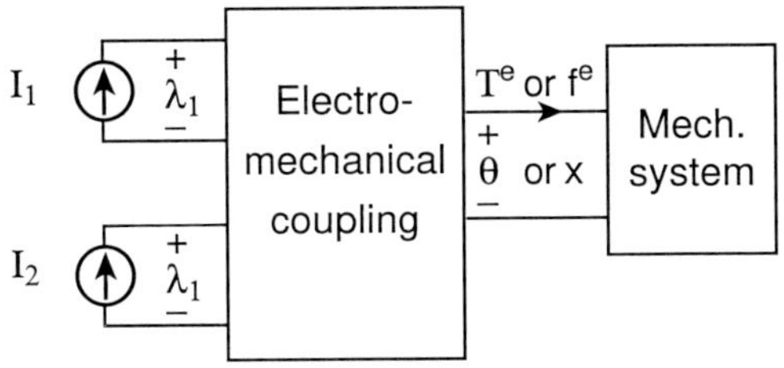

Figure 6.6: Electromechanical system

nor the mechanical system contains dissipative elements. On the electrical side, if we can constrain so that either λ or i at each of the ports is held constant, then we can expect to find a constant of motion for the electromechanical system. Thus, either $\lambda_i di_i$ or $i_i d\lambda_i$ is zero at the electrical port. Hence, there is no energy or co-energy flow into the electrical port. On the mechanical side, we will assume no dissipative elements.

The torque of electric origin is $T^e = \frac{\partial W'_m}{\partial \theta}(i_1, i_2, \theta)$. If i_1 and i_2 are held constant, then T^e is a function of θ only, i.e., $T^e = \frac{\partial W'_m}{\partial \theta}(I_1, I_2, \theta)$. Suppose the mechanical forces acting on the system are expressible as

$$T^m = \frac{-\partial U(\theta)}{\partial \theta} \tag{6.42}$$

Then

$$
\begin{aligned}
J\frac{d^2\theta}{dt^2} &= T^e + T^m \\
&= -\frac{\partial}{\partial\theta}(U(\theta) - W'_m(I_1, I_2, \theta))
\end{aligned} \tag{6.43}
$$

We call $U(\theta) - W'_m(I_1, I_2, \theta) \triangleq V(\theta)$ as the generalized potential energy.

Suppose λ_1, λ_2 are constrained to be constant as Λ_1, Λ_2, then

$$
\begin{aligned}
T^e &= -\frac{\partial W_m}{\partial\theta}(\lambda_1, \lambda_2, \theta) \\
&= -\frac{\partial W_m}{\partial\theta}(\Lambda_1, \Lambda_2, \theta)
\end{aligned} \tag{6.44}
$$

$V(\theta)$ now becomes $U(\theta) + W_m(\Lambda_1, \Lambda_2, \theta)$. Thus, for an electromechanical system, we can also use the energy function concept. $V(\theta)$ describes, therefore, a potential *well* in θ space with the stable equilibrium point at the bottom of the well.

Relationship of linearized stability and potential energy

The torque equation is

$$
J\frac{d^2\theta}{dt^2} + \frac{\partial V(\theta)}{\partial\theta} = 0 \tag{6.45}
$$

Equilibrium points are obtained by solving

$$
\frac{\partial V(\theta)}{\partial\theta} = 0
$$

Suppose θ^e is an equilibrium point. Let $\theta = \theta^e + \Delta\theta$. Then linearizing, we get

$$
J\frac{d^2\Delta\theta}{dt^2} + \left.\frac{\partial^2\, \mathrm{V}(\theta)}{\partial\theta^2}\right|_{\theta=\theta^e} \Delta\theta = 0 \tag{6.46}
$$

The stability of (6.46) depends on the sign of $\left.\frac{\partial^2 V}{\partial\theta}\right|_{\theta=\theta^e}$ since $J > 0$. θ^e is stable if $\left.\frac{\partial^2\, \mathrm{V}(\theta)}{\partial\theta}\right|_{\theta=\theta^e} > 0$ and unstable if $\left.\frac{\partial^2 V(\theta)}{\partial\theta}\right|_{\theta=\theta^e} < 0$. This property is easily verified in the case of the pendulum for $\theta^e = +\pi$, $-\pi$, and 0.

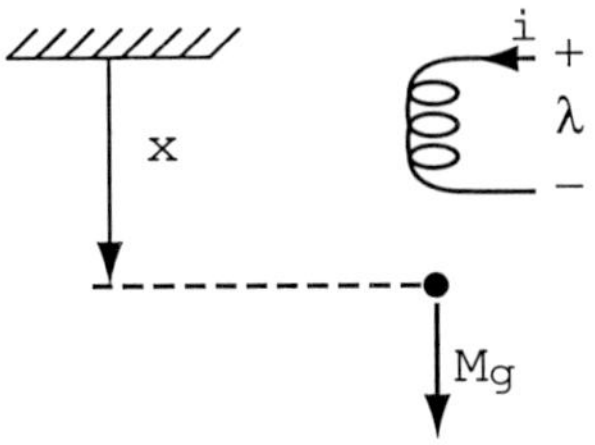

Figure 6.7: Electromechanical system

Example 6.3 The flux current relationship for the system in Figure 6.7 is $i = \lambda^2 + 2\lambda(1-x)^2$.

(a) Write the mechanical equation of motion.

(b) For the following numerical values $\lambda = \Lambda = 1$, $M = 1$, and $Mg = 2$ in a certain system of consistent units, find the equilibrium point.

(c) Write the potential energy function for the system and find the stability of the equilibrium point found in (b).

Solution:

$$\begin{aligned} W_m &= \int_0^\lambda i(\lambda', x)d\lambda' \\ &= \int_0^\lambda \left[\lambda'^2 + 2\lambda'(1-x)^2\right] d\lambda' \\ W_m &= \frac{\lambda^3}{3} + \lambda^2(1-x)^2 \\ f^e &= -\frac{\partial W_m}{\partial x} = -\lambda^2(2)(1-x)(-1) \\ &= 2\lambda^2(1-x) \end{aligned}$$

The equation of motion is

$$\begin{aligned} M\frac{d^2x}{dt^2} &= f^e + Mg \\ &= 2\lambda^2(1-x) + Mg \end{aligned}$$

Mg is the mechanical force acting other than the force of electric origin. For $\lambda = 1$ and $Mg = 2$, the equilibrium is given by

$$2(1-x) + 2 = 0, \quad x^e = 2$$
$$W_m(\lambda, x)\mid_{\lambda=1} = \tfrac{1}{3} + (1-x)^2$$

and

$$Mg = -\frac{\partial U(x)}{\partial x}$$

Hence,

$$U(x) = -Mgx$$

Hence, the potential energy $V(x)$ is

$$\begin{aligned} V(x) &= U(x) + W_m(\lambda, x) \mid_{\lambda=1} \\ &= -2x + \frac{1}{3} + (1-x)^2 \\ \frac{\partial V}{\partial x} &= -2 - 2(1-x) \text{ and } \frac{\partial^2 V}{\partial x^2} = 2 > 0 \end{aligned}$$

Hence, $x^e = 2$ is stable.

Example 6.4 The co-energy of a certain electromechanical system is given by $W'_m = \frac{16i^2}{(1+x)^2}$. The mechanical equation of motion is $\frac{d^2x}{dt^2} = 4 + f^e$.

(a) For constant current $i = 1$ A, find the equilibrium point.

(b) Find $V(x)$ and sketch it. Also find the stability of equilibrium point from $\frac{\partial^2 V}{\partial x^2}$.

Solution:

(a)

$$\begin{aligned} W'_m &= \frac{16i^2}{(1+x)^2} \\ f^e &= \frac{\partial W'_m}{\partial x} = -\frac{32i^2}{(1+x)^3} \\ \frac{d^2x}{dt^2} &= 4 - \frac{32i^2}{(1+x)^3} \end{aligned}$$

With $i = 1$ A, equilibrium point is when $4 = \frac{32}{(1+x)^3}$.

$$(1+x)^3 = 8, \quad 1 + x = 2$$

Hence, $x^e = 1$ is the equilibrium point.

(b) The potential energy is

$$V(x) = U(x) - W'_m(I, x)$$

Since

$$\frac{-\partial U(x)}{\partial x} = 4 \;\; , \;\; U(x) = -4x$$

Since $I = 1A$, $W'_m(I, x) = \frac{16}{(1+x)^2}$. Hence

$$V(x) = -4x - \frac{16}{(1+x)^2}$$

$$\begin{aligned}
\frac{\partial V}{\partial x} &= -4 + \frac{32}{(1+x)^3} \\
\frac{\partial^2 V}{\partial x^2} &= \frac{(-3)(32)}{(1+x)^4} = \frac{-96}{(1+x)^4} \\
\left.\frac{\partial^2 V}{\partial x^2}\right|_{x=1} &= \frac{-96}{(1+1)^2} = -16 \text{ which is } < 0
\end{aligned}$$

Hence, $x^e = 1$ is unstable. The potential energy curve $V(x)$ is sketched in Figure 6.8.

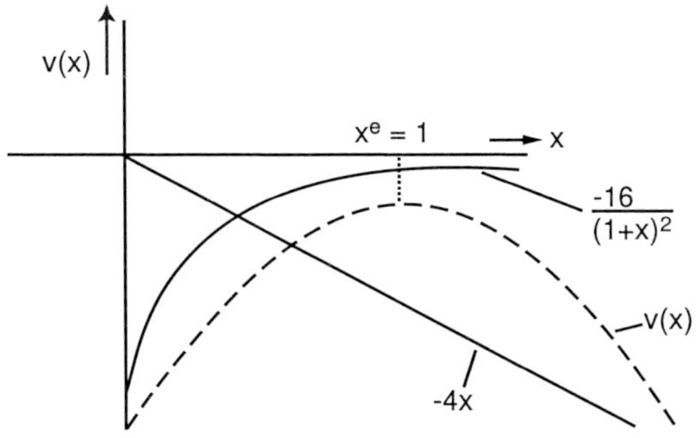

Figure 6.8: Potential energy curve

6.7 Conclusion

In this chapter, we have discussed how to evaluate the stability of an equilibrium point both for small and big disturbances. For the former, we used the linearization method, and for the latter we used the energy function method. The energy function method was also extended to the case of electromechanical systems.

6.8 Problems

6.1 A synchronous machine connected to a large system has the following nonlinear model:

$$0.04\frac{d^2\delta}{dt^2} + 0.004\frac{d\delta}{dt} = 1 - 2\sin\delta$$

where δ is the torque angle.

(a) Defining $\delta = x_1$ and $\dot{\delta} = x_2$, write the state space model.

(b) Find equilibrium points in an interval around $+\pi$ and $-\pi$. (Answer: $30°, 150°$)

(c) Linearize the state space model around each equilibrium and find eigenvalues. (Answer: $-0.05 \pm j6.58, -6.63, 6.53$)

6.2 An inverted pendulum shown in Figure 6.9

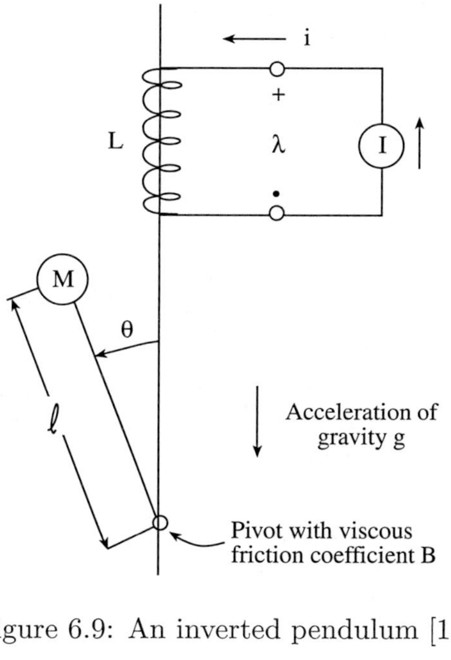

Figure 6.9: An inverted pendulum [14]

has the following equation of motion:

$$M\ell^2\frac{d^2\theta}{dt^2} = T^m + T^e$$

where $T^m = Mg\ell \sin\theta$ is the torque due to gravity and T^e is the torque of electric origin:

$$W'_m = \frac{1}{2} L_0 I^2 (1 + 0.2\cos\theta + 0.05\cos 2\theta)$$

I is the constant current source and is such that $L_0 I^2 = 6Mg\ell$.

(a) Find equilibrium points of the system in an interval of $|\theta| < \pi/2$.

(b) Find stability of *at least* one of the equilibrium points of (a) by linearization.

(c) Find the energy function E.

(d) Sketch the potential *well*. What is the height of the *well*?

(Answer: (a) $\pm 48.19°$, 0; (b) $\theta = \pm 48.19°$ (unstable); (c) $V(\theta) = Mg\ell\cos\theta - 3Mg\ell(1 + 0.2\cos\theta + 0.05\cos 2\theta) +$ constant; and (d) $0.033Mg\ell$)

6.3 The co-energy of a certain electromechanical system is given by $W'_m = \frac{3}{2} A x^3 i^2$. The equation of motion is given by

$$M\frac{d^2x}{dt^2} + Kx = f^e$$

where f^e is the force of electric origin. i is kept constant and equal to $\left(\frac{4}{9}\frac{K}{A}\right)^{1/2}$.

(a) Find the equilibrium points and their stability by linearization.

(b) Verify the results of (a) by constructing $V(x)$ and sketching it.

6.4 (a) Put the following equation in state space form

$$\ddot{x} + \dot{x} - x + x^3 = 0$$

(b) Find equilibrium points and the eigenvalues at each of them after linearization.

(c) Assuming $\Delta x(0) = 0.5$, and $\Delta\dot{x}(0) = 0$, find the solution near each of the equilibrium points for this perturbation.

(Partial answer: (b) (0,0), (1,0), (−1,0); (c) $\Delta x(t) = 0.362e^{0.618t} + 0.138e^{-1.62t}$ at (0,0); $\Delta x(t) = e^{-0.5t}(0.5\cos 1.323t + 0.189\sin 1.323t)$ at (1,0))

6.5 An electromechanical system has the following equation of motion:

$$0.1\frac{d^2\theta}{dt^2} + 0.5\frac{d\theta}{dt} = 2 - 4\sin\theta$$

(a) Find the equilibrium points in the interval $(0, \pi)$.

(b) Linearize around the equilibrium points and find the eigenvalues.

(Answer: $\theta = 30°$, $-2.5 \pm j5.33$, $\theta = 150°$, 3.89, -8.89)

6.6 For the electromechanical system shown in Figure 6.10,

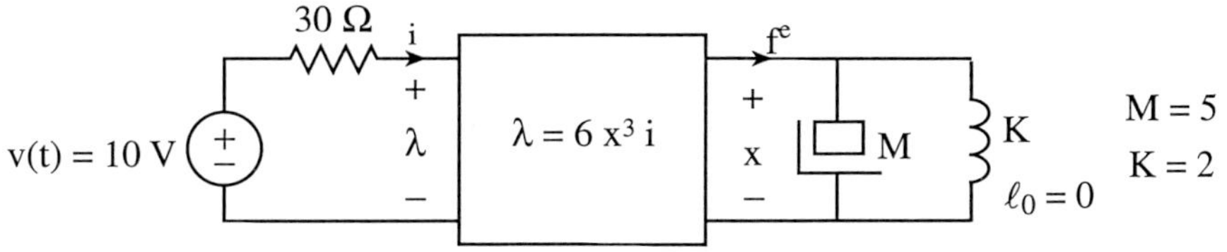

Figure 6.10: Electromechanical system [14]

(a) Find the equilibrium points.

(b) Linearize the equations around each equilibrium point. Assume the input voltage is perturbed as Δv.

6.7 An electromechanical system with one electrical and one mechanical terminal pair is shown in Figure 6.11. The electrical terminal relation is

$$\lambda = \frac{L_0 i}{(1 - x/a)^4} x$$

where L_0 and a are given constants. The system is driven by a voltage $V_0 + \Delta v(t)$ where V_0 is constant. The mass of the plunger can be ignored. Gravity acts on M as shown.

(a) Write the complete equations of motion for the system. There should be two equations in the unknowns i and x.

(b) With $\Delta v(t) = 0$, the current produced by v_0 holds for the mass M in static equilibrium at $x = x^e$. Write the linearized model in matrix form.

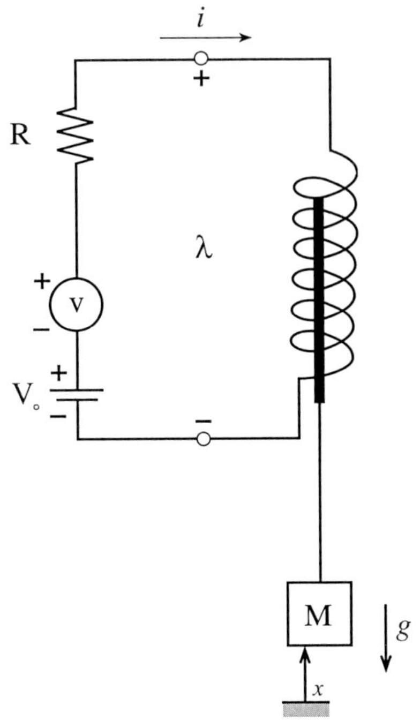

Figure 6.11: Electromechanical system [14]

Chapter 7

SYNCHRONOUS MACHINES

7.1 Introduction

Synchronous machines play an important role in power systems in terms of generation of power. The generating sets range all the way from a few MVA to 1100 MVA (1 MVA $= 10^6$ VA). The machines can be operated as either a generator or motor. The large majority of applications are as generators. The three-phase generators have an AC winding on the stator with a wye-connection. The rotor is excited by a DC field winding. It is connected to the prime movers, such as a steam or a hydro-turbine. They generate power at 60 Hz. As motors, synchronous machines are less used except at low power levels such as *permanent magnet synchronous motors* (PMSM). In certain cases, synchronous machines at a high rating are operated to act as power factor correcting devices. We will discuss the fundamental concepts of deriving torque expressions and only the sinusoidal steady-state operation using the equivalent circuit. For a proper understanding of a three-phase machine, we will motivate it via the single- and two-phase machines. The concept of co-energy is shown to be fundamental to the analysis.

7.2 Single-phase Machine

It has been shown that with a single winding on rotor and stator in Example 5.2, the flux linkages are given by

$$\lambda_s = N_s^2 L_0 i_s + N_s N_r L_0 \left(1 - \frac{2\theta}{\pi}\right) i_r \tag{7.1}$$

$$\lambda_r = N_s N_r L_0 \left(1 - \frac{2\theta}{\pi}\right) i_s + N_r^2 L_0 i_r \tag{7.2}$$

The fundamental component of mutual inductance $N_s N_r L_0 \left(1 - \frac{2\theta}{\pi}\right)$ will be $M \cos\theta$. In practical machines, the number of turns are so positioned on stator and rotor that the higher harmonics are minimized and the mutual inductance is largely due to this fundamental component. The winding, instead of being concentrated, is distributed as shown in Figure 7.1.

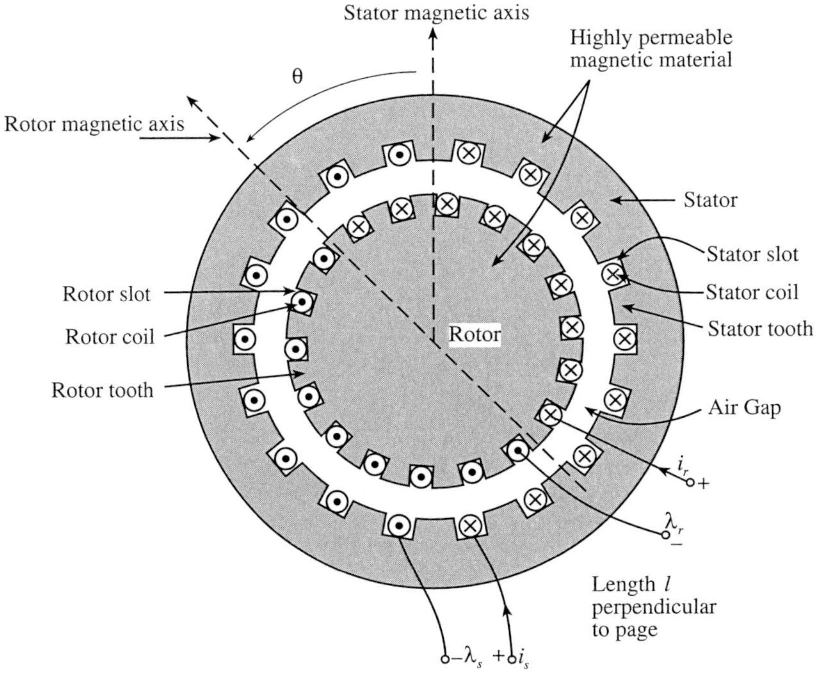

Figure 7.1: Geometry of smooth-air-gap rotating machine showing distributed windings on stator and rotor of a single-phase machine [14]

Functionally,

$$\lambda_s = L_s i_s + L_{sr}(\theta) i_r \tag{7.3}$$
$$\lambda_r = L_{sr}(\theta) i_s + L_r i_r \tag{7.4}$$
$$L_{sr}(\theta) = M\cos\theta \tag{7.5}$$

We can visualize the rotor and stator windings as two coupled coils (Figure 7.2). The co-energy W'_m is given by

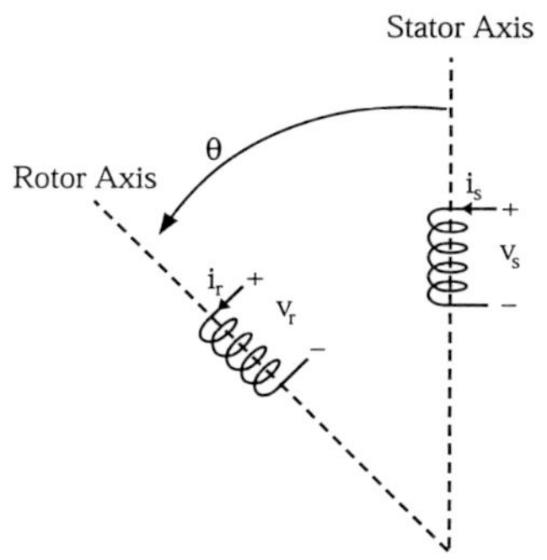

Figure 7.2: Windings in Figure 6.1 viewed as coupled coil

$$\begin{aligned} W'_m &= \int_0^{i_s} \lambda_s(i'_s, 0, \theta)\, di'_s + \int_0^{i_r} \lambda_r(i_s, i'_r, \theta)\, di'_r \\ &= \frac{1}{2} L_s i_s^2 + \frac{1}{2} L_r i_r^2 + L_{sr}(\theta) i_s i_r \end{aligned} \tag{7.6}$$

The torque of electric origin is given by

$$T^e = \frac{\partial W'_m}{\partial \theta} = i_s i_r \frac{dL_{sr}(\theta)}{d\theta} \tag{7.7}$$

Since $L_{sr}(\theta) = M\cos\theta$, the torque of the electric origin is

$$T^e = -i_s i_r M \sin\theta \tag{7.8}$$

Let the mechanical inertia be J. The electrical and mechanical equations are:

Electrical

$$v_s = i_s R_s + \frac{d\lambda_s}{dt} \tag{7.9}$$

$$v_r = i_r R_r + \frac{d\lambda_r}{dt} \tag{7.10}$$

Mechanical

$$J\frac{d^2\theta}{dt^2} + K\theta + B\frac{d\theta}{dt} = T^e + T^m \tag{7.11}$$

where T^m is some externally applied torque in the direction of positive θ. θ is measured from an equilibrium position. Equations (7.9–7.11) can be represented as a schematic diagram as in Figure 7.3.

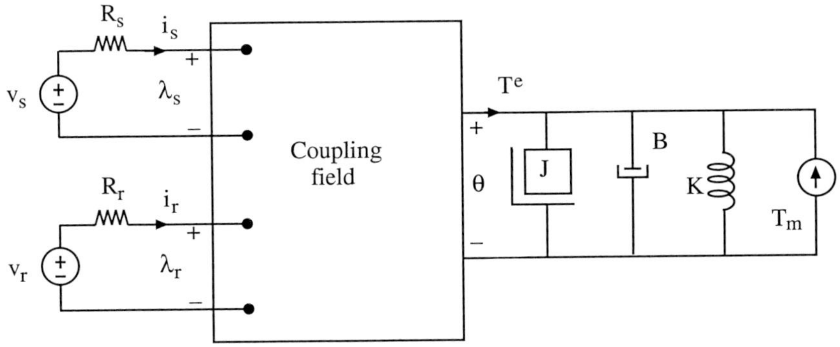

Figure 7.3: Schematic diagram

In the case of synchronous and induction machines, we are interested only in the *sinusoidal steady state*. We therefore proceed to find conditions for average power transfer. Let

$$i_s = I_s \cos \omega_s t \tag{7.12}$$

$$i_r = I_r \cos \omega_r t \tag{7.13}$$

In steady state, the mechanical power is

$$\begin{aligned} p_m &= T^e\frac{d\theta}{dt} = T^e\omega_m \\ &= -\omega_m I_s I_r M \cos \omega_s t \cos \omega_r t \sin \theta \end{aligned} \tag{7.14}$$

Let

$$\theta = \omega_m t + \gamma \tag{7.15}$$

where γ is some arbitrary constant. Hence,

$$p_m = -\omega_m I_s I_r M \cos\omega_s t \cos\omega_r t \sin(\omega_m t + \gamma) \tag{7.16}$$

Now we use the following trigonometric identities.

$$\cos\omega_s t \cos\omega_r t = \frac{1}{2}[\cos(\omega_s + \omega_r)t + \cos(\omega_s - \omega_r)t] \tag{7.17}$$

$$\begin{aligned}\cos(\omega_s + \omega_r)t \sin(\omega_m t + \gamma) &= \frac{1}{2}\{\sin[(\omega_m + \omega_s + \omega_r)t + \gamma] \\ &\quad + \sin[(\omega_m - \omega_s - \omega_r)t + \gamma]\} \qquad (7.18)\\ \cos(\omega_s - \omega_r)t \sin(\omega_m t + \gamma) &= \frac{1}{2}\{\sin[(\omega_m + \omega_s - \omega_r)t + \gamma] \\ &\quad + \sin[(\omega_m - \omega_s + \omega_r)t + \gamma]\} \qquad (7.19)\end{aligned}$$

Using the trigonometric identities (7.17–7.19), we simplify (7.16) as

$$\begin{aligned}p_m &= -\omega_m I_s I_r M\{\sin(\omega_1 t + \gamma) + \sin(\omega_2 t + \gamma) \\ &\quad + \sin(\omega_3 t + \gamma) + \sin(\omega_4 t + \gamma)]\}/4 \qquad (7.20)\end{aligned}$$

where

$$\begin{aligned}\omega_1 &= \omega_m + \omega_s - \omega_r \quad \omega_3 = \omega_m + \omega_s + \omega_r \\ \omega_2 &= \omega_m - \omega_s + \omega_r \quad \omega_4 = \omega_m - \omega_s - \omega_r \qquad (7.21)\end{aligned}$$

Since a sinusoidal function can have no average value,(7.20) can have an average value only if $\omega_i = 0$ for $i = 1, 2, 3,$ or 4, i.e.,

$$\omega_m = \pm\omega_s \pm \omega_r \tag{7.22}$$

One can say that all of the smooth-air-gap machines are constructed so as to satisfy this *frequency condition.* Suppose $\omega_2 = 0$, i.e., $\omega_m = \omega_s - \omega_r$, then the average power is

$$p_{m(av)} = \frac{-\omega_m I_s I_r M \sin\gamma}{4} \tag{7.23}$$

Thus a necessary condition for average power is that one of the ω_i's is zero, and a sufficient condition is that $\sin\gamma \neq 0$. Although there is average power when $\omega_2 = 0$, there is pulsating torque due to other ω_i's. To eliminate this, we can have a two-phase machine. In the two-phase machine, there is an additional winding on both the stator and the rotor. The two-phase machine (although not in common use) creates a rotating magnetic field.

7.3 Two-phase Machine

Consider the two-phase machine (Figure 7.4). The flux linkage equations

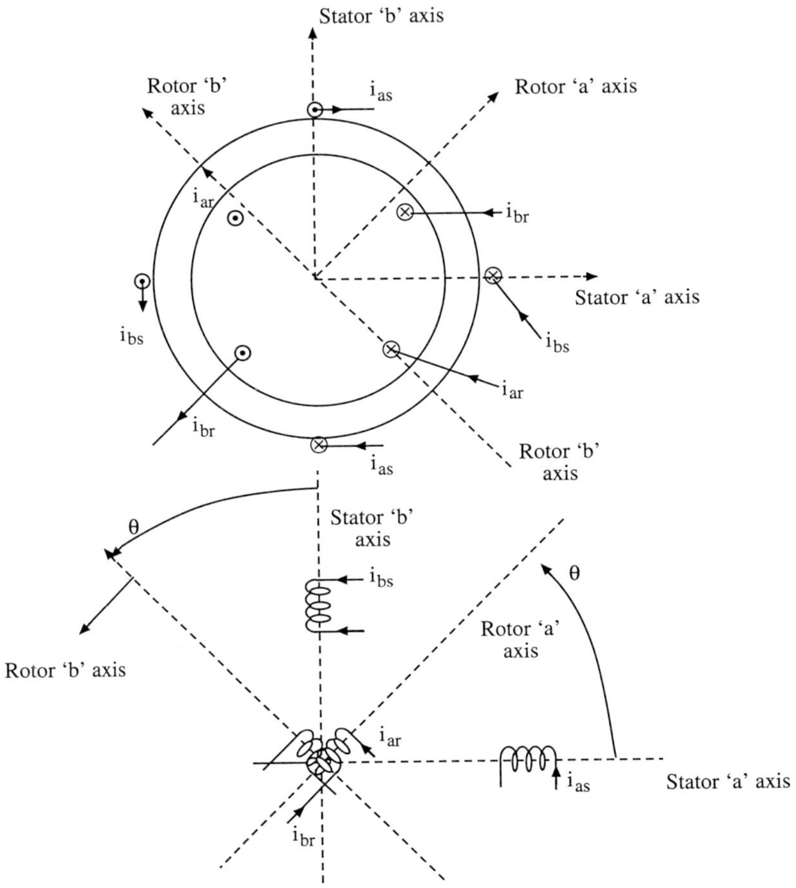

Figure 7.4: Two-phase machine

can be written by considering each coil coupled with two other coils. For simplicity of illustration, the coils are shown as being concentrated in single slots. In actual machines, the windings are distributed as in Figure 7.1, and a single slot can carry conductors from both windings. In Figure 7.4, windings a on stator and rotor represent the original windings of Figure 7.1. Windings b on rotor and stator are identical to the respective windings a in every respect except that they are displaced mechanically by 90° in positive θ direction. For example, the two stator coils a and b have no mutual coupling, similarly rotor coils a and b do not have mutual coupling. Thus while considering stator coil a, we consider mutual coupling with the two rotor coils a and b only. Similarly, we consider the mutual coupling for other coils.

Thus, we can write

$$\begin{aligned}
\lambda_{as} &= L_s i_{as} + M i_{ar}\cos\theta + M i_{br}\cos\left(\theta + \frac{\pi}{2}\right) \\
&= L_s i_{as} + M i_{ar}\cos\theta - M i_{br}\sin\theta \qquad (7.24)\\
\lambda_{bs} &= L_s i_{bs} + M i_{ar}\cos\left[-\left(\frac{\pi}{2} - \theta\right)\right] + M i_{br}\cos\theta \\
&= L_s i_{bs} + M i_{ar}\sin\theta + M i_{br}\cos\theta \qquad (7.25)\\
\lambda_{ar} &= L_r i_{ar} + M i_{as}\cos(-\theta) + M i_{bs}\cos\left(\frac{\pi}{2} - \theta\right) \\
&= L_r i_{ar} + M i_{as}\cos\theta + M i_{bs}\sin\theta \qquad (7.26)\\
\lambda_{br} &= L_r i_{br} + M i_{as}\cos\left[-\left(\frac{\pi}{2} + \theta\right)\right] + M i_{bs}\cos[-\theta] \\
&= L_r i_{br} - M i_{as}\sin\theta + M i_{bs}\cos\theta \qquad (7.27)
\end{aligned}$$

In matrix form,

$$\begin{bmatrix} \lambda_{as} \\ \lambda_{bs} \\ \lambda_{ar} \\ \lambda_{br} \end{bmatrix} = \begin{bmatrix} L_s & 0 & M\cos\theta & -M\sin\theta \\ 0 & L_s & M\sin\theta & M\cos\theta \\ M\cos\theta & M\sin\theta & L_r & 0 \\ -M\sin\theta & M\cos\theta & 0 & L_r \end{bmatrix} \begin{bmatrix} i_{as} \\ i_{bs} \\ i_{ar} \\ i_{br} \end{bmatrix} \qquad (7.28)$$

Note that the inductance matrix in (7.28) is symmetric. Equations (7.24–7.27) can be used to derive co-energy as

$$\begin{aligned}
W'_m &= \frac{1}{2}L_s i_{as}^2 + \frac{1}{2}L_s i_{bs}^2 + \frac{1}{2}L_r i_{ar}^2 + \frac{1}{2}L_r i_{br}^2 \\
&\quad + M i_{as} i_{ar}\cos\theta + M i_{bs} i_{ar}\sin\theta - M i_{as} i_{br}\sin\theta \\
&\quad + M i_{bs} i_{br}\cos\theta \qquad (7.29)
\end{aligned}$$

and

$$T^e = \frac{\partial W'_m}{\partial\theta} = M[(i_{ar} i_{bs} - i_{as} i_{br})\cos\theta - (i_{as} i_{ar} + i_{br} i_{bs})\sin\theta] \qquad (7.30)$$

Average power conditions

The windings in Figure 7.4 are called *balanced two-phase windings* since excitation with balanced two-phase currents will result in constant power conversion with no alternating components. To show this, the terminal variables of the machine are constrained by the *balanced two-phase* current sources:

$$\begin{aligned} i_{as} &= I_s \cos\omega_s t, \quad i_{bs} = I_s \sin\omega_s t \\ i_{ar} &= I_r \cos\omega_r t, \quad i_{br} = I_r \sin\omega_r t \\ \theta &= \omega_m t + \gamma \end{aligned} \tag{7.31}$$

Substitute (7.31) in (7.29) and compute p_m. After lengthy trigonometric manipulation,

$$\begin{aligned} p_m &= T^e \omega_m \\ &= -\omega_m I_r I_s M \sin[(\omega_m - \omega_s + \omega_r)t + \gamma] \end{aligned} \tag{7.32}$$

There is average power transfer if the coefficient of t is zero, i.e., $\omega_m = \omega_s - \omega_r$. This is the *frequency condition.* There are no pulsating components as in the case of a single-phase machine. Thus, by having two coils 90° apart and phase currents 90° (electrical) apart, *we have a single-frequency condition* and no pulsating torques. The torque is given by

$$p_m = -\omega_m M I_s I_r \sin\gamma \tag{7.33}$$

Concept of a rotating field

It is helpful for qualitative reasoning and a more thorough understanding of the coupling mechanism, to map the magnetic field in the air gap. Consider the two-phase machine in Figure 7.4 but with only the stator excited by current sources, as shown in Figure 7.5. The current in each winding produces

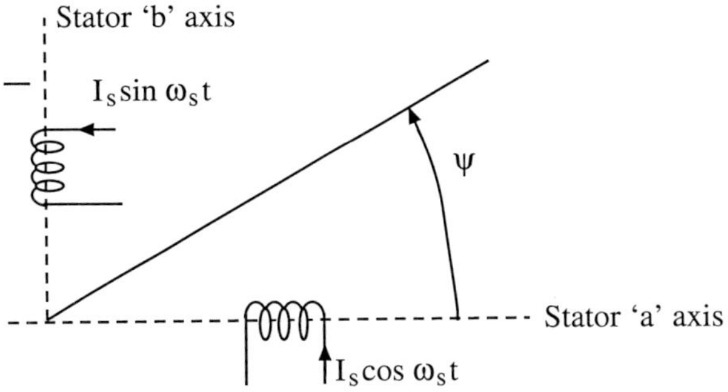

Figure 7.5: The stator windings excited by sinusoidal currents

a flux density that is maximum along its magnetic axis at some time instant and varies sinusoidally between the positive and negative directions of the axis. For the current excitations shown in Figure 7.5, we observe

(a) At $t = 0$, the flux is maximum along the positive stator a axis direction.

(b) At $t = \frac{\pi}{2\omega_s}$ seconds, it is maximum along the positive stator b axis direction.

(c) At $t = \frac{\pi}{\omega_s}$ seconds, it is maximum along the negative stator a axis direction.

(d) At $t = \frac{3\pi}{2\omega_s}$ seconds, it is maximum along the negative stator b axis direction.

We assume in each of the above cases that the flux in a spatial sense is sinusoidal in the air gap. Figure 7.6 shows the spatial distribution at five instants of time. At $t = 0$, it is maximum along the positive a axis and varies sinusoidally along the air gap. At $t = \frac{\pi}{2\omega_s}$ seconds, it is maximum along the positive b axis. It is again maximum along the positive stator a axis direction after a complete period $T = \frac{2\pi}{\omega_s}$ seconds. At an arbitrary time instant t, the magnetic field in the air gap at angle ψ is the sum of $B_{rm} \cos\omega_s t \cos\psi$ and $B_{rm} \sin\omega_s t \sin\psi$, i.e.,

$$\begin{aligned} B_r(t, \psi) &= B_{rm}[\cos\omega_s t \cos\psi + \sin\omega_s t \sin\psi] \\ &= B_{rm}(\cos(\omega_s t - \psi)) \end{aligned}$$

This is the magnetic field rotating at speed ω_s (traveling wave around the circumference).

Rotor currents excited with frequency ω_r also create a rotating field similarly. This rotating field rotates at an angular velocity of ω_r *with respect to the rotor*. From the elementary theory of magnets, the two revolving fields tend to align themselves, and power conversion occurs when these two fields are fixed in space relative to each other and the rotor is turning at a constant angular velocity. To accomplish this, a constant mechanical speed given by $\omega_m = \omega_s - \omega_r$ is required. The torque (and average power) is a function of the constant angle of separation γ. Thus, the condition for average power conversion is equivalent to the condition under which both rotor and stator fields rotate with respect to the members carrying the excitation current but are fixed in space relative to each other. In the case of a synchronous machine, $\omega_s = 2\pi 60$ rads/sec and $\omega_r = 0$.

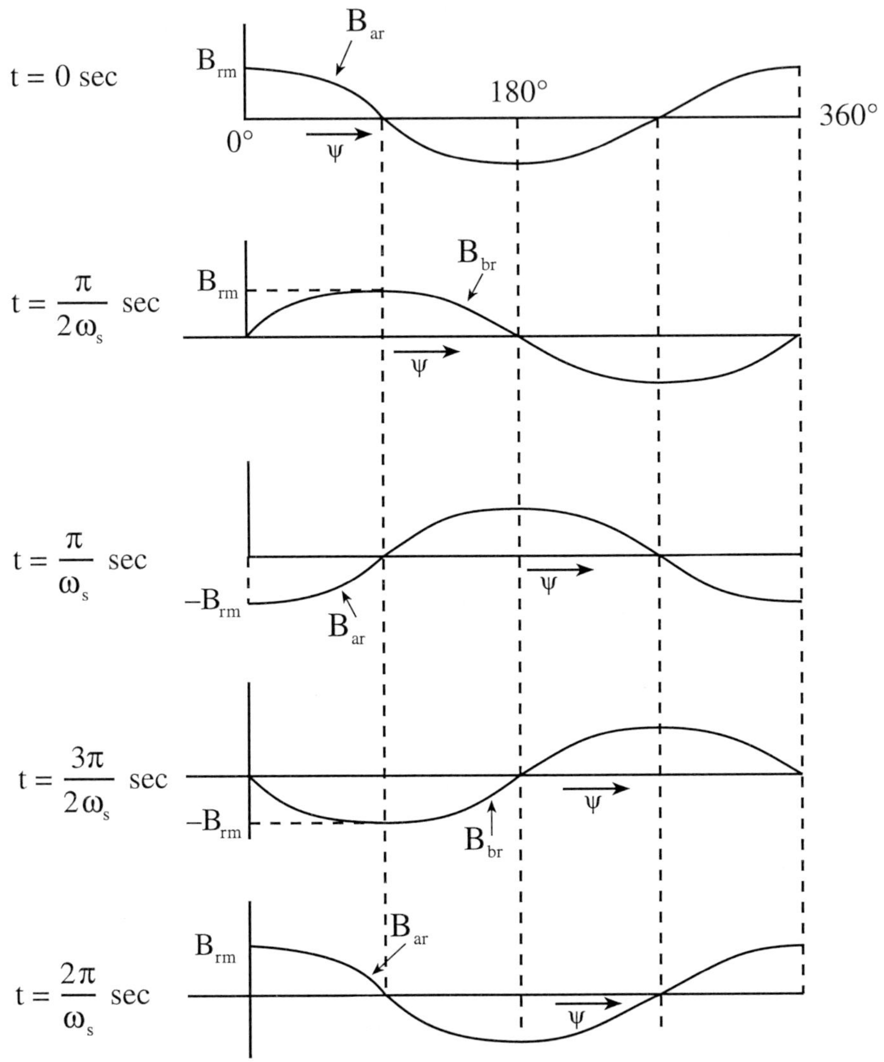

Figure 7.6: Spatial distribution of flux B_r at different time instants

7.4 Three-phase Synchronous Machine

First we consider a *salient pole* three-phase machine as opposed to a *round rotor* machine as shown in Figure (7.7). Salient pole machines are used in hydro-generators and low-power single-phase synchronous motors. We consider a two-pole machine.

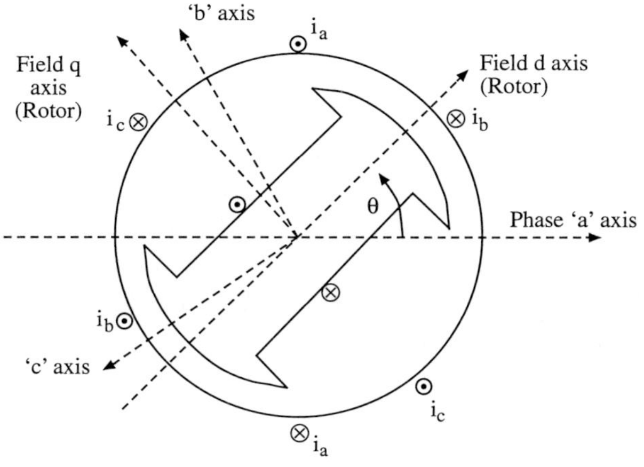

(a)

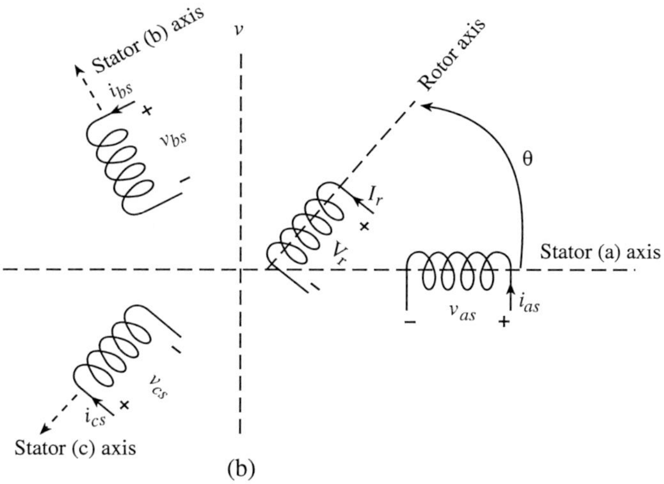

(b)

Figure 7.7: Three-phase two-pole synchronous machine (a), schematic (b)

There are three stator coils distributed so that each coil creates a sinusoidal mmf around the periphery. The rotor has a field coil that carries a constant current i_r. The stator coils are spaced 120° apart mechanically.

The flux current relations can be derived as

$$\begin{bmatrix} \lambda_a \\ \lambda_b \\ \lambda_c \\ \lambda_r \end{bmatrix} = \begin{bmatrix} L_0 + L_2\cos 2\theta & -M_0 + M_2\cos 2(\theta - 60^\circ) \\ -M_0 + M_2\cos 2(\theta - 60^\circ) & L_0 + L_2\cos 2(\theta - 120^\circ) \\ -M_0 + M_2\cos 2(\theta + 60^\circ) & -M_0 + M_2\cos 2(\theta - 180^\circ) \\ M\cos\theta & M\cos(\theta - 120^\circ) \end{bmatrix}$$

$$\begin{bmatrix} -M_0 + M_2\cos 2(\theta + 60^\circ) & M\cos\theta \\ -M_0 + M_2\cos 2(\theta - 180^\circ) & M\cos(\theta - 120^\circ) \\ L_0 + L_2\cos 2(\theta + 120^\circ) & M\cos(\theta + 120^\circ) \\ M\cos(\theta + 120^\circ) & L_r \end{bmatrix} \begin{bmatrix} i_a \\ i_b \\ i_c \\ i_r \end{bmatrix}$$

$$= \begin{bmatrix} L_a & M_{ab} & M_{ac} & M_{ar} \\ M_{ab} & L_b & M_{bc} & M_{br} \\ M_{ac} & M_{bc} & L_c & M_{cr} \\ M_{ar} & M_{br} & M_r & L_r \end{bmatrix} \begin{bmatrix} i_a \\ i_b \\ i_c \\ i_r \end{bmatrix} \tag{7.34}$$

Note that the inductance matrix is symmetric. In the case of a salient pole machine, all inductances except for L_r are functions of θ. Also $M_0 = \frac{L_0}{2}$ and $L_2 = M_2$, assuming all windings have the same number of turns and there is no leakage flux. For the *round rotor case*, $L_2 = M_2 = 0$, which simplifies the analysis. We proceed to find under *sinusoidal steady state* the two quantities T^e and v_a. Only M_{ar}, M_{br}, and M_{cr} are functions of θ.

7.4.1 Salient pole case

For this, we use (7.34):

$$\begin{aligned} W'_m &= \frac{1}{2}L_a i_a^2 + \frac{1}{2}L_b i_b^2 + \frac{1}{2}L_c i_c^2 + \frac{1}{2}L_r i_r^2 \\ &\quad + i_a i_b M_{ab} + i_a i_c M_{ac} + i_b i_c M_{bc} \\ &\quad + i_a i_r M_{ar} + i_b i_r M_{br} + i_c i_r M_{cr} \end{aligned} \tag{7.35}$$

$$\begin{aligned} T^e &= \frac{\partial W'_m}{\partial\theta} \\ &= \frac{i_a^2}{2}\frac{dL_a}{d\theta} + \frac{i_b^2}{2}\frac{dL_b}{d\theta} + \frac{i_c^2}{2}\frac{dL_c}{d\theta} \\ &\quad + i_a i_b \frac{dM_{ab}}{d\theta} + i_a i_c \frac{dM_{ac}}{d\theta} + i_b i_c \frac{dM_{bc}}{d\theta} \\ &\quad + i_a i_r \frac{dM_{ar}}{d\theta} + i_b i_r \frac{dM_{br}}{d\theta} + i_c i_r \frac{dM_{cr}}{d\theta} \end{aligned} \tag{7.36}$$

This is a symbolic expression, and the derivatives can be calculated. It is, however, quite complicated and not discussed here.

7.4.2 Round rotor case

For the simpler round rotor case, $L_2 = M_2 = 0$ and $M_0 = \frac{L_0}{2}$. Hence,

$$L(\theta) = \begin{bmatrix} L_0 & -M_0 & -M_0 & M\cos\theta \\ -M_0 & L_0 & -M_0 & M\cos(\theta - 120^\circ) \\ -M_0 & -M_0 & L_0 & M\cos(\theta + 120^\circ) \\ M\cos\theta & M\cos(\theta - 120^\circ) & M\cos(\theta + 120^\circ) & L_r \end{bmatrix} \tag{7.37}$$

Note that there is a considerable simplification of the inductance matrix. From (7.35) and (7.37)

$$\begin{aligned} W'_m &= \frac{1}{2}L_0 i_a^2 - M_0 i_a i_b + \frac{1}{2}L_0 i_b^2 - M_0 i_a i_c \\ &\quad - M_0 i_b i_c + \frac{1}{2}L_0 i_c^2 + i_a i_r M\cos\theta \\ &\quad + i_b i_r M\cos(\theta - 120^\circ) + i_c i_r M\cos(\theta + 120^\circ) + \frac{1}{2}L_r i_r^2 \end{aligned} \tag{7.38}$$

From (7.36) since $L_a = L_b = L_c = M_{ab} = M_{ac} = M_{bc} =$ constant, we have

$$\begin{aligned} T^e = \frac{\partial W'_m}{\partial \theta} &= i_a i_r \frac{dM_{ar}}{d\theta} + i_b i_r \frac{dM_{br}}{d\theta} + i_c i_r \frac{dM_{cr}}{d\theta} \\ &= -i_a i_r M\sin\theta - i_b i_r M\sin(\theta - 120^\circ) \\ &\quad - i_c i_r M\sin(\theta + 120^\circ) \end{aligned} \tag{7.39}$$

We are interested in steady-state AC conditions and in terminal conditions. Hence, $p_m = T^e\omega_m$ and v_a are of interest. Whatever happens in phase a, happens in phase b, 120° later and in phase c, 240° later. Assume a balanced set of currents in the stator:

$$\begin{aligned} i_a &= I_m \cos\omega_s t \\ i_b &= I_m \cos(\omega_s t - 120^\circ) \\ i_c &= I_m \cos(\omega_s t + 120^\circ) \\ i_r &= I_r = \text{constant} \end{aligned} \tag{7.40}$$

This is called a *balanced three-phase current source excitation.* Hence, substituting i_a, i_b, and i_c in (7.39) and evaluating, we get

$$\begin{aligned} T^e &= -I_m I_r M[\sin\theta\cos\omega_s t + \sin(\theta - 120°)\cos(\omega_s t - 120°) \\ &\quad + \sin(\theta + 120°)\cos(\omega_s t + 120°)] \\ &= -I_m I_r M\left[\frac{\sin(\theta+\omega_s t) + \sin(\theta - \omega_s t)}{2}\right. \\ &\quad + \frac{\sin(\theta + \omega_s t - 240°) + \sin(\theta - \omega_s t)}{2} \\ &\quad \left. + \frac{\sin(\theta + \omega_s t + 240°) + \sin(\theta - \omega_s t)}{2}\right] \end{aligned} \tag{7.41}$$

Now the following identity can be verified easily:

$$\sin(\theta + \omega_s t) + \sin(\theta + \omega_s t - 240°) + \sin(\theta + \omega_s t + 240°) = 0$$

Therefore, using this identity in (7.41), we get

$$T^e = \frac{-I_m I_r M 3\sin(\theta - \omega_s t)}{2} \tag{7.42}$$

Let $I_m = \sqrt{2}I_a$ where $I_a = rms$ value. Also let $\theta = \omega_m t + \gamma$ where γ is arbitrary. Then the expression in (7.42) becomes

$$T^e = -\sqrt{2}\frac{I_a I_r}{2} 3M\sin(\omega_m t + \gamma - \omega_s t) \tag{7.43}$$

For T^e to have an average value, $\omega_m = \omega_s$, which is called the *synchronous speed.*

$$T^e = \frac{-3}{\sqrt{2}} I_a I_r M\sin\gamma \tag{7.44}$$

Since the synchronous speed ω_m is equal to the electrical frequency ω_s in radians per second, we have $\omega_m = \frac{2\pi N_s}{60} = 2\pi f$ where N_s = synchronous speed in rpm (revolutions per minute). Hence for a two-pole machine $N_s = f \times 60 = 3600$ rpm.

7.4.3 Voltage in steady state (round rotor case)

Now let us compute v_a in steady state. Note $\theta = \omega_m t + \gamma = \omega_s t + \gamma$. In steady state $v_a = \frac{d\lambda_a}{dt} = L_a\frac{di_a}{dt} + M_{ab}\frac{di_b}{dt} + M_{ac}\frac{di_c}{dt} + I_r\frac{dM_{ar}}{dt}$. Using the values of inductances from (7.37),

$$\begin{aligned} v_a &= \frac{d\lambda_a}{dt} = -L_0 I_m \omega_s \sin\omega_s t + \frac{L_0}{2} I_m \omega_s \sin(\omega_s t - 120^\circ) \\ &+\frac{L_0}{2} I_m \omega_s \sin(\omega_s t + 120^\circ) - I_r M \omega_s \sin(\omega_s t + \gamma) \end{aligned} \tag{7.45}$$

Adding and subtracting $-\frac{L_0}{2} I_m \omega_s \sin\omega_s t$, we get from the right side of (7.45)

$$\begin{aligned} v_a &= -\frac{3L_0}{2} I_m \omega_s \sin\omega_s t + \frac{L_0}{2} I_m \omega_s [\sin\omega_s t + \sin(\omega_s t - 120^\circ) \\ &+ \sin(\omega_s t + 120^\circ)] - I_r M \omega_s \sin(\omega_s t + \gamma) \end{aligned} \tag{7.46}$$

The second term on the right side of (7.46) is zero. Hence,

$$v_a = -\frac{3}{2} L_0 \sqrt{2} I_a \omega_s \sin\omega_s t - \sqrt{2}\frac{M I_r}{\sqrt{2}} \omega_s \sin(\omega_s t + \gamma) \tag{7.47}$$

Let

$$v_a = \mathcal{R}e[\sqrt{2}\overline{V}_a e^{j\omega_s t}]$$

where $\overline{V}_a$ is the rms phasor. From (7.40), the rms phasor for current is $\overline{I}_a = I_a \angle 0^\circ$ where $I_a = \frac{I_m}{\sqrt{2}}$. Hence, from (7.47)

$$\begin{aligned} \mathcal{R}e\left[\sqrt{2}\,\overline{V}_a e^{j\omega_s t}\right] &= \mathcal{R}e\left[\sqrt{2}\frac{3}{2} L_0 \omega_s \overline{I}_a e^{j\pi/2} e^{j\omega_s t}\right] \\ &+\mathcal{R}e\left[\sqrt{2}\frac{M I_r}{\sqrt{2}} \omega_s e^{j\gamma} e^{j\omega_s t} e^{j\frac{\pi}{2}}\right] \end{aligned} \tag{7.48}$$

where $\overline{I}_a = I_a \angle 0^\circ$. Since $e^{j\frac{\pi}{2}} = j$ and equating terms in the brackets on both sides of (7.48), we get

$$\overline{V}_a = j\frac{3}{2} L_0 \omega_s \overline{I}_a + j\frac{M I_r}{\sqrt{2}} \omega_s e^{j\gamma} \tag{7.49}$$

In (7.49), we define

$$\begin{aligned} \frac{3}{2} L_0 \omega_s &\triangleq x_s \text{ (synchronous reactance)} \\ j\frac{\omega_s M I_r e^{j\gamma}}{\sqrt{2}} &= \frac{\omega_s M I_r}{\sqrt{2}} \angle\left(\frac{\pi}{2} + \gamma\right) \\ &\triangleq \overline{E}_{ar} \text{ (voltage phasor proportional to field (rotor) current} I_r) \end{aligned}$$

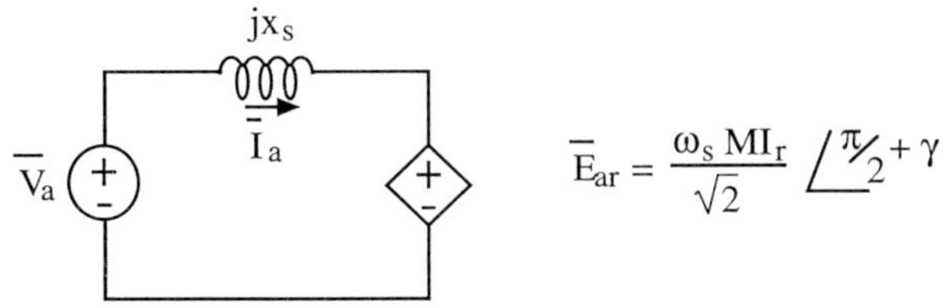

Figure 7.8: Equivalent circuit for phase a

Hence,

$$\overline{V}_a = jx_s\overline{I}_a + \overline{E}_{ar} \tag{7.50}$$

The equivalent circuit for the phase a is shown in Figure 7.8. Note that E_{ar}, the magnitude of $\overline{E}_{ar}$, is proportional to I_r the field current and hence $\overline{E}_{ar}$ is denoted as a dependent source.

For phases b and c, we have the equivalent circuits with all phasors shifted by 120° and 240°, respectively, as shown in Figure 7.9. It is therefore enough

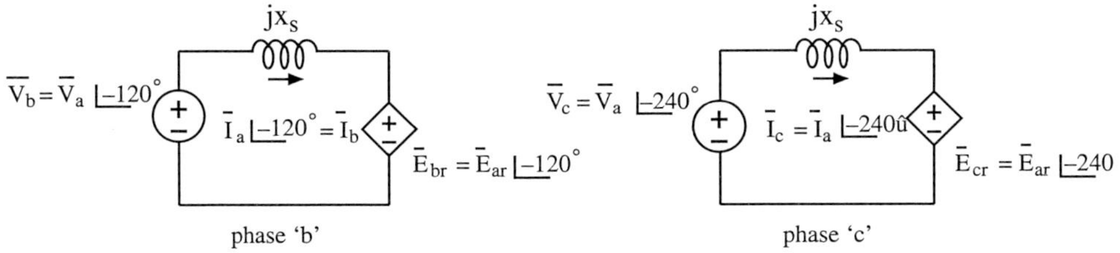

Figure 7.9: Equivalent circuit for phases b and c

to focus attention on only one phase, i.e., phase a as shown in Figure 7.8.

Since $i_a = I_m \cos\omega_s t = \sqrt{2}I_a \cos\omega_s t$, $\overline{I}_a = I_a\angle 0°$ is our reference phasor. Real power into the controlled source shown in Figure 7.8 is

$$\begin{aligned} P_a &= \mathcal{R}e[\overline{E}_{ar}\overline{I}_a^*] \\ &= \mathcal{R}e\left[\left(\frac{\omega_s M I_r}{\sqrt{2}}\angle\left(\frac{\pi}{2}+\gamma\right)\right) I_a\right] \\ &= -\frac{\omega_s M I_r}{\sqrt{2}} I_a \sin\gamma \end{aligned} \tag{7.51}$$

Total power in three phases is $P_T = 3P_a$

$$P_T = \frac{-3}{\sqrt{2}}\omega_s M I_r I_a \sin\gamma \tag{7.52}$$

The mechanical power output is

$$P_m = T^e \omega_m$$

Since we have DC field excitation, $\omega_r = 0$. Hence, from the frequency condition $\omega_m = \omega_s - \omega_r$, we have $\omega_m = \omega_s$. For a conservative system, $P_T = P_m$, and hence,

$$T^e = \frac{-3}{\sqrt{2}} M I_a I_r \sin\gamma \tag{7.53}$$

which agrees with (7.44). A multi-port representation of the three-phase machine is shown in Figure 7.10. The orientations shown are for motor

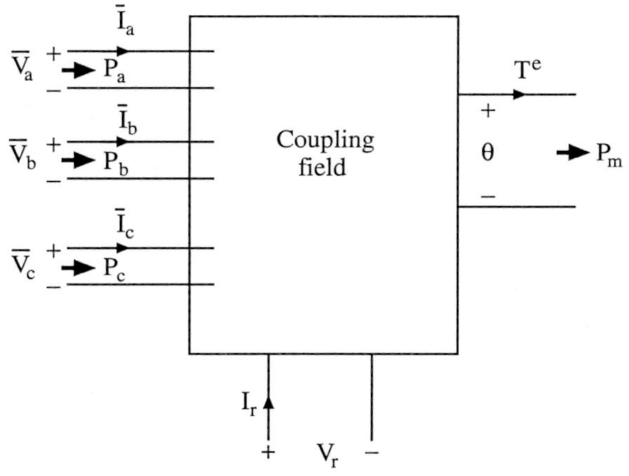

Figure 7.10: Multi-port representation

operation, i.e., electrical power "in" and mechanical power "out."

7.4.4 Phasor diagram

From (7.50)

$$\overline{V}_a = jx_s\overline{I}_a + \overline{E}_{ar} \tag{7.54}$$

For motor operation from (7.52), we must have $P_T = P_m > 0$. Hence, we must have $\gamma < 0$. Since the phasor $\overline{E}_{ar}$ has the phase angle $(\pi/2 + \gamma)$ and $\gamma < 0$, the phasor $\overline{E}_{ar}$ is in the first quadrant. The phasor diagram is shown in Figure 7.11. The PF angle is θ, and we have shown the case of lagging PF. We define torque angle as the angle δ *from* $\overline{V}_a$ *to* $\overline{E}_{ar}$. Therefore, the

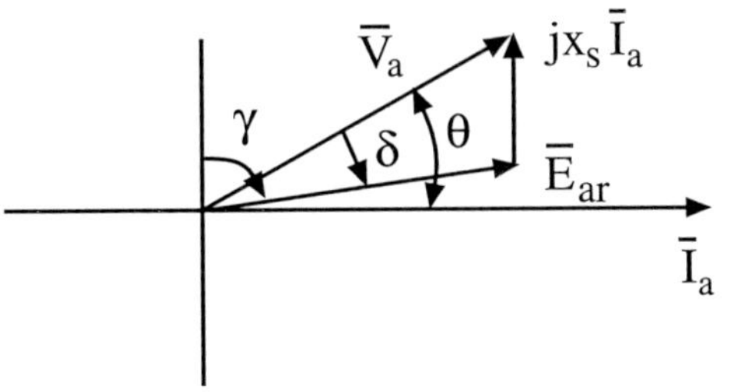

Figure 7.11: Phasor diagram (motor operation)

angle from $\overline{E}_{ar}$ to $\overline{V}_a$ is $-\delta$. As shown in Figure 7.11, the phasor $\overline{V}_a$ is

$$\begin{aligned}\overline{V}_a &= V_a e^{j(\frac{\pi}{2}+\gamma-\delta)} \\ &= V_a e^{j\theta} \qquad (7.55)\end{aligned}$$

θ is the power factor angle. $\theta > 0$ implies lagging PF and $\theta < 0$ leading PF as discussed in Chapter 2. Note from Figure 7.11, that γ and δ are < 0 for motor action.

In Figure 7.11, $\overline{I}_a$ was taken as reference since we defined $i_a = I_m \cos \omega_s t$ in (7.40) for torque calculations. In actual practice, it is convenient to take the terminal voltage $\overline{V}_a$ as reference. This can be done easily by subtracting the angle $\frac{\pi}{2} + \gamma - \delta$ from the phasors of $\overline{V}_a$, $\overline{I}_a$, and $\overline{E}_{ar}$. This is the same as adding $-(\pi/2 + \gamma - \delta)$ to these phasors. This will make $\overline{V}_a$ align with the positive real axis in Figure 7.12.

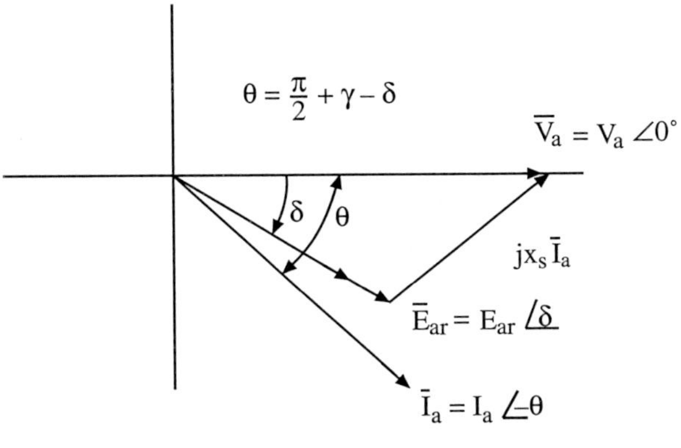

Figure 7.12: Phasor diagram for motor with voltage as reference and lagging PF

Thus,

$$\begin{aligned}\overline{V}_a &= V_a\angle 0^\circ \\ \overline{I}_a &= I_a\angle(-(\pi/2+\gamma-\delta)) = I_a\angle -\theta \\ \overline{E}_{ar} &= \frac{\omega_s M I_r}{\sqrt{2}}\angle(\pi/2+\gamma-(\pi/2+\gamma-\delta)) \\ &= \frac{\omega_s M I_r}{\sqrt{2}}\angle\delta \;=\; E_{ar}\angle\delta \qquad (7.56)\end{aligned}$$

With this, the phasor diagram becomes (according to (7.56)) as in Figure 7.12. In actual calculations, γ will play no part, as we shall see in the numerical examples below.

7.4.5 Power in terms of voltages (motor)

The current is given by

$$\overline{I}_a = \frac{\overline{V}_a - \overline{E}_{ar}}{jx_s}$$

Total power is

$$\begin{aligned}P_T &= 3P_a \\ &= 3\mathcal{R}e[\overline{E}_{ar}\overline{I}_a^*] \\ &= 3\mathcal{R}e\left[\overline{E}_{ar}\frac{(\overline{V}_a^* - \overline{E}_{ar}^*)}{-jx_s}\right] \\ &= 3\mathcal{R}e\left[\frac{j\overline{E}_{ar}\overline{V}_a^*}{x_s} - \frac{jE_{ar}^2}{x_s}\right] \\ &= 3\mathcal{R}e\left[\frac{j\overline{E}_{ar}\overline{V}_a^*}{x_s}\right] \qquad (7.57)\end{aligned}$$

Now from (7.56)

$$\overline{E}_{ar} = \frac{\omega_s M I_r}{\sqrt{2}}\angle\delta \text{ where } \delta < 0$$

Since

$$\overline{E}_{ar} = E_{ar}\angle\delta \text{ and } \overline{V}_a = V_a\angle 0^\circ \qquad (7.58)$$

we have by substituting (7.58) into (7.57)

$$\begin{aligned} P_T &= 3\mathcal{R}e\left[jE_{ar}\angle\delta\ V_a\angle 0^\circ\right]\Big/x_s \\ &= 3\mathcal{R}e\left[E_{ar}V_a\angle\left(\frac{\pi}{2}+\delta\right)\right]\Big/x_s \\ &= \frac{-3E_{ar}V_a\sin\delta}{x_s} \end{aligned} \tag{7.59}$$

The expression for torque is

$$\begin{aligned} T^e &= \frac{P_T}{\omega_m} = \frac{P_T}{\omega_s} \\ &= \frac{-3E_{ar}V_a\sin\delta}{\omega_s x_s} \end{aligned} \tag{7.60}$$

Because with motor operation $P_T > 0$, we must have $\delta < 0$. This is consistent with Figures 7.11 and 7.12. This convention is convenient for motor operation. However, it is not convenient for generator operation. We discuss next a consistent set of notations for motor and generator analysis.

7.5 Convenient Phasor Notations

7.5.1 Synchronous motor

As discussed in Section 7.4.4 the orientations of the electrical and mechanical variables are such that $P_T = P_m > 0$ implies motor operation, which requires that $\gamma < 0$ and $\delta < 0$. The formula for P_T is

$$P_T = -3E_{ar}V_a\frac{\sin\delta}{x_s} = P_m \tag{7.61}$$

The equivalent circuit is given in Figure 7.13(a), and the phasor diagram is shown in Figure 7.13(c) with voltage as reference phasor.

7.5.2 Synchronous generator

If we use the motor convention, then from Figure 7.13(b), $P_T = P_m < 0$ for the generator operation, which requires $\gamma > 0$ and $\delta > 0$ in (7.53) and (7.59), respectively. The phasor diagram will be in the second quadrant, which is inconvenient. It is convenient to have the phasor diagram for generator

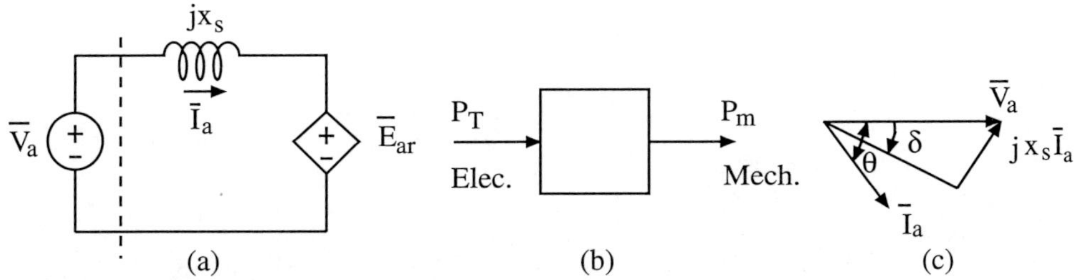

Figure 7.13: Equivalent circuit for synchronous motor operation. (The dotted line represents the physical terminal of the machine)

operation in the first and fourth quadrants for both leading and lagging PF's. Thus, we define currents such that $P_T = P_m > 0$. This leads to defining the assumed positive direction for current as shown in Figure 7.14 where the direction of T^e has also been reversed.

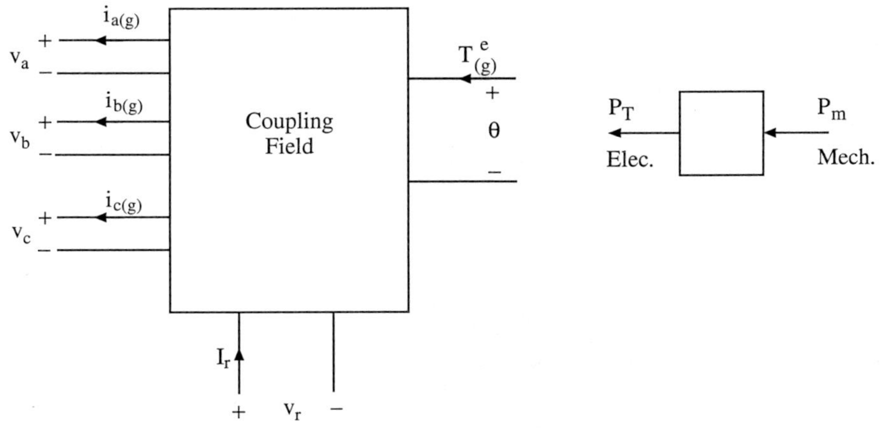

Figure 7.14: Generator orientation

$$i_{a(g)} = -i_a$$

Now both P_T and P_m will have positive orientations. The equivalent circuit (after dropping the subscript g) becomes as in Figure 7.15. Unlike the case of the motor, $\overline{V}_a$ is not a source, but it is the voltage across the terminals where load is connected. The KVL in Figure 7.15 becomes

$$\overline{V}_a = -jx_s\overline{I}_a + \overline{E}_{ar} \tag{7.62}$$

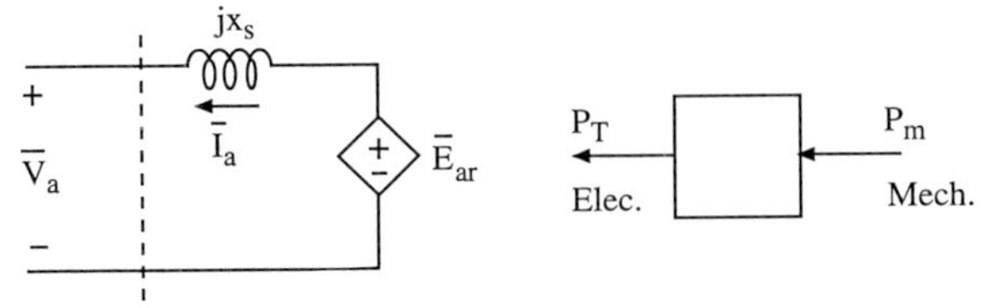

Figure 7.15: Equivalent circuit for generator

or equivalently $\overline{E}_{ar} = jx_s\overline{I}_a + \overline{V}_a$. Power delivered from the dependent source is derived as follows:

$$\overline{I}_a = \frac{\overline{E}_{ar} - \overline{V}_a}{jx_s} \tag{7.63}$$

$$P_T = 3Re[\overline{V}_a\overline{I}_a^*] = 3 \text{ Re}\left[\frac{\overline{V}_a\overline{E}_{ar}^* - V_a^2}{-jx_s}\right] \tag{7.64}$$

Using $\overline{E}_{ar}$ from (7.56) and $\overline{V}_a = V_a\angle 0°$, we get

$$\begin{aligned} P_T &= 3\mathcal{R}e\left(\frac{E_{ar}V_a e^{-j\delta}}{-jx_s}\right) \\ &= -3\mathcal{R}e\left(\frac{V_a E_{ar}(\cos\delta - j\sin\delta)}{jx_s}\right) \\ &= \frac{3V_a E_{ar}\sin\delta}{x_s} \end{aligned} \tag{7.65}$$

With this convention, for generator operation, we have $P_r = P_m > 0$ so that $\delta > 0$ and $\gamma < 0$. The phasor diagram is shown in Figure 7.16. We

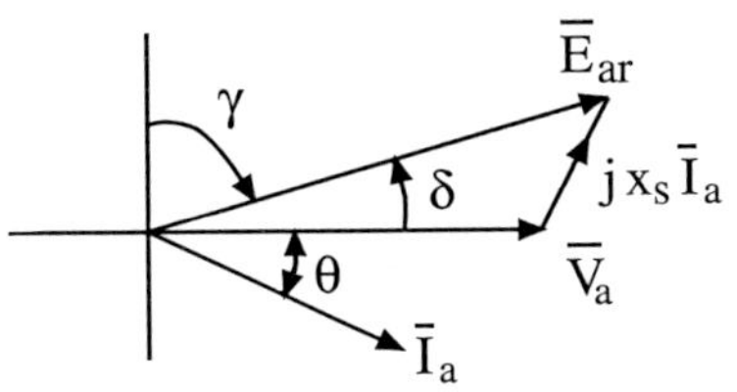

Figure 7.16: Phasor diagram for generator with lagging PF

thus resolve the difficulty of a motor or generator operation of a machine by giving proper direction for the current in the generator case. Figure 7.16

shows operation of the generator for a lagging PF case. Figures 7.12 and 7.16 show the phasor diagrams for motor and generator, respectively, for lagging PF. Figure 7.17 shows the phasor diagram for a leading PF case.

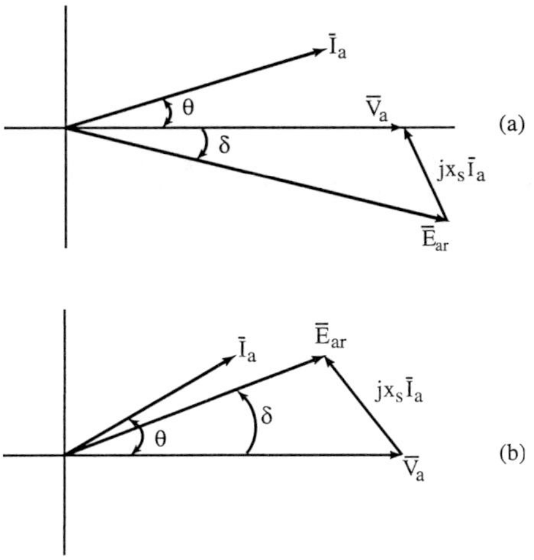

Figure 7.17: Motor operation (a) and generator operation (b) with leading PF

Example 7.1 A three-phase wye-connected 60 Hz synchronous machine with two-poles has synchronous reactance $x_s = 5.0\Omega$/phase. The machine, while operating as a motor, takes a current of is 30 A and the voltage per phase is 254 V at a PF of 0.8 leading.

Find E_{ar} and T^e. If the machine has windage, friction, and core losses of 400 W, what is the useful shaft torque? What is the efficiency?

Solution: The equivalent circuit is shown in Figure 7.18. $V_a = 254$ V, and $I_a = 30$ A

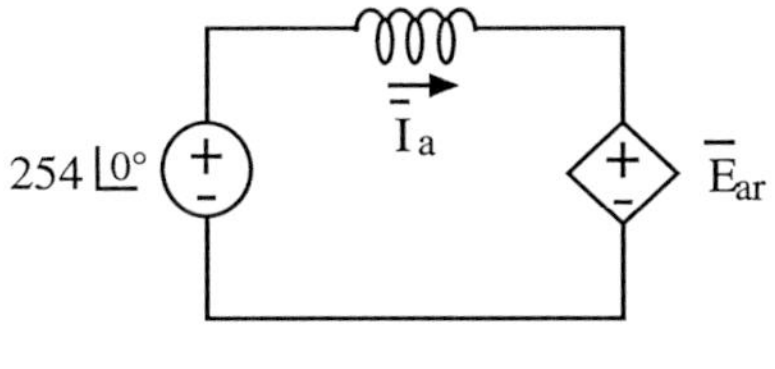

Figure 7.18: Equivalent circuit for Example 7.1

Take $\overline{V}_a$ as reference so that $\overline{V}_a = 254\angle 0°$. Power factor is the cosine of the angle θ between $\overline{V}_a$ and $\overline{I}_a$ so that $\cos\theta = 0.8$ and $\theta = 36.87°$. Since it is a leading PF,

$$\overline{I}_a = 30\angle 36.87° \text{ A}$$

From (7.53),

$$\begin{aligned}\overline{E}_{ar} &= 254\angle 0° - (j5)30\angle 36.87° \\ &= 364.3\angle -19.23° \text{ V}\end{aligned}$$

The phasor diagram is shown in Figure 7.19 Hence, the torque angle is

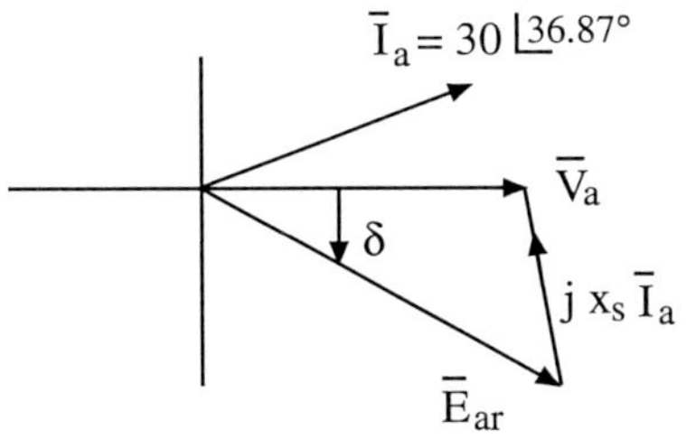

Figure 7.19: Phasor diagram for Example 7.1

$$\begin{aligned}\delta &= -19.23° \\ \omega_m &= \omega_s = 377 \text{ rad/sec}\end{aligned}$$

From (7.58),

$$\begin{aligned}P_T &= \frac{-3(364.3)(254)\sin(-19.23°)}{5} \\ &= 18286 \text{ W}\end{aligned}$$

The torque of electric origin is

$$T^e = \frac{P_T}{\omega_m} = \frac{18286}{377} = 48.5 \text{ N-m}$$

Since mechanical rotational + core losses are 400W, overall efficiency = $\frac{18286-400}{18286} = \frac{17886}{18286} = 97.8\%$. Useful shaft torque = $\frac{17886}{377} = 47.44$ N-m.

Example 7.2 Suppose in Example 7.1, the motor is operating at a PF 0.8 lagging. Repeat the calculations.

Solution:

$$\overline{V}_a = 254\angle 0°, \overline{I}_a = 30\angle -36.87°$$

From (7.54),

$$\begin{aligned} \overline{E}_{ar} &= 254\angle 0° - j5(30\angle -36.87°) \\ &= 203.2\angle -36.19° \text{ V} \\ \delta &= -36.19° \end{aligned}$$

The phasor diagram is shown in Figure 7.20.

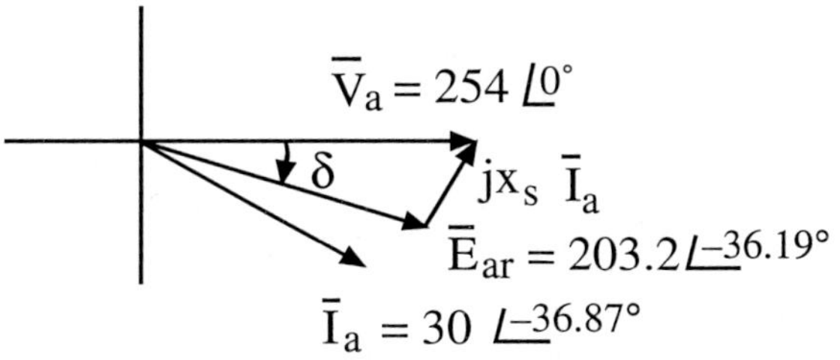

Figure 7.20: Phasor diagram for Example 7.2

$$\begin{aligned} P_T &= \frac{-3(203.2)(254)\sin(-36.19°)}{5} \\ &= 18286 \text{ W} \end{aligned}$$

Other calculations are the same as before.

Example 7.3 Suppose the field current I_r is adjusted in Example 6.2 so that the total power is the same and the PF is unity at the same input. Find the new stator current and E_{ar}.

Solution: Power per phase $= \frac{18286}{3} = 6095.3$ W

$$I_a = \frac{6095.3}{254} = 24 \text{ A}$$

Since PF is unity $\overline{I}_a = 24\angle 0°$ A.

$$
\begin{aligned}
jx_s\overline{I}_a &= j5(24\angle 0°) = j120 \text{ A} \\
\overline{E}_{ar} &= 254\angle 0° - j120 \text{ V} \\
E_{ar} &= \sqrt{254 + 120^2} = 280.9, \delta = \tan^{-1}\left(\frac{-120}{254}\right) = -25.8°
\end{aligned}
$$

The phasor diagram is shown in Figure 7.21.

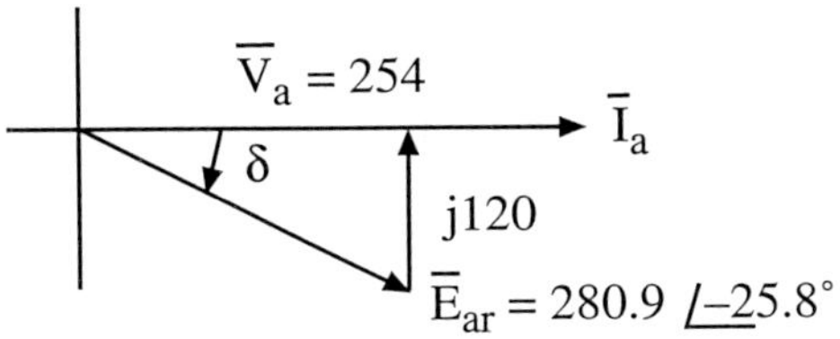

Figure 7.21: Phasor diagram for Example 7.3

Example 7.4 A two-pole, three-phase, 60 Hz, wye-connected synchronous machine has synchronous reactance $x_s = 2\Omega$ per phase. The machine is operating as a generator delivering power at a voltage of 1905 V per phase. The current is 350 A and the PF of the load is 0.8 lagging. Find E_{ar}, δ, and the torque of electric origin.

Solution: The per-phase equivalent circuit is given in Figure 7.22. Note

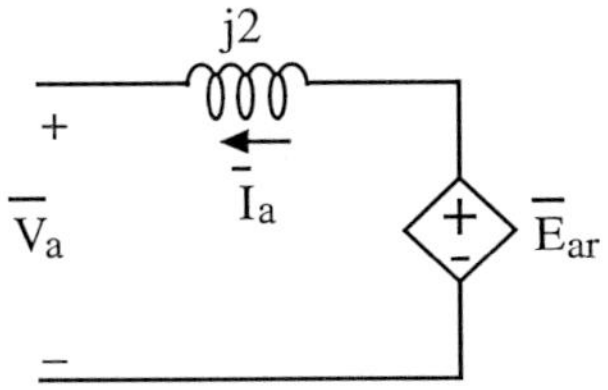

Figure 7.22: Equivalent circuit for Example 6.4

we are using the generator convention as discussed in Section 7.5.2. Take $\overline{V}_a = 1905\angle 0°$ V. Since the PF is 0.8 lagging, $\cos\theta = 0.8$, $\theta = 36.87°$ A, and $\overline{I}_a = 350\angle - 36.87°$ A.

From the equivalent circuit of Figure 7.23,

$$\begin{aligned}\overline{E}_{ar} &= \overline{V}_a + jx_s\overline{I}_a \\ &= 1905\angle 0^\circ + j2(350\angle -36.87^\circ) \\ &= 2325 + j560 = 2391\angle 13.54^\circ \text{ V}\end{aligned}$$

δ is the angle from $\overline{V}_a$ to $\overline{E}_{ar}$. Hence, $\delta = 13.54^\circ$ and $E_{ar} = 2391$ V. The power P_T is given by

$$\begin{aligned}P_T &= \frac{3E_{ar}V_a \sin\delta}{x_s} = \frac{3(2391)(1905)(.23416)}{2} \\ &= 1600,000 \text{ W} \quad = \quad 1.6 \text{ MW} \\ P_m &= T^e\omega_m = 1600,000 \text{ W} \\ \omega_m &= \omega_s \text{ since it is a two-pole machine} \\ &= 377 \text{ rads/sec}\end{aligned}$$

Therefore,

$$T^e = \frac{1600,000}{377} = 42440 \text{ N-m}$$

The phasor diagram is shown in Figure 7.23.

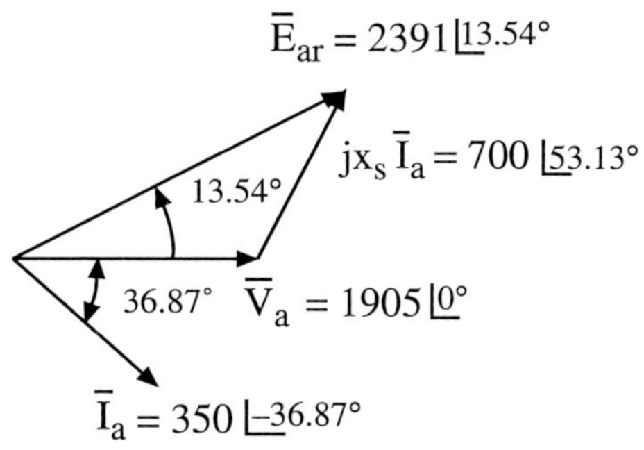

Figure 7.23: Phasor diagram for Example 7.4

7.6 Multi-pole Connections

7.6.1 Multi-pole machines

The number of poles in a machine is defined by the configuration of the magnetic field pattern. Consider the rotor of Figure 7.24(a) with a single

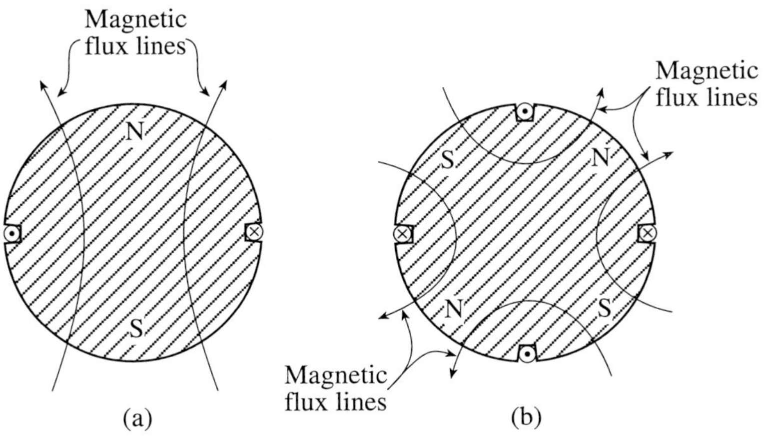

Figure 7.24: Definition of number of poles in a machine [14]

winding. When the instantaneous current is in the direction indicated, the resulting flux lines effectively create an electromagnetic field with the North (N) and the South (S) poles as shown. Going round the periphery we pass two poles. Hence, it is a two-pole machine. Consider the rotor in Figure 7.24(b). It has four slots carrying coils connected in series with polarities indicated by dots and crosses. It can be one phase of a three-phase winding. In this, we effectively have a four-pole machine. This discussion applies equally well to the windings on the stator. A four-pole three-phase machine is shown in Figure 7.25. Both the rotor and stator windings are shown. We see that we have an equivalent of four poles (N, S, N, S) on both the rotor and the stator. Only phase a is shown.

The same distribution holds for other phases as well. We thus have a rotating field that completes two cycles (720° electrical) in one mechanical revolution of 360°. This means

$$\theta_{\text{elec}} = 2\theta_{\text{mech}} \tag{7.66}$$

Since $\frac{d\theta}{dt} = \omega$, we have from (7.66) by differentiating

$$\omega_{\text{elec}} = 2\omega_{\text{mech}} \tag{7.67}$$

In general if there are p poles,

$$\omega_{\text{elec}} = \frac{p\omega_{\text{mech}}}{2} \tag{7.68}$$

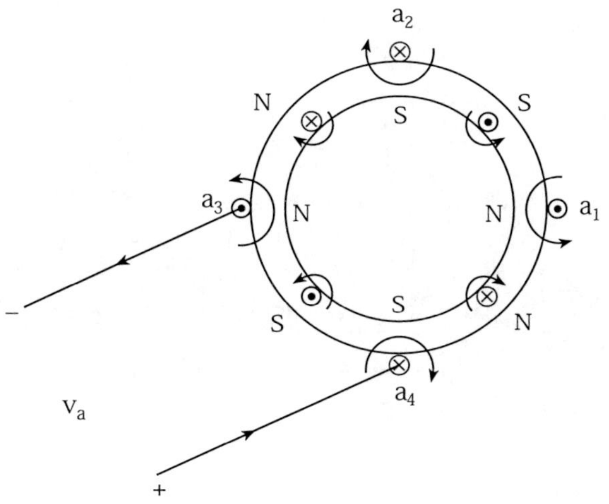

Figure 7.25: Three-phase four-pole winding (Only stator phase a is shown.)

At synchronous speed, ω_{elec} is the synchronous frequency ω_s in radians per second. Therefore,

$$\omega_s = \frac{p\omega_m}{2} \tag{7.69}$$

where ω_s is the synchronous frequency in electrical radians per second, and ω_m is the synchronous speed in mechanical radians per second. This is the fundamental relation in a synchronous machines. Now

$$\begin{aligned} \omega_s &= 2\pi f \\ &= 120\pi \end{aligned} \tag{7.70}$$

and

$$\omega_m = \frac{2\pi N_s}{60} \tag{7.71}$$

where N_s is the synchronous mechanical speed in rev/min (rpm). From (7.70) and (7.71), it follows using $\omega_s = \frac{p\omega_m}{2}$, that

$$f = \frac{pN_s}{120} \tag{7.72}$$

The following table shows the relationship between the number of poles and synchronous speed for $f = 60$ Hz.

Poles (p)	ω_m (mech.)	Speed
2	120π rad/sec	3600 rpm
4	60π rad/sec	1800 rpm
6	40π rad/sec	1200 rpm
8	30π rad/sec	900 rpm
10	24π rad/sec	720 rpm

In mathematical terms, this relationship is reflected in the mutual inductance, which instead of being $M\cos\theta$ becomes $M\cos\frac{p\theta}{2}$. The frequency condition (7.22) becomes $\omega_m = (\pm\omega_s \pm \omega_r)/p$. In the case of the synchronous machine $\omega_r = 0$, the frequency condition becomes

$$\frac{p\omega_m}{2} = \omega_s \tag{7.73}$$

i.e.,

$$\omega_s = \frac{p\omega_m}{2} \tag{7.74}$$

Example 5.5 A three-phase, 60 Hz, six-pole, wye-connected synchronous generator is driven by a turbine that delivers 16910 W to the shaft. The friction and windage losses are 500 W. The field current is adjusted so that the voltage E_{ar} proportional to field current on a per phase basis is $E_{ar} =$ 355 V. The generator supplies a load at 440 V (line-line). Find speed, $\overline{E}_{ar}$, $\overline{I}_a$, and real and reactive power output of the generator. Synchronous reactance $x_s = 5\Omega$.

Solution: From (7.72) $f = \frac{pN_s}{120}$. Since $p = 6$,

$$N_s = \frac{120 \times 60}{6} = 1200$$

$$\omega_m = \frac{2\pi(1200)}{60} = 40\pi \text{ mech. rad/sec} = 125.7 \text{ mech. rad/sec}$$

All the calculations are done on a per-phase basis. Terminal voltage per phase $= \frac{440}{\sqrt{3}} = 254$ V. Let us take terminal voltage as reference phasor so

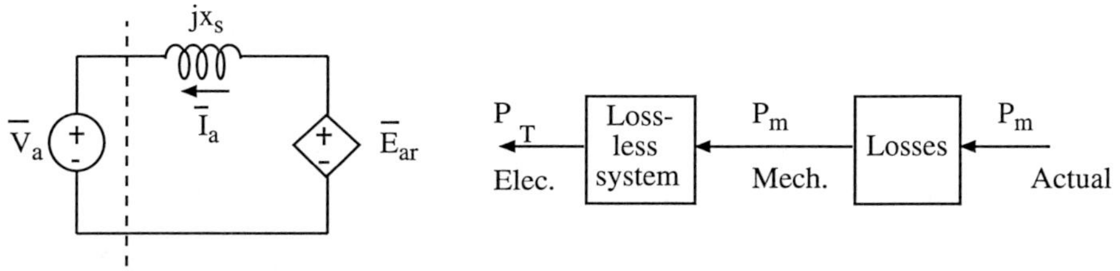

Figure 7.26: Equivalent circuit for Example 7.5

that $\overline{V}_a = 254\angle 0°$. The mechanical input P_m to the lossless system is the shaft $P_{m(\text{shaft})}$ minus the losses. Therefore, from Figure 7.26,

$$\begin{aligned} P_T = P_m &= (16910 - 500)\ \text{W} \\ &= 16410\ \text{W} \end{aligned}$$

From (7.65)

$$\begin{aligned} 16410 &= 3\frac{E_{ar}V_a}{x_s}\sin\delta \\ 16410 &= \frac{3(355)(254)\sin\delta}{5} \\ \sin\delta &= 0.3033, \quad \delta = 17.66° \end{aligned}$$

The phasor diagram is shown in Figure 7.27.

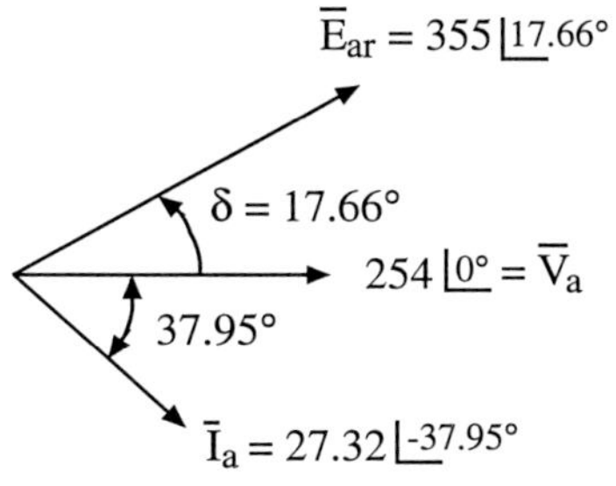

Figure 7.27: Phasor diagram for Example 7.5

$$\begin{aligned}
\overline{E}_{ar} &= 355\angle 17.66^\circ \text{ V} \\
\overline{I}_a = \frac{\overline{E}_{ar} - \overline{V}_a}{j5} &= \frac{355\angle 17.66 - 254\angle 0^\circ}{j5} \\
&= \frac{338.3 + j107.7 - 254}{j5} \\
&= \frac{84 + j107.7}{j5} = 21.54 - j16.8 \\
&= 27.32\angle -37.95^\circ \text{ A}
\end{aligned}$$

The complex power supplied by the generator is

$$\begin{aligned}
\overline{S}_T &= 3\overline{V}_a\overline{I}_a^* \\
&= 3(254\angle 0^\circ)(27.32)\angle 37.95^\circ \\
&= 16415 + j12802
\end{aligned}$$

Hence,

$$P_T = 16415W, \quad Q_T = 12802 \text{ VARs}$$

The value of P_T checks with 16410 W within engineering accuracy!

Example 7.6 A three-phase, six-pole, 60 Hz synchronous motor has $x_s = 2\Omega$ (you may neglect stator resistance). Assume the motor is operated with a voltage of 230 V (line-line), a power factor of 0.85 leading, and a per-phase current of 50 A.

(a) Find: (i) the synchronous speed of the machine in rpm; (ii) E_{ar}; (iii) torque angle δ; (iv) total complex power consumed by the motor; and (v) T^e, the torque of electric origin.

(b) Assume the field current is currently 20 A. If the real power consumption of the motor remains constant, what value of field current is needed to change the motor's power factor to unity?

Solution:

(a) From $f = \frac{pN_s}{120}$,

$$N_s = \frac{120 \times 60}{6} = 1200 \text{ rpm}$$

$$\overline{V}_a = \frac{230}{\sqrt{3}}\angle 0°V, \quad jX_s = j2\Omega \quad PF = 0.85$$

Hence $\theta = \cos^{-1}(0.85) = 31.8°$. Since the PF is leading, $\overline{I}_a = 50\angle 31.7°$ A. The equivalent circuit is shown in Figure 7.28.

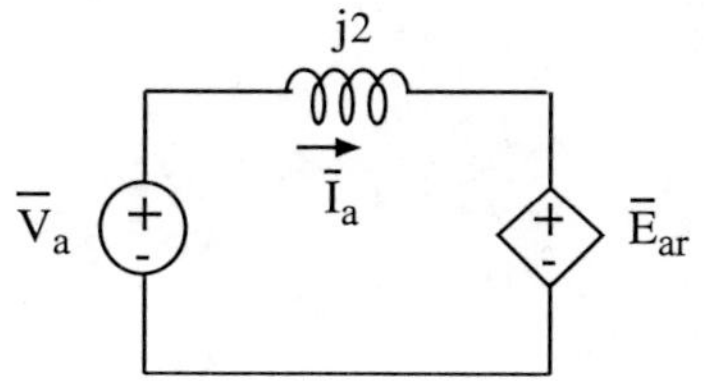

Figure 7.28: Equivalent circuit for Example 7.6

Using KVL,

$$\begin{aligned}
\overline{E}_{ar} &= \overline{V}_a - (j2)(50\angle 31.8°) \\
&= \frac{230}{\sqrt{3}}\angle 0° - j100\angle 31.8° = 204\angle -24.6° \text{ V} \\
E_{ar} &= 204 \text{ V}, \delta = -24.6°
\end{aligned}$$

Total complex power

$$\begin{aligned}
S_T &= 3\overline{V}_a\overline{I}_a^* \\
&= 3\left(\frac{230}{\sqrt{3}}\angle 0°\right)(50\angle -31.8°) \\
&= 19.92\angle -31.8° \text{ kVA} \\
&= 16.93 - j10.5 \text{ kVA}
\end{aligned}$$

The torque of electric origin T^e is

$$T^e = \frac{P_T}{\omega_m}$$

since

$$\begin{aligned}
\omega_m &= \frac{2\pi N_s}{60} = \frac{(2\pi)(1200)}{60} = 40\pi \text{ rad/sec} \\
T^e &= \frac{(16.93)10^3}{40\pi} = 134.7 \text{ N-m}
\end{aligned}$$

(b) I_a changes when I_r changes. Since P_T is constant,

$$\begin{aligned} 3V_a I_{a(\text{old})}(\text{PF})_{\text{old}} &= 3V_a I_{a(\text{new})}(\text{PF})_{\text{new}} \\ I_{a(\text{old})}(\text{PF})_{\text{old}} &= I_{a(\text{new})}(PF)_{\text{new}} \\ (50)(0.85) &= I_{a(\text{new})}(1.0)_{\text{new}} \end{aligned}$$

Therefore,

$$\begin{aligned} I_{a(\text{new})} &= 42.5 \text{ A} \\ \overline{I}_{a(\text{new})} &= 42.5\angle 0^\circ \text{ A} \end{aligned}$$

since the PF is unity.

$$\begin{aligned} \overline{E}_{ar(\text{new})} &= \overline{V}_a - (j2)I_{a(\text{new})} \\ &= \frac{230}{\sqrt{3}}\angle 0^\circ - (j2)(42.5\angle 0^\circ) \\ &= 157.7\angle -32.6^\circ \text{ V} \end{aligned}$$

E_{ar} is proportional to the field current I_r,

$$\frac{E_{ar(\text{new})}}{I_{r(\text{new})}} = \frac{E_{ar(\text{old})}}{I_{r(\text{old})}}$$

Substituting

$$\begin{aligned} I_{r(\text{new})} &= \frac{(E_{ar(\text{new})})I_{r(\text{old})}}{E_{ar(\text{old})}} \\ &= \frac{(157.7)(20)}{204} = 15.45 \text{ A} \end{aligned}$$

7.6.2 Summary for p pole machine

Generator action

With the generator convention for current, $P_T = P_m > 0$, as shown in Figure 7.29.

$$P_T = \frac{3E_{ar}V_a \sin\delta}{x_s} = P_m \tag{7.75}$$

$$\overline{E}_{ar} = jX_s\overline{I}_a + \overline{V}_a \tag{7.76}$$

In the phasor diagram, $\delta > 0$ as shown in Figure 7.16.

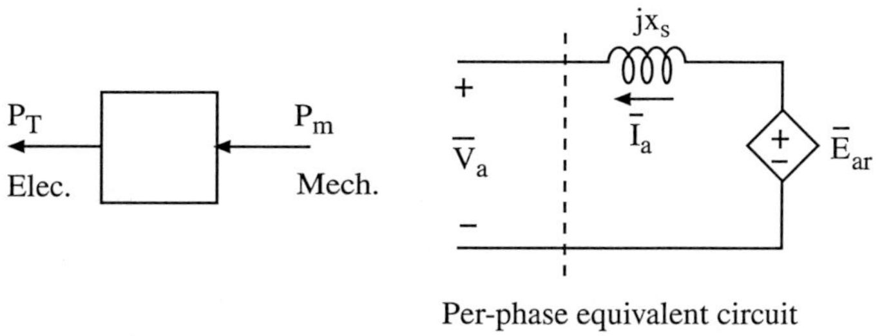

Figure 7.29: Generator

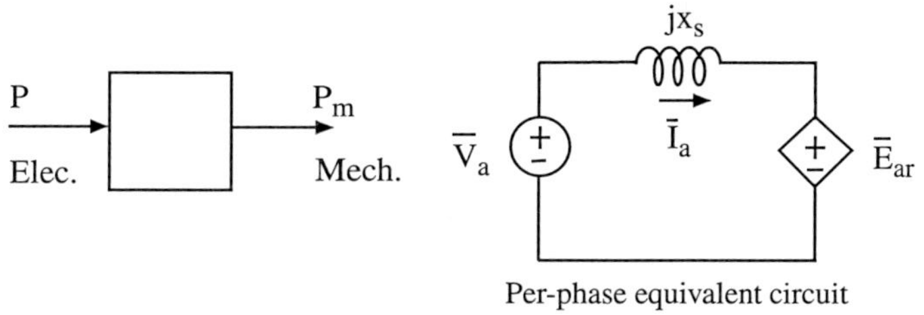

Figure 7.30: Motor

Motor action

With the motor convention $P_T = P_m > 0$, as shown in Figure 7.30.

$$P_T = \frac{-3E_{ar}V_a \sin\delta}{x_s} = P_m \quad (7.77)$$

$$\overline{V}_a = jx_s\overline{I}_a + \overline{E}_{ar} \quad (7.78)$$

In the phasor diagram, $\delta < 0$, as shown in Figure 7.13.

Both for motor and generator, $\omega_s = \frac{p}{2}\omega_m$, where ω_s is the supply frequency and ω_m is the synchronous speed in mechanical radians per second. Also, the synchronous speed N_s in rpm is $120f/p$.

The power factor is cosine of the angle between terminal voltage $\overline{V}_a$ and terminal current $\overline{I}_a$. It is called *leading PF* when $\overline{I}_a$ leads $\overline{V}_a$. It is called *lagging PF* when $\overline{I}_a$ lags $\overline{V}_a$. δ is called the *torque angle* and is the angle from $\overline{V}_a$ to $\overline{E}_{ar}$. It is *positive* for generator action and *negative* for motor action. $\overline{E}_{ar}$ is a dependent voltage source proportional to field current I_r.

7.7 Power Factor Improvement

A synchronous motor can supply mechanical power while at the same time receiving electrical power at a leading or lagging PF. It can therefore act either in a capacitive or inductive mode. Keeping shaft output power constant and input voltage constant, we can vary the field current I_r and make $\overline{I}_a$ vary from a lagging to a leading PF. This fact is used for improving the PF of a load using the synchronous motor. We explain this phenomenon through phasor diagrams.

Consider the equivalent circuit of the motor in Figure 7.31. The total

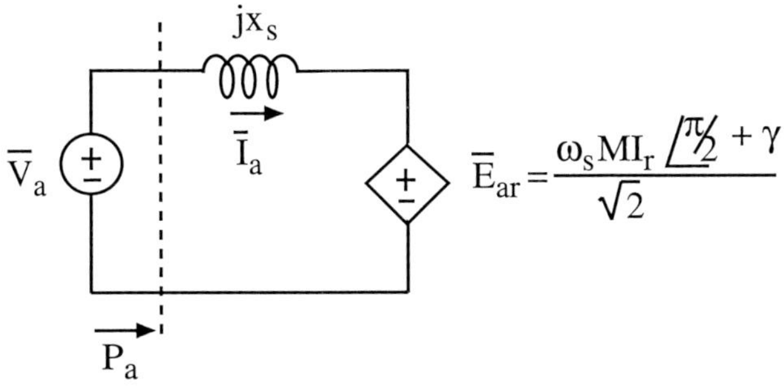

Figure 7.31: Per-phase equivalent of a synchronous motor

power $P_T = 3P_a = 3V_aI_a\cos\theta = \frac{-3E_{ar}V_a\sin\delta}{x_s}$ where θ is the power factor and δ is the torque angle. We have P_T, V_a as constant. Hence, $I_a\cos\theta =$ constant and $E_{ar}\sin\delta =$ constant. δ is the angle from $\overline{V}_a$ to $\overline{E}_{ar}$ and is negative for a motor. Let $\overline{V}_a = V_a\angle 0°$ be chosen as reference phasor, shown in Figure 7.31. We draw the loci of $\overline{I}_a$ and $\overline{E}_{ar}$ on the phasor diagram as I_r is varied. Now $\overline{V}_a = jx_s\overline{I}_a + \overline{E}_{ar}$. Vary $\overline{I}_a$ keeping $I_a\cos\theta =$ constant. The locus of $\overline{I}_a$ is the line B on which $I_a\cos\theta =$ constant. The phasor diagram for three values of I_a, namely, I_{a1}, I_{a2}, and I_{a3} are shown. The corresponding locus of $\overline{E}_{ar}$ is the line A on which $E_{ar}\sin\delta =$ constant as shown in Figure 7.32.

In Figure 7.32, the lines A and B provide the locus of $\overline{E}_{ar}$ and $\overline{I}_a$ for three typical positions from a lagging PF, through unity PF, to a leading PF. The lengths of phasors $\overline{I}_a$ and $\overline{E}_{ar}$ are proportional to I_a and I_r, respectively. For small values of I_a (small value of E_{ar}), $\overline{I}_a$ lags $\overline{V}_a$ ($\overline{I}_{a1}$, $\overline{E}_{ar1}$). As I_r increases, I_a initially decreases and the PF increases to unity until a certain point. After that, the current increases again and the PF becomes leading ($\overline{I}_{a3}, \overline{E}_{ar3}$). Thus, as I_r increases we go from a lagging to a leading PF.

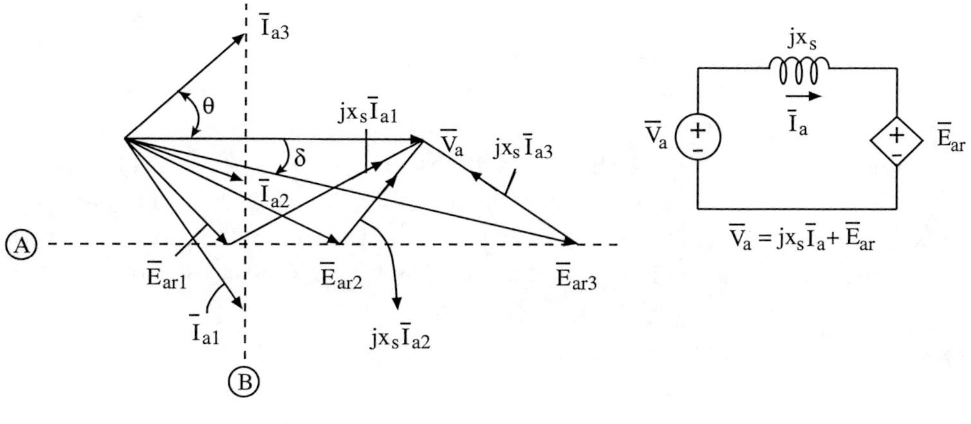

Figure 7.32: Phasor diagram

Qualitatively, the curves of I_a versus I_r look like those in Figure 7.33 for

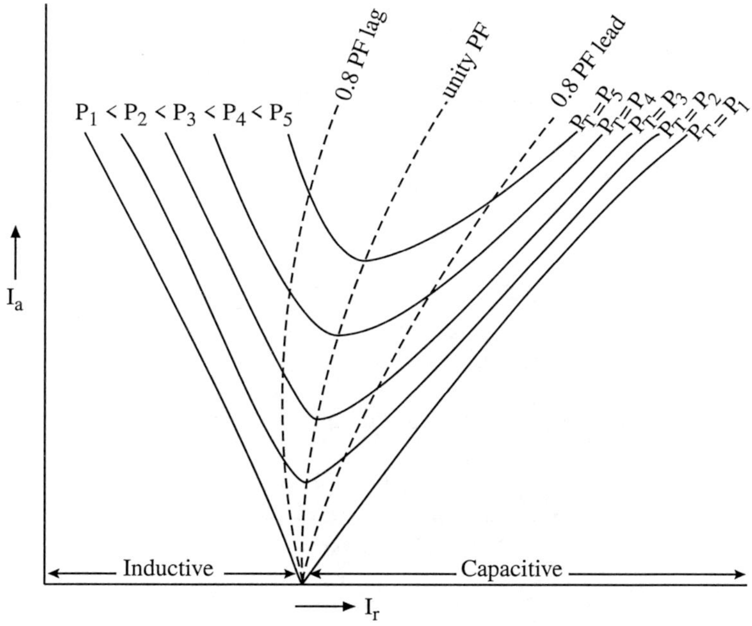

Figure 7.33: "V" curves for synchronous motor

different values of P_T. These are called V curves. These curves are nonlinear. The dotted line (center) represents the unity PF line. To the right of that line, the motor behaves like a capacitor (over-excited) supplying reactive power to the external system, and to the left, like an inductor (under-excited) absorbing reactive power from the external system. The phrase *over-* and *under*-excited refers to the field excitation or current. A high value of I_r is *over excitation*, meaning thereby a high value of E_{ar}. Similarly, *under*-excited refers to a low value of I_r and hence, a low value of E_{ar}. This unique capacity of the device to look inductive or capacitive, depending on the modulation of I_r, makes it an ideal device for power factor correction.

Example 7.7 A three-phase wye-connected load of 1500 kW at 0.8 PF lagging is connected to a three-phase supply with a line-line voltage of 1732 V. A synchronous motor with no load is connected across the load to improve the PF to unity. Find the current taken by the synchronous motor.

Solution: We work on a per-phase basis and rms values.

$$V_a = \frac{1732}{\sqrt{3}} = 1000 \text{ V}$$

$\overline{V}_a$ is taken as reference so that $\overline{V}_a = 1000\angle 0°$. The PF is

$$\cos\theta = 0.8$$

Hence,

$$\theta = 36.87°$$

$$\overline{V}_a = 1000\angle 0° \text{ V}$$

The per-phase circuit diagram is shown in Figure 7.34.

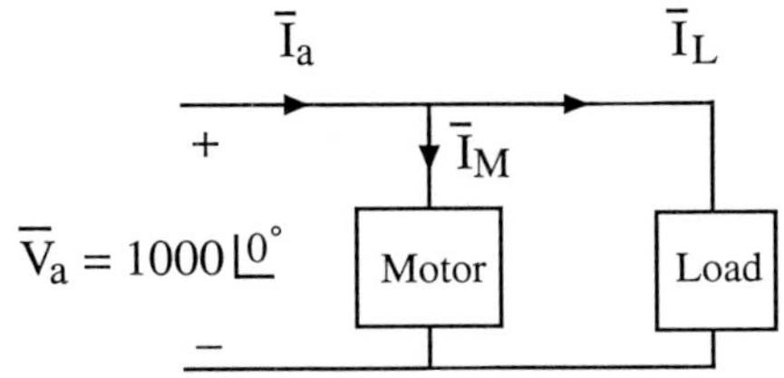

Per-phase diagram

Figure 7.34: Power factor improvement

The real power of the load per phase is

$$\begin{aligned} V_a I_L \cos\theta &= \frac{1500}{3} \times 10^3 \text{ W} \\ &= 500 \times 10^3 \text{ W} \end{aligned}$$

Therefore,

$$I_L = \frac{500 \times 1000}{1000 \times 0.8} = 625 \text{ A}$$

Since it is a lagging PF load,

$$\overline{I}_L = 625\angle -36.87^\circ \text{ A}$$

After connecting the motor, the overall PF is the motor unity. Hence

$$\overline{I}_a = I_a \angle 0^\circ$$

Since a synchronous motor has no load, using the conservation of power principle

$$V_a I_a = 500 \times 10^3 + 0$$

Hence,

$$\begin{aligned} I_a &= \frac{500 \times 10^3}{1000} \\ &= 500 \text{ A} \end{aligned}$$

In phasor notation,

$$\overline{I}_a = 500\angle 0^\circ \text{ A}$$

since $\overline{I}_a$ is in phase with $\overline{V}_a$. Let

$$\overline{I}_M = I_M \angle\phi = I_M \cos\phi + jI_M \sin\phi$$

By KCL,

$$\begin{aligned} 500\angle 0^\circ &= \overline{I}_M + \overline{I}_L \\ &= I_M \cos\phi + jI_M \sin\phi + 625\angle -36.87^\circ \\ 500 &= I_M \cos\phi + jI_M \sin\phi + 500 - j375 \end{aligned}$$

Equating real and imaginary parts,

$$\begin{aligned} I_M \cos\phi &= 0 \\ I_M \sin\phi &= +375 \end{aligned}$$

Hence,

$$\phi = \frac{+\pi}{2} \text{ and } I_M = 375\text{A}$$

Therefore,

$$\overline{I}_M = 375\angle 90^\circ \text{ A} = \jmath 375 \text{ A}$$

$\overline{I}_M$ leads $\overline{V}_a$ by 90°. The synchronous motor acts like an ideal capacitor.

Example 7.8 If in Example 7.7 the synchronous motor drives a load of 300 kW, find the current taken by the motor.

Solution: The per-phase load of the synchronous motor is

$$\frac{300 \times 10^3}{3} = 100 \text{ kW}$$

Hence, using the conservation of real power principle

$$\begin{aligned} V_a I_a &= 500 \times 10^3 + 100 \times 10^3 \\ &= 600 \times 10^3 \text{ VA} \end{aligned}$$

Hence,

$$\begin{aligned} I_a &= \frac{600 \times 10^3}{1000} \\ &= 600 \text{ A} \end{aligned}$$

Again

$$\overline{I}_a = 600\angle 0^\circ \text{ A}$$

By KCL,

$$\begin{aligned} 600\angle 0^\circ &= I_M \cos\phi + jI_M \sin\phi + 500 - \jmath 375 \\ 100 - \jmath 375 &= \overline{I}_M = 388\angle 76^\circ \end{aligned}$$

Hence,

$$\begin{aligned} I_M \cos\phi &= 100 \\ I_M \sin\phi &= 375 \\ \tan\phi &= 3.75, \phi = 75^\circ \\ I_M &= \sqrt{(375)^2 + (100)^2} \\ &= 388.1 \text{ A} \end{aligned}$$

Hence, the synchronous motor current $\overline{I}_M$ is

$$\overline{I}_M = 388.1\angle 75^\circ \text{ A}$$

The motor takes a leading current from the supply, i.e., acts capacitive. In power engineering terminology, we say the motor is *over excited.*

For Examples 7.7 and 7.8, it is noted that when the synchronous motor is not driving any mechanical load, it is acting like an ideal capacitor; and when it is driving a load, it is *capacitive* with the current in the motor leading the voltage by less than 90°. In both cases it is *over excited.*

7.8 Conclusion

In this chapter, we have developed the electromechanical equations of a three-phase synchronous machine from the fundamental concept of co-energy. Expression for torque of electric origin is developed. Then assuming sinusoidal steady-state conditions and a balanced three-phase operation, we developed the per-phase equivalent circuit. Both motor and generator operation are considered. Use of a synchronous motor in PF correction is illustrated.

7.9 Problems

7.1 A three-phase, 60 Hz, 550 V (line-line), six-pole, wye-connected synchronous generator is delivering 300 kW at rated voltage. The excitation voltage E_{ar} is 460 V per phase, and the synchronous reactance $x_s = 2\Omega$.

(a) Calculate torque angle δ and torque in N-m.

(b) Find generator current, power factor, and reactive generation Q.

7.2 A three-phase, six-pole 60 Hz wye-connected synchronous *motor* with $x_s = 2\Omega$ and a three-phase wye-connected load are connected in parallel. The load current $\overline{I}_L$ is $200\angle -45^\circ$ A and the voltage across the motor and the load is 2300 V (line-line). The *total* three-phase complex power taken by both the load and the motor is $P + jQ$ where $P = 1500$ kW and $Q = 300$ kVAR.

(a) Find current $\overline{I}_{aM}$ taken by the motor.

(b) Find PF of the motor and state whether it is leading or lagging.

(c) Find $\overline{E}_{ar}$ of the motor and the torque angle δ.

7.3 A three-phase, 60 Hz, 550 V, wye-connected synchronous generator is delivering 250 kW at rated voltage. The excitation voltage E_{ar} is 460 V per phase, and the synchronous reactance $x_s = 1.2\Omega$.

(a) Calculate torque angle δ. (Answer: 43.2°)

(b) Find generator current, power factor, and reactive generation Q. (Answer: $Q = 14.095$ kVAR)

(c) For the same terminal voltage and active power generation P, determine the excitation voltage E_{ar} to achieve a power factor of 0.8 lagging. What is the new value of the torque angle δ in this case? (Answer: $\delta = 29.628°$)

7.4 A 1000 kVA, 3300 V, 60 Hz, four-pole, three-phase, wye-connected synchronous generator has synchronous reactance $x_s = 1\Omega$. The generator is delivering power to a load at a power factor of 0.8 lagging at rated voltage and at rated current. Find the speed, E_{ar}, δ, and the torque of electric origin. (Answer: $E_{ar} = 2015$ V; $\delta = 3.98°$; $T^e = 4244$ N-m)

7.5 A 2300 V, wye-connected, three-phase, 60 Hz, twenty-pole synchronous motor has $x_s = 2\Omega$. Suppose the total input power is 800 kW at rated voltage, and the field excitation is adjusted such that the PF is unity. Find the speed, torque angle δ, and the torque of electric origin. (Answer: 360 rpm, $\delta = -16.16°$, $T^e = 21,220$ N-m)

7.6 A 6600 V, ten-pole, wye-connected 60 Hz, three-phase synchronous motor has $x_s = 15\Omega$. I_r and the load are such that the stator current is 95 A at a PF of 0.8 leading with rated voltage at the terminals. Neglect all losses.

(a) Compute δ and T^e.

(b) Compute reactive power drawn from supply.

(c) Draw a phasor diagram.

(Answer: $\delta = -13.87°$; $T^e = 11,524$ N-m; $Q = 651.6 \times 10^3$ VAR)

7.7 A three-phase wye-connected load takes 50 A current at 0.707 PF lagging with 220 V between the lines. A three-phase wye-connected round rotor synchronous motor with $x_s = 1.27\Omega$ is connected in parallel. The power taken by the motor is 33 kW, and the torque angle is $\delta = 30°$. Calculate (a) current taken by the motor and PF of the motor, and (b) overall PF of the motor and load. (Hint: Take terminal voltage per phase as reference phasor. Recall, δ is negative.) (Answer: $I_a = 100\angle 30°$; PF of motor = 0.866 leading; overall PF = 0.993 leading)

7.8 A three-phase, 60 Hz, 11,000 kVA, 3400 V, 24-pole, wye-connected synchronous generator is delivering 8,000 kW at unity power factor and at rated voltage. The synchronous reactance is $x_s = 1\Omega$ per phase.

(a) Calculate the generator speed and the torque of electric origin.

(b) Find E_{ar} and the torque angle δ. (Partial answer: 2387.2V; 34.68°)

(c) Under constant terminal voltage and constant turbine input (P_m = constant), the rotor current is increased by 20%. Find the new torque angle δ, stator current I_a, and the reactive power output of the generator. (Partial answer: $\delta = 28.31°$, $Q = 3292$ kVAR)

7.9 A composite load consists of the following two components in parallel as shown in Figure 7.35:

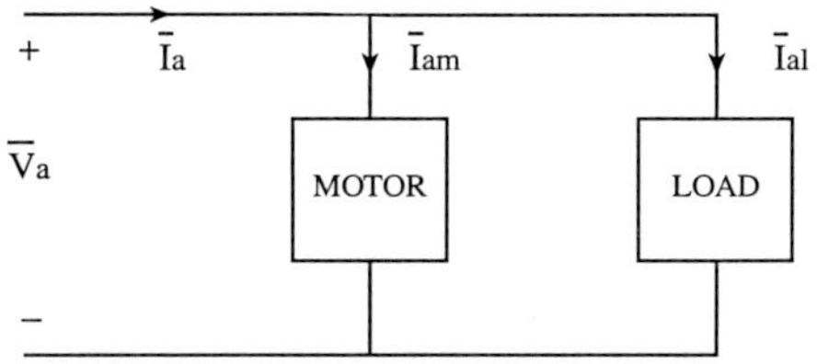

Figure 7.35: Load connection for Problem 6.9

(a) a three-phase, wye-connected, 2300 V, 60 Hz, six-pole, synchronous motor with $x_s = 10\Omega$ (The mechanical load of the motor is 87 kW and the total mechanical losses are 3 kW.) and

(b) a three-phase, wye-connected load, absorbing 20 A per phase at 0.8 lagging power factor.

The supply voltage for both loads is equal to the rated motor voltage.

(a) Find the motor speed and torque of electrical origin, as well as the torque of the mechanical load. (Partial answer: 716.2 N-m; 692.3 N-m)

(b) Calculate the motor excitation voltage E_{ar}, in order to achieve a unity power factor of the composite load. Find the torque angle δ under this condition. (Partial answer: 1465.5 V, $-8.87°$)

7.10 A three-phase, 60 Hz, 550 V, wye-connected synchronous generator is delivering 250 kW at rated voltage. The excitation voltage E_{ar} is 460 V per phase, and the synchronous reactance $x_s = 1.2\Omega$.

(a) Calculate the torque angle δ. (Answer: 43.025°)

(b) Find the generator current, power factor, and reactive generation Q. (Partial answer: $Q = 14.095$ kVAR)

(c) For the same terminal voltage and active power generation P, determine the excitation voltage E_{ar} to achieve a power factor of 0.8 lagging. What is the new value of the torque angle δ in this case? (Partial answer: $\delta = 29.628°$)

7.11 A 25 kVA, 230 V, three-phase, four-pole, 60 Hz, wye-connected synchronous machine is operating as a *generator* at *rated* voltage and rated current at 0.8 PF lagging. The synchronous reactance $x_s = 1.5\Omega$.

(a) Find E_{ar} and the torque angle δ. (Answer: $E_{ar} = 203.8$ V; $\delta = 21.7°$)

(b) Suppose the rotor current I_r is increased by 20% while keeping the terminal voltage and the mechanical input to the generator constant. Find the new torque angle, $\overline{I}_a$, and the new power factor. (Partial answer: $\delta_{\text{new}} = 17.946°$; $\overline{I}_a = 83.356\angle - 53.0°$)

7.12 An industrial plant is consuming a total load of 900 kW at 0.9 lagging power factor and *rated* voltage from a three-phase, 2 kV source. Part of this load is a three-phase, 60 Hz, 400 kVA, 2 kV, four-pole, wye-connected synchronous motor operating at its *rated* current and *rated* voltage at unity power factor. The synchronous reactance of the motor is 10Ω.

(a) Find the current $\overline{I}_L$ and the power factor of the non-motor part of the load L. (Answer: $191.5\angle - 41.08°$)

(b) Find speed, torque, excitation voltage E_{ar} and torque angle δ of the synchronous motor. (Partial answer: $\delta = -45°$)

(c) Find the value of the excitation voltage E_{ar}, in order to achieve unity power factor of the total load. (Answer: 2675 V)

(Hint: Draw the per-phase equivalent circuit as shown in Figure 7.36.)

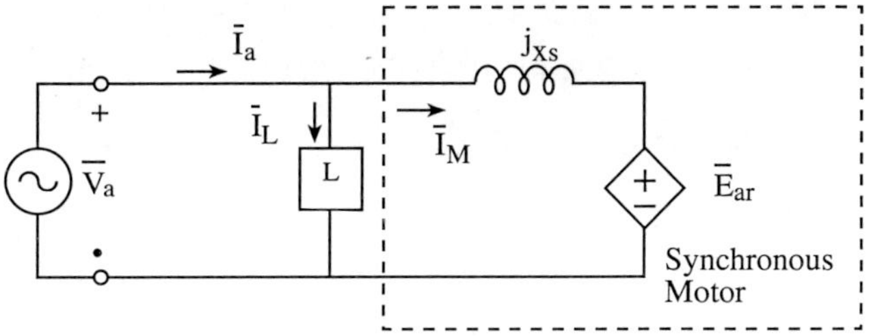

Figure 7.36: Equivalent circuit for Problem 6.12

7.13 Assume a 30 kW, three-phase, 0.8 PF lagging load is connected in parallel with a three-phase, two-pole, 60 Hz synchronous motor that has $x_s = 2\Omega$ (you may neglect stator resistance). The terminal voltage for the load and the motor is 230 V (line-line). The motor is used for power factor correction and supplies zero real power. For parts (a) to (c), assume that the motor is used to correct the power factor to unity.

(a) How much reactive power must the motor supply to correct the power factor to unity?

(b) What is the magnitude of the motor's current?

(c) What is the magnitude of the internal voltage on the motor?

(d) Now assume the motor is only used to correct the net power factor to 0.95 lagging. What is the new internal voltage?

(e) To change the net power factor from unity to 0.95 lagging, what percentage change is required to the machine's field current (i.e., how does the field current change from part (c) to part (d))?

7.14 An industrial plant is supplied from a three-phase 2300 V, 60 Hz system, and assume the load to be wye connected. The total load of the plant is 2600 kW at a PF of 0.87 lagging (each phase of the load also has PF of 0.87 lagging). A wye-connected synchronous motor is connected in parallel so as to improve the overall PF of the plant to unity ($x_s = 1.6\Omega$).

(a) Find the current taken by the plant before the synchronous motor is connected.

(b) Find the current in the motor and the new current to the parallel combination of motor and industrial plant.

(c) Find E_{ar} for the motor and the torque angle.

(Answer: (a) $I_L = 750.2$ A; (b) 369.9 A, 652.7 A; (c) 1919.7 V, 0°)

7.15 An eight-pole, synchronous motor draws 45 kW from a 208 V, 60 Hz, three-phase system at a PF of 0.8 lagging. The motor is wye connected, and $x_s = 0.6\Omega$.

(a) What is the torque angle and E_{ar}?

(b) What is the maximum torque possible?

(Answer: (a) 98.56, −49.5°; (b) 627.43 N-m)

7.16 An industrial plant uses two three-phase induction motors at 460 V. At full load one draws 232 A at 0.85 PF lagging, and the other draws 169 A at 0.82 PF lagging. Find kVA rating of the synchronous motor that can serve to bring the plant PF to 0.9 lagging. (Answer: 481 kVA)

7.17 A composite load consists of the following two components in parallel:

(i) a three-phase, wye-connected, 2300 V, 60 Hz, six-pole, synchronous motor with $x_s = 10\Omega$ (The mechanical load of the motor is 87 kW, and the total mechanical losses are 3 kW.) and

(ii) a three-phase, wye-connected load, absorbing 20 A per phase at 0.8 lagging power factor (The supply voltage for both loads is equal to the rated motor voltage.)

The supply voltage for both loads is equal to the rated motor voltage.

(a) Find the motor speed and torque of electrical origin, as well as the torque of the mechanical load. (Partial answer: 716.2 N-m; 692.3 N-m)

(b) Calculate the motor excitation voltage E_{ar}, in order to achieve a unity power factor of the composite load. Find the torque angle δ under this condition. (Partial answer: 1465.5 V; $-8.87°$)

7.18 A 3000 kVA, 3300 V, 60 Hz, four-pole, wye-connected synchronous generator has $x_s = 1.5\Omega$. The generator delivers power to a load at a PF of 0.9 lagging at *rated* voltage and *rated* current. Find δ and T^e. (Answer: $\delta = 17.5°$, $T^e = 14,326$ N-m)

7.19 (a) A three-phase, wye-connected, 2300 V, 60 Hz, six-pole synchronous motor has $x_s = 2\Omega$. Total input power to the motor is 900 kW at rated voltage, and the field current is adjusted so that PF is unity. Find speed, δ and T^e. (Answer: $\delta = -18.7°$, $T^e = 7160$ N-m)

(b) Suppose this motor is connected across a three-phase wye-connected load taking 200 A at 0.707 PF lagging, find the overall PF and the total complex power taken by the load and motor. (Answer: $P + jQ = (1463 + j564.25)$ kVA)

7.20 A three-phase, 60 Hz, 550 V, four-pole, wye-connected synchronous generator is delivering 250 kW at rated voltage. The excitation voltage E_{ar} is 460 V per phase, and the synchronous reactance $x_s = 1.2\Omega$.

(a) Calculate the torque angle δ, and the torque in N-m.

(b) Find the generator current, power factor and reactive generation Q. (Partial answer: PF = 0.998 lagging, $Q = 14.09$ kVAR)

Chapter 8

INDUCTION MACHINES

8.1 Introduction

The induction machine is the most widely used machine in the industry. It is used mainly as a motor and very rarely as a generator. Both the stator and the rotor carry alternating currents. In large sizes, three-phase induction motors are used in pumps, compressors, fans, etc. Due to the rapid advances in power electronics, the inductance motor can be made to have excellent torque-speed capabilities. In this chapter, we study the basic phenomena as well as the analysis based on equivalent circuits derived from the energy point of view.

8.2 Physical Features and Frequency Relationships

In the previous chapter, we discussed the synchronous machine with a three-phase winding on the stator and a DC field winding on the rotor. It runs at a *synchronous speed* $\omega_m = \omega_s = 2\pi f$ *determined* by $f = \frac{pN_s}{120}$ where $p =$ the number of poles and N_s is the synchronous speed in rpm. Each phase is wound for p poles, and the DC field winding also is wound for p poles.

The other class of machines, the *work horse* of the industry, is the induction machine. It has a three-phase winding on both the stator and the rotor, wound for p poles in each phase. The three-phase winding on the rotor is, however, short circuited either through external means or on the rotor itself. This gives rise to the (i) wound rotor and (ii) *squirrel cage rotor*. As a motor, it runs as a rugged device requiring very little maintenance. It was invented by Nikola Tesla in 1888. It consists of two basic components—*stator* and the *rotor*—and is analogous to a transformer with a rotating secondary.

8.2.1 Wound rotor

The rotor windings that rotate with the rotor are connected to the slip rings, as shown in Figure 8.1. The fixed brushes provide a short-circuited

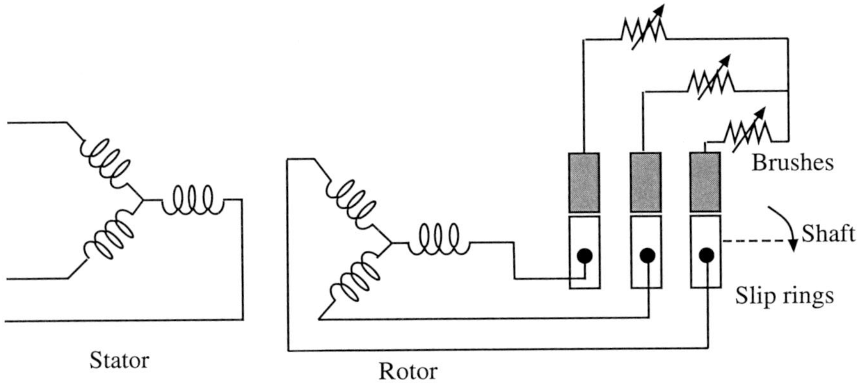

Figure 8.1: Wound-rotor induction machine

path through an external resistance. The purpose of the external resistance is to achieve a desired torque-speed characteristic, depending on the load. In particular, a higher starting torque can be achieved together with limited control over speed. A fixed-torque speed characteristic of the machine can be achieved by having the rotor windings shorted on the rotor itself. This gives rise to a *squirrel cage* motor.

8.2.2 Squirrel cage rotor

The technique of having shorted windings on the rotor is very simple. The conductors through the slots are simply shorted through a ring at the ends, as shown in Figure 8.2. Conceptually for the purpose of analysis, we can think of the rotor as having the three-phase windings shorted.

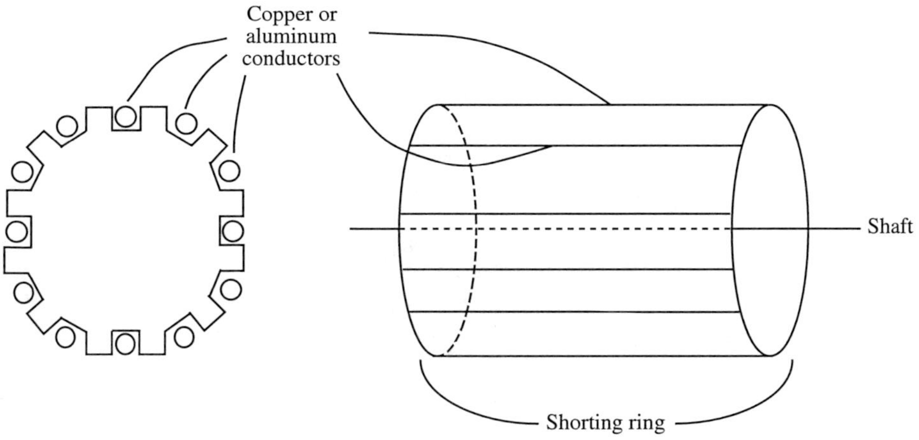

Figure 8.2: Squirrel cage induction machine

8.2.3 Operation of an Induction Machine

An induction machine is mostly run as a motor. Three-phase currents are supplied to the stator windings, resulting in a rotating field. Here again, each phase may be wound for p poles so that the rotating field revolves at the synchronous speed N_s given by $f = \frac{pN_s}{120}$ where f is the frequency in Hz and N_s is in revolutions per minute (rpm). Currents are induced in the rotor windings, which interact with the rotating field to produce torque. Hence, the rotor revolves. The induced currents on the rotor give rise to p poles on the rotor. As the rotor revolves, the relative velocity between the rotor and rotating field due to the stator decreases. The speed increases to a point where the torque developed between rotor and stator is *just enough* to balance the opposing mechanical load torque. So long as this opposing torque (due to bearing friction, etc., under no-load conditions) exists, the rotor speed can never equal the speed of the stator rotating field speed. The rotor currents have a frequency ω_r, and the mechanical speed in mechanical rads per second satisfies the relation by $\omega_s - \omega_r = \left(\frac{p}{2}\right)\omega_m$. This satisfies the frequency condition for average power, which is one of the combinations of $\frac{p\omega_m}{2} = \pm\omega_s \pm \omega_r$. The frequency of induced currents in the rotor is

$$\omega_r = \omega_s - \left(\frac{p}{2}\right)\omega_m \text{ electrical rads/sec}$$

$\omega_m = \frac{2\pi N_{\text{act}}}{60}$ where N_{act} is the actual speed of the rotor in rpm. On the mechanical side, $N_s - N_{\text{act}}$ is normalized by dividing by N_s, and we define a new dimensionless quantity called *slip* as

$$s = \frac{N_s - N_{\text{act}}}{N_s} \tag{8.1}$$

Multiply the right side of (8.1) by $\frac{2\pi p}{120}$ to get

$$\begin{aligned} s &= \frac{2\pi\left(\frac{pN_s}{120}\right) - \frac{2\pi p}{120}N_{\text{act}}}{\frac{2\pi p N_s}{120}} \\ &= \frac{2\pi f - \left(\frac{p}{2}\right)\frac{2\pi N_{\text{act}}}{60}}{2\pi f} = \frac{\omega_s - \left(\frac{p}{2}\right)\omega_m}{\omega_s} \end{aligned} \tag{8.2}$$

From (8.2), we have

$$\omega_m = \frac{\omega_s(1-s)}{p/2} \text{ mechanical rads/sec} \tag{8.3}$$

Since $\omega_r = \omega_s - \left(\frac{p}{2}\right)\omega_m$, it follows from (8.2) that

$$\omega_r = s\omega_s \tag{8.4}$$

From this, we get $f_r = sf$ where f_r is the frequency of current in the rotor. The quantity *slip* is dimensionless, and it gives an idea of the speed of the machine with respect to the synchronous speed. $s = 0$ corresponds to synchronous speed (a speed at which an induction motor can never run) and $s = 1$ corresponds to *standstill* or *starting conditions*. Generally, s is a very small number between 0.01 to 0.05.

Example 8.1 Consider a three-phase, 60 Hz, six-pole induction machine run at 1125 rpm. Find the slip and the rotor current frequency f_r.

Solution:

$$f = \frac{pN_s}{120}$$

Hence,

$$\begin{aligned} N_s &= \frac{60 \times 120}{6} = 1200 \text{ rpm} \\ s &= \frac{1200 - 1125}{1200} = 0.0625 \\ f_r &= sf \\ &= (0.0625)60 = 3.75 \text{ Hz} \end{aligned}$$

8.3 Analysis of a Two-pole Machine

Just as in the case of the synchronous machine, we first do the analysis for a two-pole machine and then generalize to a p pole machine. The stator has three-phase windings. Unlike in the synchronous machine, there are three windings on the rotor, as shown in Figure 8.3. We will label stator windings

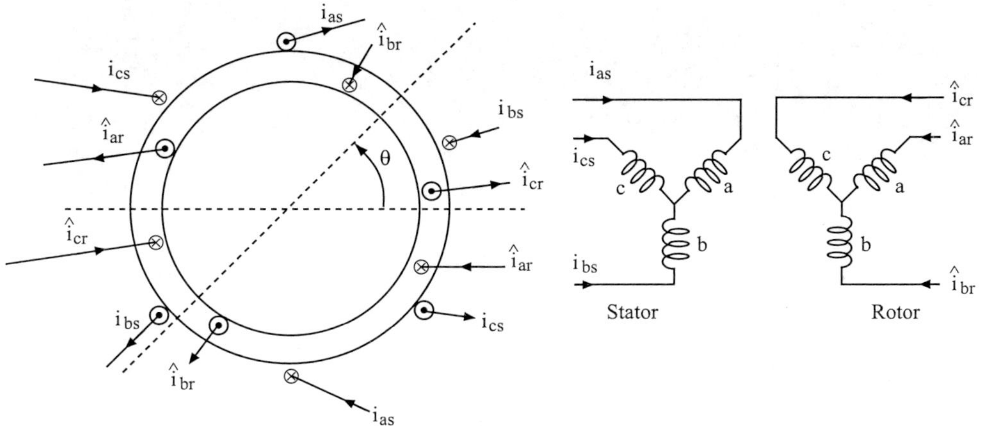

Figure 8.3: A two-pole three-phase induction machine

as s and rotor windings as r. Thus the $\lambda - i$ relations are

$$\begin{bmatrix} \lambda_{as} \\ \lambda_{bs} \\ \lambda_{cs} \\ \lambda_{ar} \\ \lambda_{br} \\ \lambda_{cr} \end{bmatrix} = \mathbf{A} \begin{bmatrix} i_{as} \\ i_{bs} \\ i_{cs} \\ \hat{i}_{ar} \\ \hat{i}_{br} \\ \hat{i}_{cr} \end{bmatrix} \tag{8.5}$$

where the matrix $\mathbf{A}$ is

$$\left[\begin{array}{ccc|ccc} L_s & \frac{-L_{ms}}{2} & \frac{-L_{ms}}{2} & M\cos\theta & M\cos(\theta+120^\circ) & M\cos(\theta-120^\circ) \\ \frac{-L_{ms}}{2} & L_s & \frac{-L_{ms}}{2} & M\cos(\theta-120^\circ) & M\cos\theta & M\cos(\theta+120^\circ) \\ \frac{-L_{ms}}{2} & \frac{-L_{ms}}{2} & L_s & M\cos(\theta+120^\circ) & M\cos(\theta-120^\circ) & M\cos\theta \\ \hline & & & L_r & \frac{-L_{mr}}{2} & \frac{-L_{mr}}{2} \\ & M_r^T & & \frac{-L_{mr}}{2} & L_r & \frac{-L_{mr}}{2} \\ & & & \frac{-L_{mr}}{2} & \frac{-L_{mr}}{2} & L_r \end{array}\right]$$

where M_r is the 3×3 matrix corresponding to mutual inductance between the stator and rotor coils. L_s is the self inductance of the stator coil, being the sum of the magnetizing inductance L_{ms} and the leakage inductance L_{ls}, i.e., $L_s = L_{ms} + L_{\ell s}$. Similarly, L_r is the self inductance of the rotor coil being the sum of magnetizing inductance L_{mr}, and leakage inductance $L_{\ell r}$, i.e., $L_r = L_{mr} + L_{\ell r}$. The formulas for L_{ms}, L_{mr}, and M in terms of the physical parameters of the machine are given by $L_{ms} = \left(\frac{N_s}{2}\right)^2 \frac{\pi\mu_0 r\ell}{g}$, $L_{mr} = \left(\frac{N_r}{2}\right)^2 \frac{\mu_0 \pi r\ell}{g}$, and $M = \frac{N_s N_r}{4}\frac{\pi\mu_0 r\ell}{g}$ where g is the air gap length, r is the radius of the air gap, ℓ is the axial length of the core, and N_s and N_r are the effective turns per phase on the stator and rotor, respectively. Note both for the stator and rotor windings, the current directions are *into* the windings. The rotor currents are denoted as $\hat{i}_{ar}$, $\hat{i}_{br}$, and $\hat{i}_{cr}$ into the coils; in the final equivalent circuit, we need to reverse their directions to conform to the standard equivalent circuit in the literature. The co-energy is given by

$$
\begin{aligned}
W'_m &= \int_0^{i_{as}} \lambda_{as}(i'_{as}, 0, \ldots, \theta)di'_{as} + \int_0^{i_{bs}} \lambda_{bs}(i_{as}, i'_{bs}, 0, \ldots, \theta)di'_{bs} + \ldots \\
&\quad + \int_0^{\hat{i}_{cr}} \lambda_{cr}(i_{as}, i_{bs}, i_{cs}, \hat{i}_{ar}, \hat{i}_{br}, \hat{i}'_{cr}, \theta)d\hat{i}'_{cr} \\
&= \frac{1}{2}L_s i_{as}^2 + \frac{1}{2}L_s i_{bs}^2 + \frac{1}{2}L_s i_{cs}^2 + \frac{1}{2}L_r \hat{i}_{ar}^2 + \frac{1}{2}L_r \hat{i}_{br}^2 + \frac{1}{2}L_r \hat{i}_{cr}^2 \\
&\quad - \frac{L_{ms}}{2} i_{as} i_{bs} - \frac{L_{ms}}{2} i_{bs} i_{cs} - \frac{L_{ms}}{2} i_{cs} i_{as} \\
&\quad - \frac{L_{mr}}{2} \hat{i}_{ar} \hat{i}_{br} - \frac{L_{mr}}{2} \hat{i}_{br} \hat{i}_{cr} - \frac{L_{mr}}{2} \hat{i}_{cr} \hat{i}_{ar} \\
&\quad + i_{as}\hat{i}_{ar} M\cos\theta + i_{bs}\hat{i}_{ar} M\cos(\theta - 120^\circ) + i_{cs}\hat{i}_{ar} M\cos(\theta + 120^\circ) \\
&\quad + i_{as}\hat{i}_{br} M\cos(\theta + 120^\circ) + i_{bs}\hat{i}_{br} M\cos\theta + i_{cs}\hat{i}_{br} M\cos(\theta - 120^\circ) \\
&\quad + i_{as}\hat{i}_{cr} M\cos(\theta - 120^\circ) + i_{bs}\hat{i}_{cr} M\cos(\theta + 120^\circ) + i_{cs}\hat{i}_{cr} M\cos\theta
\end{aligned}
\tag{8.6}
$$

The torque of electric origin is given by

$$\begin{aligned} T^e &= \frac{\partial W'_m}{\partial \theta} \\ &= i_{as}\frac{d}{d\theta}(\hat{\imath}_{ar}M\cos\theta + \hat{\imath}_{br}M\cos(\theta+120°) + \hat{\imath}_{cr}M\cos(\theta-120°)) \\ &\quad + i_{bs}\frac{d}{d\theta}(\hat{\imath}_{ar}M\cos(\theta-120°) + \hat{\imath}_{br}M\cos\theta + \hat{\imath}_{cr}M\cos(\theta+120°)) \\ &\quad + i_{cs}\frac{d}{d\theta}(\hat{\imath}_{ar}M\cos(\theta+120°) + \hat{\imath}_{br}M\cos(\theta-120°) \\ &\quad + \hat{\imath}_{cr}M\cos\theta)) \end{aligned} \tag{8.7}$$

From (8.7), the torque can be written as

$$T^e = i_{as}\frac{d\lambda_{as,r}}{d\theta} + i_{bs}\frac{d\lambda_{bs,r}}{d\theta} + i_{cs}\frac{d\lambda_{cs,r}}{d\theta} \tag{8.8}$$

where $\lambda_{as,r}$ is the flux linkage of phase a with the three rotor windings and similarly for phases b and c, as can be verified from (8.5). Assume a balanced set of rotor currents as

$$\begin{aligned} \hat{\imath}_{ar} &= \hat{I}_{mr}\cos(\omega_r t + \beta) \\ \hat{\imath}_{br} &= \hat{I}_{mr}\cos(\omega_r t + \beta - 120°) \\ \hat{\imath}_{cr} &= \hat{I}_{mr}\cos(\omega_r t + \beta + 120°) \end{aligned} \tag{8.9}$$

where β is an arbitrary angle. It can be verified from (8.5) that

$$\begin{aligned} \lambda_{as,r} &= \hat{\imath}_{ar}M\cos\theta + \hat{\imath}_{br}M\cos(\theta+120°) + \hat{\imath}_{cr}M\cos(\theta-120°) \\ &= \frac{3}{2}M\hat{I}_{mr}\cos(\theta+\omega_r t+\beta) \end{aligned} \tag{8.10}$$

Similarly,

$$\lambda_{bs,r} = \frac{3}{2}M\hat{I}_{mr}\cos(\theta+\omega_r t+\beta-120°) \tag{8.11}$$

$$\lambda_{cs,r} = \frac{3}{2}M\hat{I}_{mr}\cos(\theta+\omega_r t+\beta+120°) \tag{8.12}$$

To prove (8.10), we use the fact that $\cos x\cos y + \cos(x-120°)\cos(y+120°) + \cos(x+120°)\cos(y-120°) = \frac{3}{2}\cos(x+y)$. Let the angle between the rotor and stator axes of phase a be $\theta = \omega_m t + \gamma$ where γ is an arbitrary constant, and let the stator currents be

$$\begin{aligned} i_{as} &= I_{ms}\cos\omega_s t \\ i_{bs} &= I_{ms}\cos(\omega_s t - 120^\circ) \\ i_{cs} &= I_{ms}\cos(\omega_s t + 120^\circ) \end{aligned} \tag{8.13}$$

Substitution of (8.10–8.13) in (8.8) and simplifying yields (assuming $\omega_r = \omega_s - \omega_m$, since it is a two-pole machine)

$$T^e = -\frac{9}{4} I_{ms}\hat{I}_{mr} M \sin(\beta + \gamma) \tag{8.14}$$

The detailed derivation is omitted.

8.3.1 Derivation of an equivalent circuit

A systematic derivation of the per-phase circuit can be done along the following lines:

1. Solve the rotor circuit for the rotor current.
2. Assume a *turns ratio* between stator and rotor windings, i.e., $\frac{N_s}{N_r} = a$.
3. Refer all rotor quantities to the stator side and write the KVL for the stator side in the phasor form.
4. From the phasor equation and the equivalent circuit, identify a term associated with slip and rotor resistance that will be associated with mechanical power delivered to the rotor.

From the electrical side, we will focus on the KVL equations for phases a of the stator and the rotor. They are given by

$$v_{as} = i_{as}R_s + \frac{d\lambda_{as}}{dt} \tag{8.15}$$

and

$$v_{ar} = \hat{i}_{ar}R_r + \frac{d\lambda_{ar}}{dt} \tag{8.16}$$

To develop an equivalent circuit, we need to *refer* all the rotor quantities to the stator side, just as in the case of the transformer in Chapter 4. We will assume that the effective number of turns on the stator is a times the number of turns on the rotor. Thus, (8.16) becomes after multiplying by a

$$av_{ar} = \left(\frac{\hat{i}_{ar}}{a}\right) a^2 R_r + \frac{d}{dt}(a\lambda_{ar}) \tag{8.17}$$

Knowing λ_{ar} from (8.5), we can solve for the rotor current $\hat{i}_{ar}$ from (8.17) and substitute it in (8.15). This is done next.

Solving the rotor circuit

Referring all rotor quantities to the stator side, we get

$$av_{ar} = v'_{ar}, \frac{\hat{i}_{ar}}{a} = \hat{i}'_{ar}, a^2 R_r = R'_r, a^2 L_r = L'_r, a^2 L_{mr} = L'_{mr}$$

It follows that $a^2 L_{\ell r} = L'_{\ell r}$. From the formula in the previous section for L_{ms}, L_{mr}, and M, we can easily verify that $aM = L_{ms} = L'_{mr}$. Thus (8.17) becomes

$$\begin{aligned} v'_{ar} &= \hat{i}'_{ar} R'_r + a\frac{d\lambda_{ar}}{dt} \\ &= \hat{i}'_{ar} R'_r + a\frac{d}{dt}(i_{as} M \cos\theta + i_{bs} M \cos(\theta - 120^\circ) + i_{cs} M \cos(\theta + 120^\circ)) \\ &+ \frac{d}{dt}(L'_r \hat{i}'_{ar} - \frac{L'_{mr}}{2}\hat{i}'_{br} - \frac{L'_{mr}}{2}\hat{i}'_{cr}) \end{aligned} \tag{8.18}$$

In (8.18), we have used the relationship $a^2 L_r = L'_r$, $a^2 L_{mr} = L'_{mr}$. Substituting the expression for current from (8.9) and (8.13) in (8.18) and noting $\theta = \omega_m t + \gamma$, and $\hat{i}'_{ar} = \frac{\hat{i}_{ar}}{a}$, we get

$$\begin{aligned} v'_{ar} &= \hat{i}'_{ar} R'_r + aMI_{ms}[(-\omega_m \sin(\omega_m t + \gamma)\cos\omega_s t - \omega_s \sin\omega_s t \cos(\omega_m t + \gamma)) \\ &+(-\omega_m \sin(\omega_m t + \gamma - 120^\circ)\cos(\omega_s t - 120^\circ) \\ &-\omega_s \sin(\omega_s t - 120^\circ)\cos(\omega_m t + \gamma - 120^\circ)) \\ &+(-\omega_m \sin(\omega_m t + \gamma + 120^\circ)\cos(\omega_s t + 120^\circ) \\ &-\omega_s \sin(\omega_s t + 120^\circ)\cos(\omega_m t + \gamma + 120^\circ))] \\ &-L'_r \omega_r \hat{I}'_{mr} \sin(\omega_r t + \beta) + \frac{L'_{mr}}{2}\omega_r \hat{I}'_{mr} \sin(\omega_r t + \beta - 120^\circ) \\ &+\frac{L'_{mr}}{2}\omega_r \hat{I}'_{mr} \sin(\omega_r t + \beta + 120^\circ) \end{aligned} \tag{8.19}$$

We use the trigonometric identity

$$\sin x \cos y + \sin(x - 120°)\cos(y - 120°) + \sin(x + 120°)\cos(y + 120°)$$
$$= \frac{3}{2}\sin(x - y) \tag{8.20}$$

for simplifying the terms in the brackets [ ] in (8.19). Also add and subtract $\frac{L'_{mr}}{2}\omega_r \hat{I}'_{mr}\sin(\omega_r t + \beta)$ and use the identity $\sin x + \sin(x + 120°) + \sin(x + 120°) = 0$. The result is

$$\begin{aligned} v'_{ar} &= \hat{i}'_{ar}R'_r - aMI_{ms}\frac{3}{2}\left[(\omega_m \sin((\omega_m - \omega_s)t + \gamma) + \omega_s \sin((\omega_s - \omega_m)t - \gamma))\right] \\ &\quad - L'_{rr}\omega_r \hat{I}'_{mr}\sin(\omega_r t + \beta) \\ &= \hat{i}'_{ar}R'_r - aMI_{ms}(\omega_s - \omega_m)\sin((\omega_s - \omega_m)t - \gamma) \\ &\quad - L'_{rr}\omega_r \hat{I}'_{mr}\sin(\omega_r t + \beta) \end{aligned} \tag{8.21}$$

where $L'_{rr} = L'_{\ell r} + \frac{3}{2}L'_{mr}$. Since the rotor winding is short circuited, $v'_{ar} = 0$. Therefore, (8.21) becomes (since $\omega_s - \omega_m = \omega_r$)

$$\hat{i}'_{ar}R'_r + L'_{rr}\frac{d}{dt}(\hat{I}'_{mr}\cos(\omega_r t + \beta)) = \frac{3}{2}aM\omega_r I_{ms}\sin(\omega_r t - \gamma) \tag{8.22}$$

Equation (8.22) represents an $R - L$ circuit of the type $R + L\frac{di}{dt} = v(t)$ where both i and v have the same frequency ω_r, as shown in Figure 8.4. Note from (8.9) that $\hat{I}'_{mr}\cos(\omega_r t + \beta) = \hat{i}'_{ar}$. In the phasor circuit as shown

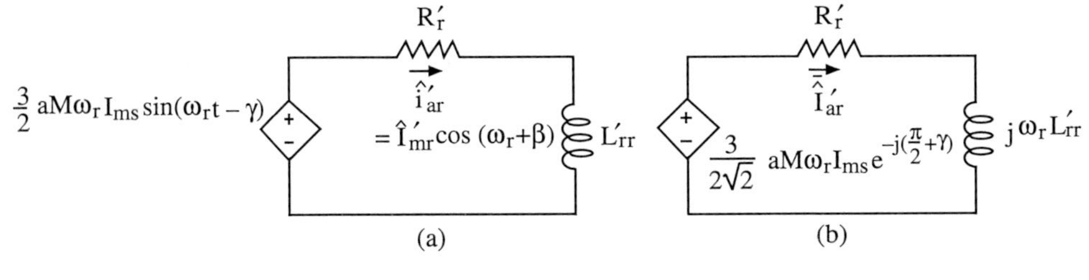

Figure 8.4: Rotor circuit in (a) time domain and (b) phasor domain

in Figure 8.4, the voltage and current magnitudes are rms values. Solving for the steady-state value of $\hat{i}'_{ar}$ using phasor algebra, we get

$$\hat{i}'_{ar(s.s)} = \frac{3}{2}\frac{aM\omega_r I_{ms}\sin(\omega_r t - \gamma - \psi)}{Z}$$

where

$$Z = \sqrt{(R_r')^2 + (\omega_r L_{rr}')^2} \text{ and } \tan\psi = \frac{\omega_r L_{rr}'}{R_r'} \tag{8.23}$$

ψ is the impedance angle of the rotor circuit. $\hat{i}_{ar(s.s)}$ can also be written as

$$\hat{i}'_{ar(s.s)} = \frac{\frac{3}{2}aM\omega_r I_{ms}}{\sqrt{(R_r')^2 + (\omega_r L_{rr}')^2}} \cos(\omega_r t - \frac{\pi}{2} - \gamma - \psi) \tag{8.24}$$

This form is needed to identify β. Therefore, comparing (8.24) and (8.9) referred to stator side, we have

$$\hat{I}'_{mr} = \frac{\frac{3}{2}aM\omega_r I_{ms}}{\sqrt{(R_r')^2 + (\omega_r L_{rr}')^2}} \text{ and } \beta = \frac{-\pi}{2} - \gamma - \psi \tag{8.25}$$

$\hat{i}'_{br}$ and $\hat{i}'_{cr}$ will lag $\hat{i}'_{ar}$ by 120° and 240°, respectively. This completes the rotor circuit computations.

Stator circuit

Let us now compute $v_{as} = \frac{d\lambda_{as}}{dt}$ from (8.15) assuming $R_s = 0$. Substitute for i_{as}, i_{bs}, i_{cs}, $\hat{i}_{ar}$, $\hat{i}_{br}$, and $\hat{i}_{cr}$ from (8.9) and (8.13) in the equation for λ_{as} in (8.5) as follows.

In (8.5), we change $\hat{i}_{ar}$, $\hat{i}_{br}$, and $\hat{i}_{cr}$ to $\hat{i}'_{ar}$, $\hat{i}'_{br}$, and $\hat{i}'_{cr}$ and M to aM in the first three equations. Thus,

$$\begin{aligned}
v_{as} &= L_s\frac{di_{as}}{dt} - \frac{L_{ms}}{2}\frac{di_{bs}}{dt} - \frac{L_{ms}}{2}\frac{di_{cs}}{dt} \\
&\quad + \frac{d}{dt}(\hat{i}'_{ar} aM\cos\theta) + \frac{d}{dt}(aM\cos(\theta + 120°)\hat{i}'_{br}) \\
&\quad + \frac{d}{dt}(aM\cos(\theta - 120°)\hat{i}'_{cr}) \\
&= L_s\frac{d}{dt}(I_{ms}\cos\omega_s t) - \frac{L_{ms}}{2}\frac{d}{dt}(I_{ms}\cos(\omega_s t - 120°)) \\
&\quad - \frac{L_s}{2}\frac{d}{dt}(I_{ms}\cos(\omega_s t + 120°)) \\
&\quad + \frac{d}{dt}[aM\cos(\omega_m t + \gamma)\hat{I}'_{mr}\cos(\omega_r t + \beta) \\
&\quad + aM\cos(\omega_m t + \gamma + 120°)\hat{I}'_{mr}\cos(\omega_r t + \beta - 120°) \\
&\quad + (aM\cos(\omega_m t + \gamma - 120°)\hat{I}'_{mr}\cos(\omega_r t + \beta + 120°)]
\end{aligned} \tag{8.26}$$

For the terms in the brackets [ ], use the identity $\cos x \cos y + \cos(x + 120°)\cos(y - 120°) + \cos(x - 120°)\cos(y + 120°) = \frac{3}{2}\cos(x + y)$. For the first three terms in (8.26), we add and subtract $\frac{L_{ms}}{2}\frac{d}{dt}(I_m \cos\omega_s t)$. The result is

$$\begin{aligned} v_{as} &= L_{ss}\frac{d}{dt}(I_{ms}\cos\omega_s t) + \frac{3}{2}aM\hat{I}'_{mr}\frac{d}{dt}(\cos(\omega_m + \omega_r)t + \gamma + \beta) \\ &= L_{ss}\frac{d}{dt}(I_{ms}\cos\omega_s t) + \frac{3}{2}aM\hat{I}'_{mr}\frac{d}{dt}(\cos\omega_s t + \gamma + \beta) \qquad (8.27) \end{aligned}$$

since $\omega_s = \omega_m + \omega_r$ for a two-pole machine. In (8.27), $L_{ss} = L_s + \frac{L_{ms}}{2} = L_{\ell s} + \frac{3}{2}L_{ms}$. This is the KVL equation for the stator phase a, and (8.22) is the KVL equation for the rotor phase a. Note, however, that each equation has the respective frequencies, namely ω_s for the stator and ω_r for the rotor.

Per-phase equivalent circuit

Our objective is to combine both (8.22) and (8.27) so that we can construct an equivalent circuit for phase a. We do this by transforming (8.27) in phasor form as follows. (Note $I_a = I_{ms}/\sqrt{2}$.)

$$\overline{V}_a = j\omega_s L_{ss}\overline{I}_a + \frac{3}{2}\omega_s aM\frac{\hat{I}'_{mr}}{\sqrt{2}}e^{j(\pi/2+\beta+\gamma)} \qquad (8.28)$$

where

$$\overline{I}_a = \frac{I_{ms}}{\sqrt{2}}\angle 0°$$

Substituting from (8.25) for $\beta = -\frac{\pi}{2} - \gamma - \psi$, (8.28) becomes

$$\overline{V}_a = j\omega_s L_{ss}\overline{I}_a + \frac{3}{2}\omega_s aM\frac{\hat{I}'_{mr}}{\sqrt{2}}e^{-j\psi} \qquad (8.29)$$

Substituting for $\hat{I}'_{mr}$ from (8.25), we get

$$\overline{V}_a = j\omega_s L_{ss}\overline{I}_a + \frac{(\frac{3}{2}\omega_s aM)\frac{3}{2}aM\omega_r\frac{I_{ms}}{\sqrt{2}}e^{-j\psi}}{\sqrt{(R'_r)^2 + (\omega_r L'_{rr})^2}} \qquad (8.30)$$

Now $\frac{I_{ms}}{\sqrt{2}} = I_a = \overline{I}_a\angle 0°$ and for a two-pole machine $\omega_s - \omega_m = \omega_r = s\omega_s$. Substituting for $\omega_r = s\omega_s$ in (8.30) and rearranging terms, we get

$$\overline{V}_a = j\omega_s L_{ss}\overline{I}_a + \frac{(\frac{3}{2}\omega_s aM)^2\overline{I}_a e^{-j\psi}}{\sqrt{(\frac{R'_r}{s})^2 + (\omega_r L'_{rr})^2}} \qquad (8.31)$$

$$\overline{V}_a = j\omega_s L_{ss}\overline{I}_a + \frac{(\frac{3}{2}\omega_s aM)^2\overline{I}_a}{\frac{R'_r}{s} + j\omega_r L'_{rr}} \qquad (8.32)$$

since

$$\tan\psi = \frac{\omega_r L'_{rr}}{R'_r} = \frac{\omega_s L'_{rr}}{(R'_r/s)}$$

We can rearrange (8.32) by adding and subtracting $j\frac{3}{2}\omega_s aM\overline{I}_a$ from the right side. We then get

$$\overline{V}_a = j\omega_s(L_{ss} - \frac{3}{2}aM)\overline{I}_a + \frac{(j\frac{3}{2}\omega_s aM)(-j\frac{3}{2}\omega_s aM)}{(\frac{R'_r}{s} + j\omega_s L'_{rr})}\overline{I}_a + j\frac{3}{2}\omega_s aM\overline{I}_a$$

Rearranging terms,

$$\overline{V}_a = j\omega_s(L_{ss} - \frac{3}{2}aM)\overline{I}_a + j\frac{3}{2}\omega_s aM\overline{I}_a\left(\frac{\frac{R'_r}{s} + j\omega_s(L'_{rr} - \frac{3}{2}aM)}{\frac{R'_r}{s} + j\omega_s L'_{rr}}\right) \quad (8.33)$$

We now recall that

$$\begin{aligned} L_{ss} - \frac{3}{2}aM &= L_{\ell s} + \frac{3}{2}L_{ms} - \frac{3}{2}aM \\ &= L_{\ell s} \end{aligned}$$

since $L_{ms} = aM$. $L_{\ell s}$ is the leakage inductance of the stator winding. Similarly, $L'_{rr} - \frac{3}{2}aM = L'_{\ell r} + \frac{3}{2}L'_{mr} - \frac{3}{2}aM = L'_{\ell r}$ since $L'_{mr} = aM$. Thus $L'_{\ell r}$ is the leakage inductance of the rotor winding referred to the stator side. Equation (8.33) can be written as

$$\overline{V}_a = j\omega_s L_{\ell s}\overline{I}_a + j\frac{3}{2}\left(\frac{\omega_s aM\overline{I}_a\left(\frac{R'_r}{s} + j\omega_s L'_{\ell r}\right)}{\frac{R'_r}{s} + j\omega_s\left(L'_{\ell r} + \frac{3}{2}aM\right)}\right) \quad (8.34)$$

The equivalent circuit is shown in Figure 8.5. Equation (8.34) can be given an equivalent circuit interpretation by recognizing that two impedances $Z_1 = j\frac{3}{2}\omega_s aM$ and $Z_2 = \frac{R'_r}{s} + j\omega_s(L'_{\ell r})$ are in parallel as shown in Figure 8.5.

Thus, (8.34) is equal to $\overline{V}_a = j\omega_s(L_{\ell s})\overline{I}_a + \overline{I}_a\frac{Z_1 Z_2}{Z_1+Z_2}$ where $Z_1 = j\frac{3}{2}\omega_s aM$ and $Z_2 = \frac{R'_r}{s} + j\frac{3}{2}\omega_s L'_{\ell r}$. $\overline{\hat{I}}'_r$ is obtained from Figure 8.5 by current division as

$$\overline{\hat{I}}'_r = \frac{-\overline{I}_a j\frac{3}{2}\omega_s aM}{\frac{R'_r}{s} + j\omega_s L'_{rr}} = \frac{-j\frac{3}{2}\omega_s aM\overline{I}_a}{\sqrt{(\frac{R'_r}{s})^2 + (\omega_s L'_{rr})^2}}e^{-j\psi} \quad (8.35)$$

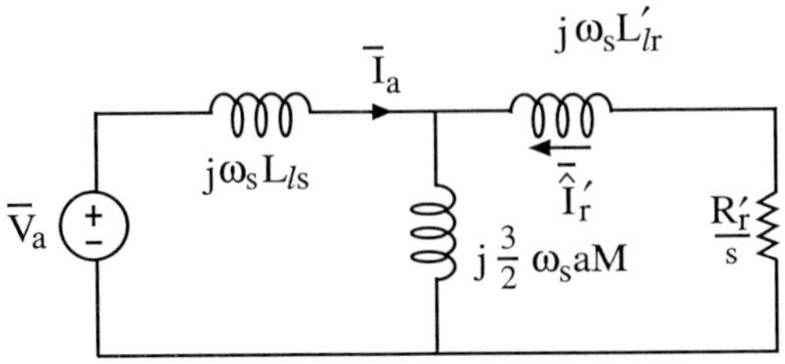

Figure 8.5: Per-phase equation circuit (all impedances referred to stator side)

since $L'_{rr} = L'_{\ell r} + \frac{3}{2}aM$. $\overline{\hat{I}}'_r$ is shown in Figure 8.5. It can be verified from (8.35) that the magnitude of $\overline{\hat{I}}'_r$ is

$$\hat{I}'_r = \frac{\frac{3}{2}\omega_s aMI_a}{\sqrt{(\frac{R'_r}{s})^2 + (\omega_s L'_{rr})^2}} \tag{8.36}$$

This is the rms value of the rotor current. In (8.25), we have the peak value of the rotor current. Noting that $\omega_s = \frac{\omega_r}{s}$ and $I_a = \frac{I_{ms}}{\sqrt{2}}$, (8.25) and (8.36) are consistent.

8.4 Torque-speed Characteristic (Two-pole Machine)

8.4.1 Power relationships

Consider the per-phase equivalent of Figure 8.5. Note that $\frac{R'_r}{s} = R'_r + \frac{R'_r(1-s)}{s}$. Thus the rotor resistance R'_r is separated from $\frac{R'_r(1-s)}{s}$, which involves the mechanical speed through the slip term. We will also denote $\overline{\hat{I}}'_r = -\overline{I}'_r$ to conform to the usual convention in textbooks. This results in the circuit as shown in Figure 8.6(a). If in addition we introduce the ideal transformer with the turns ratio of a, we get Figure 8.6(b), which is analogous to the transformer equivalent circuit of Section 4.3. A further simplification can result by moving the magnetizing inductance aM in Figure 8.6(a) to the left as shown in Figure 8.7. The hysteresis and eddy current losses in the machine can be represented by a shunt resistance R_c. We finally represent the equivalent circuit as in Figure 8.8, where all inductances

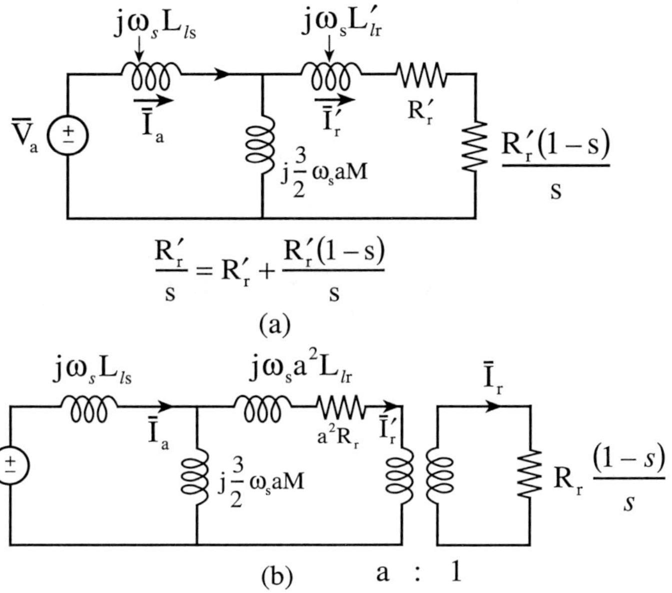

Figure 8.6: Per-phase equivalent circuit (a) without an ideal transformer; (b) with an ideal transformer

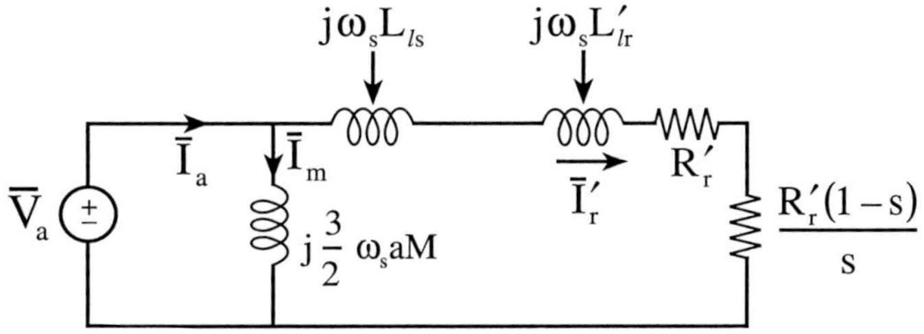

Figure 8.7: Approximate equivalent circuit

are replaced by reactances and a stator resistance R_a is introduced. This is similar to the transformer equivalent circuit in Section 4.3. $x_{\ell s}$ and $x'_{\ell r}$ are the leakage reactances of the stator and the rotor referred to the stator side, respectively. R'_r is the rotor resistance referred to the stator side. From Figure 8.8, the total input power is given by

$$P_T = 3V_a I_a \cos\phi$$

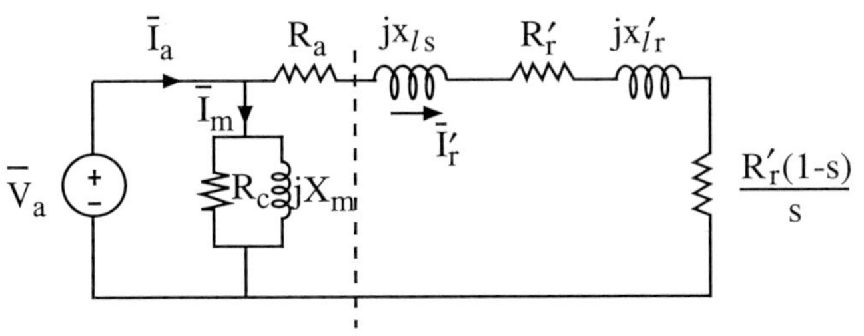

Figure 8.8: Approximate equivalent circuit

where $\cos\phi$ is the power factor between $\overline{V}_a$ and $\overline{I}_a$. P_T is also equal to

$$\begin{aligned} P_T &= 3I_r'^2\frac{R_r'}{s} + 3(I_r')^2R_a + 3V_a^2/R_c & (8.37)\\ &= P_{ag} + P_{scl} + P_c & (8.38)\end{aligned}$$

where

$$\begin{aligned} P_c &= \text{core loss}\\ &= 3V_a^2/R_c\\ P_{scl} &= \text{stator copper loss}\\ &= 3(I_r')^2R_a\\ P_{ag} &= \text{power across the air gap} = 3I_r'^2\frac{R_r'}{s}\end{aligned}$$

P_{ag} is the power transferred across the air gap. $\frac{R_r'}{s}$ is the sum of the two resistances R_r' and $\frac{R_r'(1-s)}{s}$ as shown in Figure 8.8. Thus,

$$P_{ag} = 3I_r'^2(R_r' + \frac{R_r'(1-s)}{s}) \qquad (8.39)$$

This power transformed across the air gap is the sum of the mechanical power P_m developed and the rotor I^2R loss. The rotor copper loss, or I^2R loss in the rotor denoted by P_r, is given by $3I_r'^2R_r'$. $3I_r'^2R_r'\frac{(1-s)}{s}$ is the mechanical power developed P_m. It is easy to verify that

$$P_m = 3I_r'^2\frac{R_r'(1-s)}{s} = P_{ag}(1-s) \qquad (8.40)$$

Also, the rotor copper loss $3I_r'^2R_r'$ can be expressed in terms of P_{ag} as

$$\begin{aligned} P_r &= 3I_r'^2R_r'\\ &= sP_{ag} & (8.41)\end{aligned}$$

Thus, from (8.40) and (8.41), $P_{ag} = P_m + P_r$. We thus have the relations

$$P_{ag} = \frac{3I_r'^2 R_r'}{s}, P_m = \frac{3I_r'^2 R_r'(1-s)}{s}, \text{ and } P_r = 3I_r'^2 R_r' \tag{8.42}$$

Also,

$$P_m = P_{ag}(1-s) \text{ and } P_r = sP_{ag} \tag{8.43}$$

Efficiency

Taking all losses into account, the efficiency is

$$\eta = \frac{P_m}{P_T} = \frac{P_T - (P_{scl} + P_c + P_r)}{P_T} \tag{8.44}$$

If we neglect the stator copper loss P_{scl} and the core losses P_c, then the efficiency of the induction motor is given by

$$\begin{aligned} \eta &= \frac{\text{mechanical power developed}}{\text{input power}} \\ &= \frac{P_m}{P_{ag}} = 1 - s \end{aligned} \tag{8.45}$$

This is a simplified formula since we are neglecting stator copper loss and core loss. Given the actual speed, it gives a quick idea about the efficiency of the motor. If rotational losses P_{rot} are considered, they should be subtracted from P_m to get P_{shaft}, which is the available power at the shaft, i.e., $P_{\text{shaft}} = P_m - P_{\text{rot}}$. The overall efficiency is therefore $\eta = \frac{P_{\text{shaft}}}{P_T}$.

8.4.2 Expression for T^e in terms of V_a (Use approximate equivalent circuit)

We use the approximate per-phase equivalent circuit from Figure 8.8 for calculations. The current $\overline{I}_r'$ is given by

$$\overline{I}_r' = \frac{\overline{V}_a}{\left(R_a + \frac{R_r'}{s}\right) + j(x_{\ell s} + x_{\ell r}')}$$

The magnitude of the current $\overline{I}_r'$ is

$$I_r' = \frac{V_a}{\sqrt{\left(R_a + \frac{R_r'}{s}\right)^2 + \left(x_{\ell s} + x_{\ell r}'\right)^2}} \tag{8.46}$$

From (8.42),

$$\begin{aligned} P_m &= 3(I_r')^2 R_r' \frac{(1-s)}{s} \\ &= \frac{3V_a^2 R_r' \frac{(1-s)}{s}}{\left(R_a + \frac{R_r'}{s}\right)^2 + \left(x_{\ell s} + x_{\ell r}'\right)^2} \end{aligned} \quad (8.47)$$

For a two-pole machine $\omega_m = \omega_s(1-s)$, and hence,

$$P_m = T^e \omega_m = T^e \omega_s (1-s) \quad (8.48)$$

Equating (8.47) and (8.48) and solving for T^e, we get the expression for torque as

$$\begin{aligned} T^e &= \frac{\frac{3V_a^2}{s\omega_s} R_r'}{\left(R_a + \frac{R_r'}{s}\right)^2 + (x_{\ell s} + x_{\ell r}')^2} \\ &= f(V_a, s) \end{aligned} \quad (8.49)$$

This is an important formula. We have now completed the theoretical discussion of the two-pole induction motor.

Example 8.2 A three-phase, 866 V, wye-connected, 60 Hz, two-pole induction motor has $\omega_s L_{\ell s} = 0.5\Omega$, $\frac{3}{2}\omega_s a M = 50\Omega$, $\omega_s L_{\ell r}' = 0.5\Omega$, and $R_r' = 0.1\Omega$. Find the torque at $s = 0.05$ and the three-phase input complex power. Neglect R_a and R_c. Use approximate as well as the exact equivalent circuit.

Approximate equivalent circuit

The per-phase equivalent circuit is shown in Figure 8.9. Voltage per phase

$$\begin{aligned} V_a &= \frac{866}{\sqrt{3}} = 500 \text{ V} \\ \frac{R_r'}{s} &= \frac{0.1}{0.05} = 2\Omega \end{aligned}$$

The formula for T^e in (8.49) is used:

$$\begin{aligned} T^e &= \frac{\frac{3(500)^2(0.1)}{(0.05)(377)}}{(2)^2 + (0.5+0.5)^2} = \frac{3978.98}{5} \\ &= 795.76 \text{ N-m} \end{aligned}$$

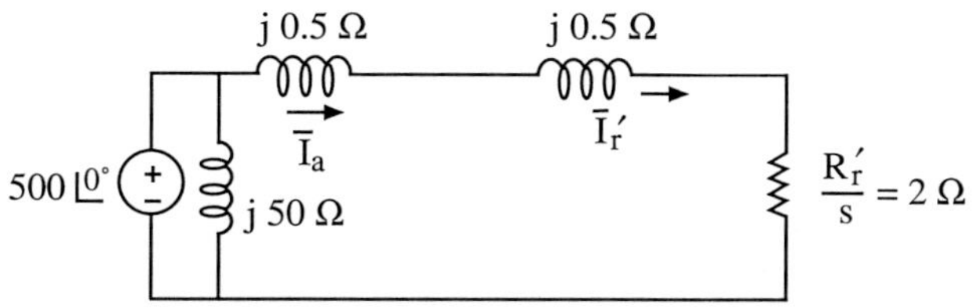

Figure 8.9: Per-phase approximate equivalent circuit

Exact equivalent circuit

If we use the exact equivalent circuit as shown in Figure 8.10(a), we can

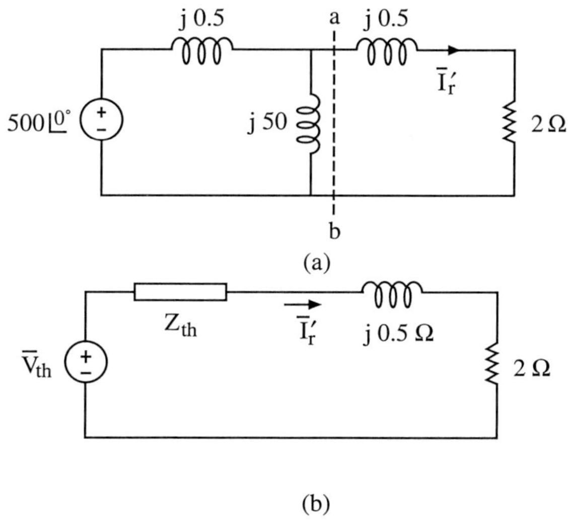

Figure 8.10: Exact equivalent circuit; (a) before simplification; (b) after simplification

use the Thevenin equivalent to the left of the dotted line *ab*. This results in Figure 8.10(b). The Thevenin equivalent to the left of the dotted line is given by

$$\begin{aligned}
\overline{V}_{th} &= \frac{500(j50)}{j(50+0.5)} = 495.05 \text{ V} \\
Z_{th} &= \frac{(j50)(j.5)}{j(50.5)} = j0.495\Omega \\
\overline{I}'_r &= \frac{495.05}{2+j(0.5+0.495)} = 221.61\angle-26.45^\circ A \\
P_{ag} &= 3(I'_r)^2\frac{R'_r}{s} = 3(0.495)^2 2 = 294.68 \text{ kW} \\
P_m &= P_{ag}(1-s) = 279.94 \text{ kW} \\
T^e &= \frac{P_m}{\omega_m} = \frac{P_m}{\omega_s(1-s)} = 781.64 \text{ N-m}
\end{aligned}$$

The error between the approximate and equivalent calculation is about 1.8%. One can also calculate T^e using the formula (8.49) by replacing V_a by V_{th} and $x_{\ell s}$ by $|Z_{th}|$.

Example 8.3 Using the approximate equivalent circuit for Example 8.2 and shown in Figure 8.9, compute $\overline{I}'_r$, P_{ag}, P_m, P_r, and the torque.

Solution:

$$\begin{aligned}
\overline{I}'_r &= \frac{500\angle 0^\circ}{2+j(0.5+0.5)} = \frac{500\angle 0^\circ}{2+j1} = 223.61\angle-26.565^\circ \text{ A} \\
P_{ag} &= 3I'^2_r\frac{R'_r}{s} = 3(223.61)^2(2) = 300 \text{ kW} \\
P_r &= 3I'^2_r R'_r = 3(223.61)^2(0.1) = 15 \text{ kW}
\end{aligned}$$

$$\begin{aligned}
P_m = 3I'^2_r\frac{R'_r(1-s)}{s} &= 3(223.61)^2 0.2(1-0.05) \\
&= 3(223.61)^2 0.2(0.95) \\
&= 285 \text{ kW}
\end{aligned}$$

It is easily verified that $P_{ag} = P_r + P_m$. Now

$$\begin{aligned}
T^e\omega_m &= P_m \\
T^e &= \frac{P_m}{\omega_m} = \frac{P_m}{\omega_s(1-s)} = \frac{285\times 10^3}{377(0.95)} = 795.76 \text{ N-m}
\end{aligned}$$

This agrees with the result of Example 8.2.

8.5 Torque-speed Characteristic (Approximate Equivalent Circuit)

The expression for torque of electric origin T^e has been derived in (8.49) as

$$\begin{aligned} T^e &= \frac{\frac{3V_a^2 R_r'}{s\omega_s}}{\left(R_a + \frac{R_r'}{s}\right)^2 + (x_{\ell s} + x_{\ell r}')^2} \\ &= f(V_a, s) \end{aligned} \tag{8.50}$$

Suppose V_a is constant, then (8.50) can be put in the form

$$T^e = \frac{\beta(\frac{R_r'}{s})}{\propto^2 + \left(R_a + \frac{R_r'}{s}\right)^2} \tag{8.51}$$

where $\propto$ and β are constants. $\beta = \frac{3V_a^2}{\omega_s}$ and $\propto = (x_{\ell s} + x_{\ell r}')$. The shape of T^e as a function of s is given by Figure 8.11. The speed, or equivalently the

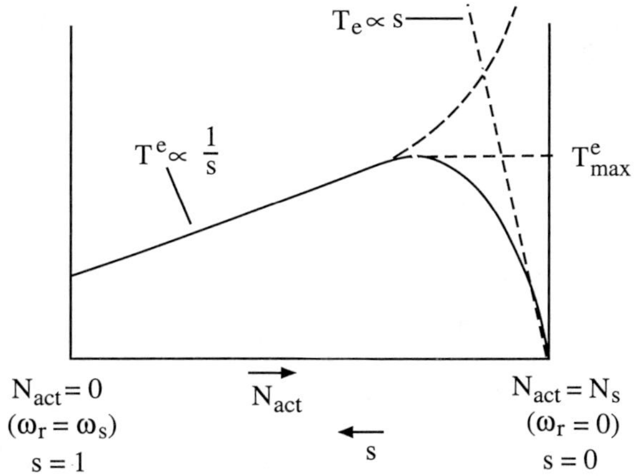

Figure 8.11: Torque-speed characteristic

slip, is on the horizontal axis, and torque is on the vertical axis. This shape of the curve can be justified as follows.

1. Consider *small values* of s and neglect R_a in (8.51). Then

$$(\frac{R_r'}{s})^2 \ll \propto^2 \tag{8.52}$$

Hence,

$$T^e \approx \frac{\beta(\frac{R_r'}{s})}{(\frac{R_r'}{s})^2} \approx \beta \frac{s}{R_r'} \tag{8.53}$$

T^e is linear in s near $s = 0$. This is shown by the dotted line in Figure 8.11.

2. Consider large values of s, with R_a being negligible

$$\propto^2 \gg (\frac{R_r'}{s})^2 \tag{8.54}$$

Then

$$T^e \approx \frac{\beta}{\propto^2} \frac{R_r'}{s} \tag{8.55}$$

Thus, T^e varies inversely with s. This variation is also shown by a dotted line in Figure 8.11.

In between $s = 0$ and $s = 1$, one can extrapolate, and the solid curve in Figure 8.11 is obtained, which shows a typical torque-speed characteristic. Note that since the rotor frequency is $\omega_r = \omega_s - \omega_m = s\omega_s$, for small values of s, i.e., low rotor frequency, the resistance dominates the equivalent circuit. For high values of s, the rotor frequency is high and the inductance dominates.

8.6 Expression for Maximum Torque

From the torque expression (8.51), we can derive the expression for maximum torque as follows:

$$T^e = \frac{\beta\left(\frac{R_r'}{s}\right)}{\propto^2 + (R_a + R_r'/s)^2} = \frac{s\beta R_r'}{s^2 \propto^2 + (sR_a + R_r')^2} \tag{8.56}$$

For T^e to have a maximum value, we must have $\frac{dT^e}{ds} = 0$.

$$\frac{dT^e}{ds} = \frac{\beta R_r'(s^2 \propto^2 + (sR_a + R_r')^2) - s\beta R_r'(\propto^2 2s + 2(sR_a + R_r')R_a)}{(D)^2} \tag{8.57}$$

where $D = s^2 \propto^2 + (sR_a + R_r')^2$. Therefore, the numerator in (8.57) is zero, i.e.,

$$s^2 \propto^2 + (sR_a + R_r')^2 = s(2s \propto^2 + 2R_a(sR_a + R_r'))$$

i.e.,

$$
\begin{aligned}
s^2 \propto^2 &= (sR_a + R_r')(sR_a + R_r' - 2R_a s) \\
&= (R_r' + sR_a)(R_r' - sR_a) \\
&= (R_r')^2 - (sR_a)^2
\end{aligned}
\tag{8.58}
$$

Thus, (8.58) simplifies to

$$
\begin{aligned}
\left(\frac{R_r'}{s}\right)^2 &= R_a^2 + \propto^2 \\
&= R_a^2 + (x_{\ell s} + x_{\ell r}')^2 \\
\frac{R_r'}{s} &= \sqrt{R_a^2 + (x_{\ell s} + x_{\ell r}')^2}
\end{aligned}
\tag{8.59}
$$

Maximum power is thus transferred to the rotor resistance $\frac{R_r'}{s}$ when this resistance is equal to the impedance looking back into the source from the resistance $\frac{R_r'}{s}$. Thus, the slip at which maximum torque occurs is obtained from (7.60) as

$$
s_{mT} = \frac{R_r'}{\sqrt{R_a^2 + (x_{\ell s} + x_{\ell r}')^2}}
\tag{8.60}
$$

When the value of s_{mT} is substituted in (8.56), we get $T_{\max}^e$. This expression is in general analytically quite complicated. In the special case where $R_a = 0$, it turns out that

$$
s_{mT} = \frac{R_r'}{(x_{\ell s} + x_{\ell r}')}
\tag{8.61}
$$

and hence $T_{\max}^e$ becomes

$$
\begin{aligned}
T_{\max}^e &= \frac{\frac{\beta R_r'}{s_{mT}}}{(x_{\ell s} + x_{\ell r}')^2 + \left(\frac{R_r'}{s_{mT}}\right)^2} \\
&= \frac{\frac{3V_a^2}{\omega_s}\frac{R_r'}{s_{mT}}}{2\left(\frac{R_r'}{s_{mT}}\right)^2} \qquad (8.62) \\
&= \frac{3}{2}\frac{V_a^2}{\omega_s} \cdot \frac{s_{mT}}{R_r'} \\
&= \frac{3}{2}\frac{V_a^2}{\omega_s(x_{\ell s} + x_{\ell r}')} \qquad (8.63)
\end{aligned}
$$

Thus, $T^e_{\max}$ is independent of the rotor resistance, but s_{mT} is dependent on the value of R'_r. The torque-speed characteristic is given in Figure 8.12 for

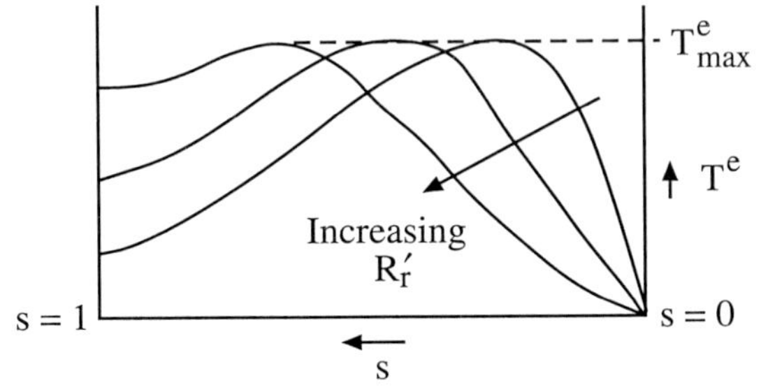

Figure 8.12: Torque-speed characteristics

various values of R'_r. The maximum value of T^e is the same, but it occurs at different values of s depending on R'_r according to (8.60) or (8.61).

Example 8.4 In Example 8.2, find at what value of s maximum torque occurs and compute the value of this torque.

Solution: From (8.61)

$$\begin{aligned} s_{mT} = \frac{R'_r}{(x_{\ell s} + x'_{\ell r})} &= \frac{0.1}{(0.5 + 0.5)} \\ &= 0.1 \end{aligned}$$

From (8.63)

$$\begin{aligned} T^e_{\max} &= \frac{3}{2}\frac{(500)^2}{377(1)^2} \\ &= 994.5 \text{ N-m} \end{aligned}$$

8.7 Multi-pole Induction Machines

How does our analysis affect the treatment of p pole machines? For a p pole machine, we can repeat the analysis with mechanical angle θ substituted by $\left(\frac{p}{2}\right)\theta$. We use the *approximate* equivalent circuit. The per-phase equivalent circuit remains the same. Since $\omega_r = \omega_s - \left(\frac{p}{2}\right)\omega_m$ from (8.1), we have

$s = \frac{\omega_s - (p/2)\omega_m}{\omega_s}$. Hence $\omega_m = \frac{\omega_s(1-s)}{p/2}$. The mechanical power P_m is given by

$$P_m = T^e \omega_m = T^e \frac{\omega_s(1-s)}{(p/2)} \tag{8.64}$$

Using P_m from (8.47) in (8.64), we get

$$\begin{aligned} T^e &= \left(\frac{p}{2}\right) 3V_a^2 \, \frac{R_r'}{s\omega_s} \Big/ \left[\left(R_a + \frac{R_r'}{s} \right)^2 + (x_{\ell s} + x_{\ell r}')^2 \right] \\ &= f(V_a, s) \end{aligned} \tag{8.65}$$

The slip for maximum torque can be calculated from (8.60) as before. The expression s_{mT} for the special case $R_a = 0$ remains the same as in (8.63). The maximum torque is computed as

$$T^e_{\max} = \left(\frac{p}{2}\right) \frac{3}{2} \frac{V_a^2}{\omega_s(x_{\ell s} + x_{\ell r}')} \tag{8.66}$$

Example 8.5 A three-phase, wye-connected, 60 Hz 400 V, four-pole induction motor has the following parameters:

$$\begin{aligned} \frac{3}{2}\omega_s a M &= 20\Omega, \qquad \omega_s L_{\ell s} = 0.5\Omega \\ \omega_s L_{\ell r}' &= 0.2\Omega, \qquad R_r' = 0.1\Omega \end{aligned}$$

Compute (i) torque at a speed of 1755 rpm and (ii) s_{mT} and $T^e_{\max}$. For (i) use the formula for the approximate equivalent circuit, and for (ii) use the Thevenin approach for the exact circuit. Assume $R_a = 0$ and ignore core and stator copper losses.

Solution:

(i) Using the formula and approximate equivalent circuit,

$$V_a = \frac{400}{\sqrt{3}} = 230.9 \text{ V} \quad ; \quad N_s = \frac{120f}{p} = 1800 \text{ rpm}$$

$$s = \frac{1800 - 1755}{1800} = 0.025 \quad ; \quad \frac{R_r'}{s} = \frac{0.1}{0.025} = 4\Omega$$

The equivalent circuit is shown in Figure 8.13(a). In the approximate

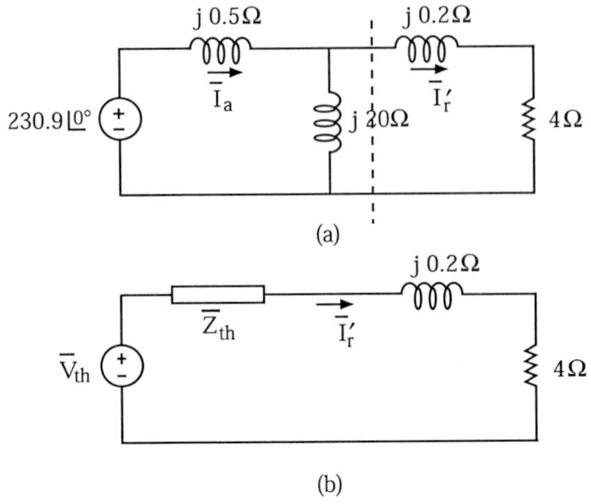

Figure 8.13: Equivalent circuit (a) and Thevenin equivalent (b)

equivalent circuit, we move the magnetizing inductance to the left as in Figure 8.7.

Using the formula (8.65), we get

$$T^e = \frac{\left(\frac{4}{2}\right) 3(230.9)^2 \frac{4}{377}}{(4)^2 + (0.2 + 0.5)^2} = 205.82 \text{ N-m}$$

Using (8.61), we get (since $R_a = 0$)

$$s_{mT} = \frac{0.1}{(0.2 + 0.5)} = 0.14286$$

Using (8.66), we obtain

$$\begin{aligned} T^e_{\max} &= \left(\frac{p}{2}\right) \frac{3}{2} \frac{V_a^2}{\omega_s (x_{\ell s} + x'_{\ell r})} \\ &= 2 \frac{3}{2} \frac{(230.9)^2}{(377)(0.2 + 0.5)} = 606.08 \text{ N-m} \end{aligned}$$

(ii) Using the exact equivalent circuit and Thevenin's theorem.

In the equivalent circuit of Figure 8.13(a), we construct the Thevenin equivalent to the left of the dotted line, as shown in Figure 8.13(b).

$$
\begin{aligned}
\overline{V}_{th} &= \frac{(230.9)(j20)}{(j20.5)} = 225.37\angle 0^\circ \text{ V} \\
\overline{Z}_{th} &= \frac{(j0.5)(j20)}{j20.5} = j0.4878\Omega \\
I'_r &= \frac{225.37}{\sqrt{4^2 + (0.4878 + 0.2)^2}} \text{ A} \\
&= \frac{225.37}{4.059} = 55.528 \text{ A} \\
P_{ag} &= \frac{3I_r'^2 R'_r}{s} \\
&= 3(55.528)^2(4) \text{ W} \\
&= 37 \text{ kW} \\
P_m &= T^e \omega_m = T^e \frac{\omega_s(1-s)}{p/2} \\
(0.975)37 \times 10^3 &= T^e \frac{(377)(0.975)}{2}
\end{aligned}
$$

Thus, $T^e = 196.29$ N-m. For s_{mT}, and $T^e_{\max}$, we use the same formula as in (8.63) and (8.66) but use V_{th} and X_{th} in place of V_a and $x_{\ell s}$, respectively. Thus

$$
\begin{aligned}
s_{mT} &= \frac{0.1}{(0.2 + 4878)} = 0.145 \\
T^e_{\max} &= \left(\frac{p}{2}\right) \frac{3}{2} \frac{V_{th}^2}{\omega_s \left(X_{th} + x'_{\ell r}\right)} \\
&= 2\left(\frac{3}{2}\right) \frac{(225.37)^2}{377(0.4878 + 0.2)} \\
&= 587.64 \text{ N-m}
\end{aligned}
$$

There is a slight discrepancy between the two methods due to the approximation involved.

Example 8.6 A three-phase, 60 Hz, 866 V, six-pole, wye-connected induction motor has the following parameters:

$$x_{\ell s} = 1.5\Omega, x'_{\ell r} = 1.15\Omega, \text{ and } R'_r = 0.6\Omega, X_m = 13.5\Omega$$

Both R_a and R_c are neglected. The motor operates at rated voltage and has torque of electric origin $T^e = 160$ N-m. Use the exact equivalent circuit.

(a) Find slip s, speed N_{act} in rpm, and f_r (the frequency of rotor current).

(b) Find $T^e_{\max}$ and T^e_{starting}.

(c) Repeat (a) and (b) using the approximate equivalent.

Solution: The equivalent circuit is shown in Figure 8.14(a).

(a)

$$\begin{aligned} V_a &= \frac{866}{\sqrt{3}} = 500 \text{ V} \\ \omega_s &= 377 \text{ rad/s}, p = 6 \\ N_s &= \frac{60 \times 120}{6} = 1200 \text{ rpm} \end{aligned}$$

The Thevenin equivalent circuit is constructed to the left of the dotted line in Figure 8.14(a), and shown in Figure 8.14(b).

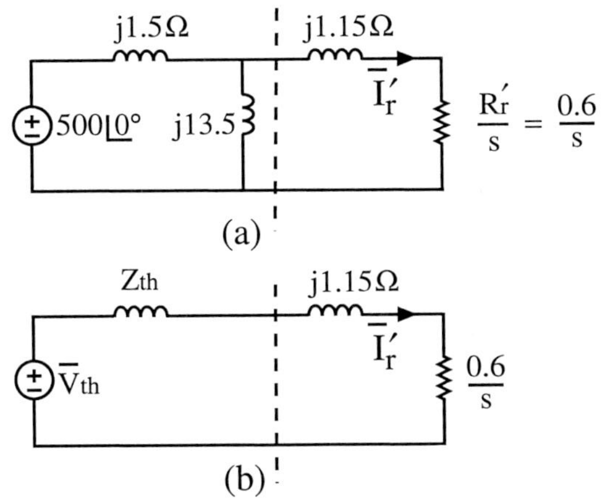

Figure 8.14: Equivalent circuit for Example 8.6

$$
\begin{aligned}
Z_{th} &= \frac{(j1.5)(j13.5)}{j1.5+j13.5} = j1.35\Omega \\
\overline{V}_{th} &= \frac{j13.5}{j13.5+j1.5} 500\angle 0^\circ = 450\angle 0^\circ \text{ V} \\
I_r' &= \frac{450}{\sqrt{(1.35+1.15)^2 + \left(\frac{0.6)}{s}\right)^2}} \text{ A} \\
P_m &= 3\frac{R_r'}{s} I_r'^2 (1-s) = 3\frac{0.6}{s}\left[\frac{450}{\sqrt{(2.5)^2+\left(\frac{0.6}{s}\right)^2}}\right]^2 (1-s) \\
T^e &= \frac{P_m}{\omega_m} = \frac{P_m}{\omega_s(1-s)/(p/2)} = \frac{p}{2}\frac{P_m}{\omega_s(1-s)}
\end{aligned}
$$

Substituting for the various values,

$$
160 = \left(\frac{6}{2}\right) 3 \left(\frac{0.6}{s}\right) \left[\frac{(450)^2}{(2.5)^2+\left(\frac{0.6}{s}\right)^2}\right] \frac{1}{377}
$$

Let

$$
\frac{0.6}{s} = y
$$

Then,

$$
160 = \frac{1822500y}{377y^2 + 2356.25}
$$

Solving

$$
y = 0.2083, 30
$$

This gives

$$
s = 0.02, 2.88
$$

Since $s < 1$, we chose $s = 0.02$.

$$
\begin{aligned}
f_r &= sf_s = 0.02 \times 60 = 1.2 \text{ Hz} \\
N_{\text{act}} &= N_s(1-s) = 1200(1-0.02) = 1176 \text{ rpm}
\end{aligned}
$$

(b)

$$
\begin{aligned}
s_{mT} &= \frac{R_r'}{X_{th} + x_{\ell r}'} = \frac{0.6}{1.35+1.15} = 0.24 \\
T_{\max}^e &= \left(\frac{p}{2}\right)\frac{3}{2}\frac{V_{th}^2}{\omega_s(X_{th}+x_{\ell r}')} = (3)\frac{3}{2}\frac{(450)^2}{(377\times 2.5)} = 966.8 \\
T_{\max}^e &= 966.8 \text{ N-m}
\end{aligned}
$$

The starting torque is obtained by evaluating T^e at $s = 1$.

$$T^e_{\text{start}} = \left(\frac{p}{2}\right) \frac{3V_{th}^2 R_r'}{s\omega_s \left[\left(\frac{R_r'}{s}\right)^2 + \left(X_{th} + x_{\ell r}'\right)^2\right]}\Bigg|_{s=1}$$

$$T^e_{\text{start}} = 3\frac{3 \times (450)^2 \times 0.6}{377[(0.6)^2 + (2.65)^2]} = 438.8 \text{ N-m}$$

(c) The approximate equivalent circuit is shown in Figure 8.15.

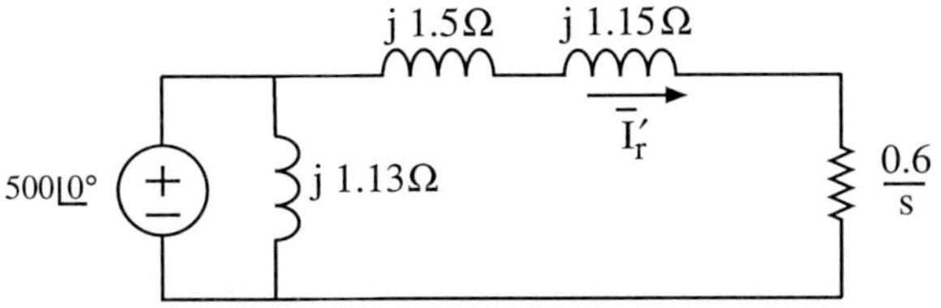

Figure 8.15: Approximate equivalent circuit for Example 8.6

The expression for T^e is given by (8.65) as

$$
\begin{aligned}
T^e &= \frac{p}{2}\frac{3V_a^2\frac{R_r'}{s}}{\omega_s\left[\left(\frac{R_r'}{s}\right)^2+\left(x_{\ell s}+x_{\ell r}'\right)^2\right]} \\
160 &= 3\frac{3\times 500^2\left(\frac{0.6}{s}\right)}{377\left[\left(\frac{0.6}{s}\right)^2+(2.65)^2\right]}
\end{aligned}
$$

Following the same procedure as in part (a), we get

$$
\begin{aligned}
s &= 0.016 \\
f_r &= 0.016\times 60 = 0.97\text{ Hz} \\
N_{\text{act}} &= 1200(1-0.016) = 1180.8\text{ rpm} \\
T_{\max}^e &= \frac{p}{2}\frac{3}{2}\frac{V_a^2}{\omega_s(x_{\ell s}+x_{\ell r}')} = 3\frac{3}{2}\frac{(500)^2}{377\times 2.65} = 1126\text{ N-m} \\
T_{\text{start}}^e &= \frac{p}{2}\cdot\left.\frac{3V_a^2R_r'}{s\omega_s\left[\left(\frac{R_r'}{s}\right)^2+\left(x_{\ell s}+x_{\ell r}'\right)^2\right]}\right|_{s=1} = 3\frac{3(500)^2\times 0.6}{377[(0.6)^2+2.65^2]} \\
&= 485\text{ N-m}
\end{aligned}
$$

Example 8.7 For an induction motor, show that

$$
\frac{T_{\text{start}}^e}{T_{\max}^e} = \frac{2s_{mT}}{1+s_{mT}^2}
$$

Assume $R_a = 0$. Use the formula (8.65) for the approximate equivalent circuit.

$$
\begin{aligned}
T_{\text{start}}^e = T^e|_{s=1} &= \frac{p}{2}\frac{3V_a^2R_r'}{\omega_s\left[(R_a+R_r')^2+\left(x_{\ell s}+x_{\ell r}'\right)^2\right]} \\
T_{\max}^e &= \frac{p}{2}\frac{3V_a^2R_r'}{s_{mT}\omega_s\left[\left(R_a+\frac{R_r'}{s_{mT}}\right)^2+\left(x_{\ell s}+x_{\ell r}'\right)^2\right]} \\
\frac{T_{\text{start}}^e}{T_{\max}^e} &= \frac{s_{mT}\left[\left(R_a+\frac{R_r'}{s_{mT}}\right)^2+(x_{\ell s}+x_{\ell r}')^2\right]}{(R_a+R_r')^2+\left(x_{\ell s}+x_{\ell r}'\right)^2}
\end{aligned}
$$

From (8.60),

$$s_{mT} = \frac{R'_r}{\sqrt{R_a^2 + (x_{\ell s} + x'_{\ell r})^2}}$$

Since

$$R_a = 0, s_{mT} = \frac{R'_r}{(x_{\ell s} + x'_{\ell r})}$$

$$1 + s_{mT}^2 = \frac{R'^2_r + (x_{\ell s} + x'_{\ell r})^2}{(x_{\ell s} + x'_{\ell r})^2}$$

Therefore, simplifying

$$\frac{T^e_{\text{start}}}{T^e_{\max}} = \frac{s_{mT}[2(x_{\ell s} + x'_{\ell r})^2]}{R'^2_r + (x_{\ell s} + x'_{\ell r})^2} = \frac{2s_{mT}}{[R'^2_r + (x_{\ell s} + x'_{\ell r})^2]/(x_{\ell s} + x'_{\ell r})^2}$$

Hence

$$\frac{T^e_{\text{start}}}{T^e_{\max}} = \frac{2s_{mT}}{1 + s_{mT}^2}$$

Example 8.8 The per-phase equivalent circuit of an induction motor is shown in Figure 8.16. Find the Thevenin equivalent at ab, $\overline{I}'_r$, P_{ag}, P_m, and

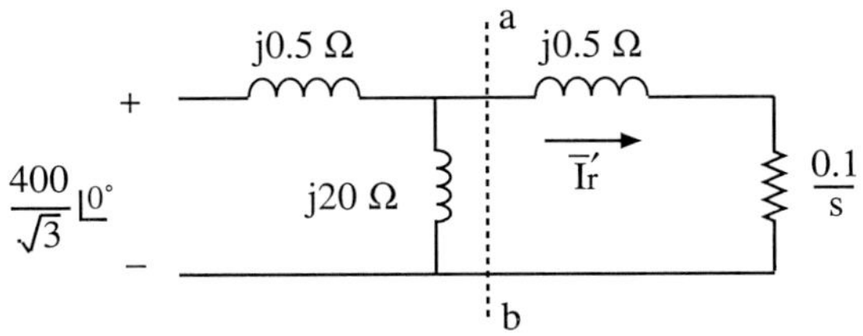

Figure 8.16: Per-phase equivalent circuit for Example 8.8

T^e. It is a three-phase, 400 V, 60 Hz, four-pole motor running at 1755 rpm.

Solution: The computation for the Thevenin equivalent circuit results in

$$Z_{th} = \frac{j20 \times j0.5}{j20 + j0.5} = j0.4878\Omega$$

$$\overline{V}_{th} = \frac{j20}{j20.5}\frac{400}{\sqrt{3}}\angle 0^\circ = 225.3\angle 0^\circ \text{ V}$$

$$N_s = \frac{60 \times 60}{4/2} = 1800 \text{ rpm}$$

The resulting equivalent circuit is shown in Figure 8.17.

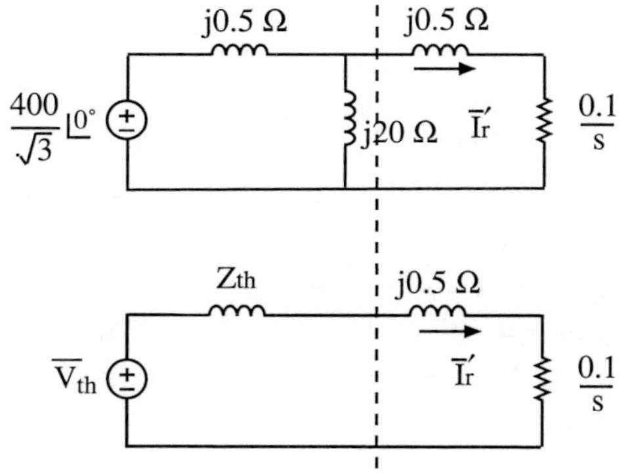

Figure 8.17: Thevenin equivalent for Example 8.8

$$
\begin{aligned}
s &= \frac{N_s - N_{\text{act}}}{N_s} = \frac{1800 - 1755}{1800} = 0.025 \\
\overline{I}_r' &= \frac{225.3\angle 0^\circ}{\frac{0.1}{0.025} + j0.9878} = 54.68\angle -13.87^\circ \text{ A} \\
P_{ag} &= 3I_r'^2 \frac{R_r'}{s} = 3(54.68)^2 \frac{0.1}{0.025} = 35878.8 \text{ W} \\
P_m &= P_{ag}(1-s) = 34981.8 \text{ W} \\
T^e &= \frac{P_m}{\omega_s(1-s)/\frac{p}{2}} = 2\frac{34981.8}{377(1-0.025)} = 190.34 \text{ N-m}
\end{aligned}
$$

8.8 Induction Generators

In the first part of this chapter, the induction machine was analyzed primarily when it was operating as a motor. In the motor case, the primary objective is to use the grid to control the shaft with the following typical applications of shaft loads: fans, pumps, compressors, actuators, position control and vehicle drives. Sometimes induction machines are used for both motoring and generating – two examples are pumped hydro storage plants and vehicle regenerative braking. In the generating case, the primary objective is to use the shaft to control the terminal voltage and current (or real and reactive power). In recent years the wind turbine industry has greatly expanded the use of induction machines as generators. This section presents the additional basic theory and practical application of the induction machine as a generator.

8.8.1 Induction Generator Basics

Fortunately, a large amount of the circuit theory and analysis for induction generator operation is the same as that for motor operation with one fundamental difference. For a squirrel-cage induction motor, the slip was typically a small positive number for stable operation. Similarly, for efficient generator operation, the slip of a squirrel-cage induction machine is typically a small negative number (it must be turned by a prime mover at a speed faster than the stator revolving magnetic field and in the same direction). It will be shown that if the rotor of the induction machine is fed with a separate controllable supply, the machine can have generator or motor operation for positive or negative slip and also perform the control of power flow between two different grids as a device called a variable frequency transformer (VFT) with slip slightly above or below 1.0. These are called doubly-fed induction machines (DFIMs). When used for generation, they are called doubly-fed induction generators (DFIGs).

Recall that if the slip is 1.0, the machine is at standstill and essentially a transformer with the same frequency on both sides. There are three sticky issues that often cause problems for people who think about the derivation of the induction machine equivalent circuit and its use to analyze operation. The first one has to do with the fact that the rotor currents have a frequency (slip times stator frequency) that is different from the stator currents. This makes people question the use of phasors for both the stator and rotor in the same diagram. The second has to do with the fact that the derivation of the equivalent circuit requires the transfer of the rotor side variables to the stator side using an ideal transformer with stator to rotor turns ratio Ns/Nr. In some ways, this can be thought to avoid any problems with the first issue. The third issue has to do with the fact that while the "transferred" rotor current phasor magnitude at stator frequency is the same as the rotor current phasor magnitude at rotor frequency, the phase angles on the two are different. In addition, the fact that when the machine is operated at a speed above stator synchronous speed, the frequency of the rotor currents is negative, creating additional issues.

For example, assume the currents for phase a in steady state are written as follows:

$$i_s = I_{ms}cos(\omega_s t + \alpha)$$
$$i_r = I_{mr}cos(s\omega_s t + \beta)$$

and the machine angle in the earlier diagram of steady state (constant speed) shaft rotation from the stator phase a axis to the rotor phase a-axis in the counter-clockwise direction is:

$$\theta = \omega_m t + \gamma$$

Notice that the following results might be different from what was given earlier in this chapter since that was for a regular squirrel-cage machine which did not have the doubly-fed feature. Note also that the examples earlier in this chapter assumed the stator current phase angle α was zero.

Using the earlier machine without the doubly-fed rotor feature, the phase angle on the slip frequency rotor current was written from the circuit as β which was given to be $-\pi/2-\gamma-\psi$ where γ is the physical angle from the stator magnetic field to the rotor magnetic field at time equal to zero (as defined above), and ψ is the angle on the slip frequency impedance of the rotor circuit (including both leakage and magnetizing inductances – $(tan\psi = \omega_s L_{rr}/(Rr/s))$ – note that the primes are omitted for simplicity). Similarly, the angle on the rotor current referred to the stator from the equivalent circuit (with all quantities at synchronous frequency) has the angle equal to $-\pi/2+0-\psi$, where the zero is the angle on the a-phase stator current assumed in the first part of this chapter. This referred rotor current phase angle preserves conservation of energy and the physical laws of the equivalent circuit. Most of these can be explained by the careful use of the quantity called slip which reflects the importance of the mechanical system in the model, the revolving magnetic fields, and the mechanical rotor angle at time zero (ψ).

From the previously discussed induction machine model:

- The speed of the rotor in rad/sec is ω_m.
- The stator revolving magnetic field speed relative to the stator in rad/sec is ω_s.
- Slip is $s = (\omega_s - \omega_m)/\omega_s$ and $\omega_m = (1-s)\omega_s$.
- The rotor's revolving magnetic field speed, relative to the rotor, is ω_r rad/sec. This makes the rotor revolving magnetic field speed relative

to the stator in rad/sec to be $\omega_r + \omega_m$. If this is to be equal to ω_s to maintain constant torque to the shaft, then $\omega_r = \omega_s - \omega_m = s\omega_s$. This means that if the shaft speed is greater than the synchronous speed, then the slip will be negative which means that the rotor currents have a negative frequency.

- Frequency of the stator currents is $f_s = \omega_s/2\pi$.
- Frequency of the rotor currents is $f_r = \omega_r/2\pi = sf_s$.
- The number of poles per phase on the stator and rotor is P. Remember that in other countries, the symbol P stands for the number of pole pairs per phase.
- The revolutions/minute of the magnetic field produced by the stator currents around the stator coils in the positive (assumed to be counter-clockwise) direction is $N_{ss} = 120f_s/P$.
- The revolutions/minute of the magnetic field produced by the rotor currents around the rotor coils is $N_{rr} = 120f_r/P$ and the revolutions/minute of the rotor is N_m (note that if the frequency is negative, N_{rr} will be negative). In order to have a constant speed with constant torque to the shaft of the machine, the stator and rotor magnetic fields must turn together at the same speed and the same direction relative to the stator so that the speed of the rotor magnetic field with respect to the stator is:

$$N_{rs} = N_m + N_{rr} = N_{ss}$$

These relationships are used in a series of examples.

Example 8.9 Consider a three-phase, 60 Hz (stator current frequency), six-pole squirrel-cage induction machine spinning at 1280 RPM (assumed to be counter-clockwise). Call this case I.

(a) What is the speed of the magnetic field produced by the stator currents (N_{ss}) relative to the stator?

(b) What is the slip and the frequency of the rotor currents?

(c) What is the speed of the magnetic field produced by the rotor currents relative to the rotor?

Solution:

(a) The speed of the magnetic field produced by the stator currents (N_{ss}) relative to the stator is $120 \cdot 60/6 = 1200RPM$. Note that the actual direction of this field will depend on the phase sequence of the stator currents, but we have assumed that to be counter-clockwise.

(b) The slip is equal to (1200–1280)/1200 = −0.0667, and the frequency of the rotor currents is about $-4Hz$. What does this mean? This means that the rotor magnetic field must revolve in the direction opposite to the direction that the rotor is turning and of speed $N_{rr} = 120f_r/P$ (which is negative in this case).

(c) As indicated above, the speed of the magnetic fields produced by the rotor currents relative to the rotor is 120*(-4)/6 = -80 RPM (which would be clockwise and therefore 1200 counter-clockwise relative to the stator to make Nrs = Nss. Clearly the two speeds of revolution of the magnetic fields produced by the stator currents and those produced by the rotor currents are the same (1200 RPM) relative to the stator and in the same direction, thus producing constant torque to the shaft.

At this point, it is prudent to say a little more about negative frequency. First consider the case of some positive frequency. If we use $i(t) = Icos(\omega t)$ (phase shift of zero) into the positive terminal of voltage across an inductor of L Henries, where ω is a positive angular frequency in rad/sec, that voltage by Faraday's law would be

$$V = \omega LI(-sin(\omega t)) = \omega LIcos(\omega t + (\pi/2))$$

which indicates that the angle on the voltage is 90° ahead of the current (current is lagging), which is consistent with having positive reactance (using V=ZI). Complex power into the inductor with phasors V and I is S = VI* (* is for conjugate), and the reactive power into the inductor is positive. This means that the inductor absorbs reactive power. But, what if the frequency ω is a negative number?

For negative frequency, the phase angle of the current is still zero. Looking at the voltage waveform above, it clearly has a negative magnitude and a +90° phase shift. When the minus sign from the magnitude (due to negative ω) is replaced by using the magnitude of the frequency (or $-\omega$ in this case) and subtracting 180 degrees to the phase angle (leaving the ωt alone for comparison with the current), the voltage is:

$$V = -\omega LIcos(\omega t + (\pi/2) - \pi)) = -\omega LIcos(\omega t - (\pi/2))$$

which has a phasor that is $-90°$ and the current phasor is at zero. Clearly for positive magnitudes, the negative frequency phase angle is -90 degrees (current is leading) which is consistent with having a negative reactance (using V=ZI).

Now recall from earlier coursework that voltage and current phasors are defined as rotating vectors and the direction of rotation is counter-clockwise for positive frequency, and clockwise for negative frequency. The above lead/lag reversal is illustrated in the Figure 8.18 below shown with phasors in the complex plane. Since the direction of rotation of the phasor changes, so does the lead/lag concept.

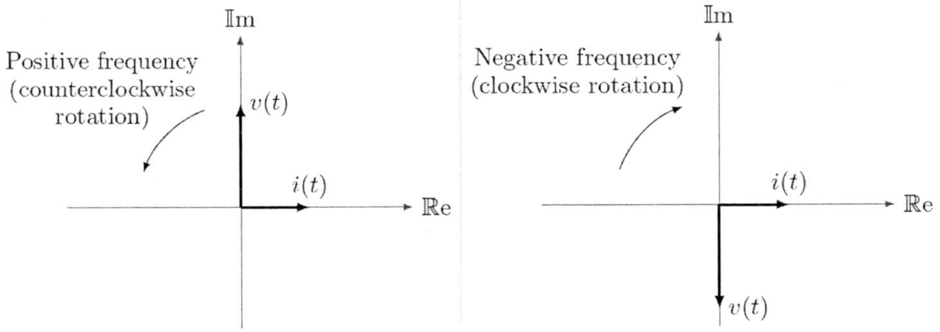

Figure 8.18: Inductors have lagging current for either positive or negative frequency

If we use complex power into the inductor with phasors V (negative frequency) and I (negative frequency), then using S=VI* (* is for conjugate) results in the reactive power into the inductor being negative. This means that the inductor produces reactive power.

We now assume that a simple inductor should absorb reactive power with sinusoidal excitation of positive or negative frequency. We propose that for negative frequency phasors, the complex power into a load should be computed as V*I (* is for conjugate). This would correctly make the inductor with negative frequency absorb reactive power if an "all-negative frequency" circuit was used. In the derivation of the induction machine equivalent circuit, this problem is essentially avoided by "transferring" the rotor variables over to the stator side where the frequency is positive and avoids the use of two different frequencies. However, the "three sticky issues" discussed near the beginning of this section still remain things to be properly explained.

Case I was for an induction machine that is either a squirrel-cage or wound-rotor shorted through an impedance. A very important alternative is to use a wound-rotor machine with a fixed balanced AC supply connected to the stator and another balanced controllable AC supply connected to the rotor through AC/DC and DC/AC converters. Such devices are traditionally called doubly-fed induction machines (DFIMs) with the following structure for a generator (called a DFIG):

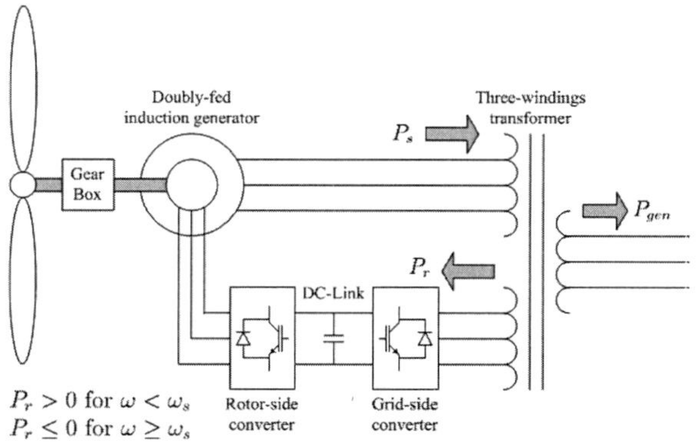

Figure 8.19: Wound rotor doubly-fed induction generator (DFIG) basic configuration (Hector Pulgar Dissertation, 2010 [15])

Note that with this convention, the power to the grid from the DFIG is $\bar{S}_{gen} = P_{gen} + jQ_{gen} = \bar{S}_s - \bar{S}_r. = (P_s - P_r) + j(Q_s - Q_r)$. The per-phase steady-state equivalent circuit of this DFIG is shown in Figure 8.20 (note that the circuit has been flipped from earlier circuits to show the stator on the right and rotor on the left). Note that the rotor source is usually obtained from converters connected to the stator grid (which also feeds the stator). All quantities are referred to the stator – the prime on the rotor parameters and variables indicating that they are referred to the stator has been omitted for convenience. Generator notation is used in this figure (current out of the positive terminal of the stator – into the grid). The rotor circuit uses motor notation (current out of the grid into the converters and then into the positive terminal of the rotor voltage) which it needs to ensure proper conservation of power. The subscript a in this circuit stands for phase

a. Other parameters are the same as defined earlier. The major difference in the doubly-fed machine is the addition of the rotor voltage source which appears in the figure divided by slip. This rotor voltage is typically supplied from the stator grid through AC/DC/AC converters as shown in Fig. 8.19.

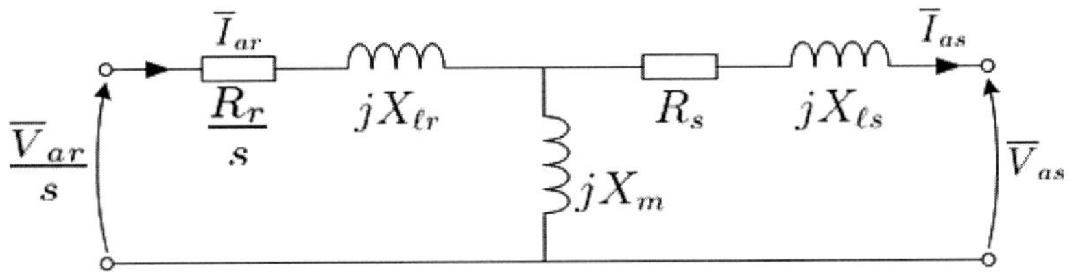

Figure 8.20: Steady-state equivalent circuit of a doubly-fed induction generator (DFIG) (Hector Pulgar Dissertation, 2010 [15])

To refresh everyone's memory, the a-phase rotor voltage and current (from the converters) shown here at synchronous frequency have root-mean-square (RMS) magnitudes equal to the actual a-phase slip-frequency rotor voltage RMS magnitude (multiplied times the N_s/N_r turns ratio) and the actual a-phase slip-frequency rotor current RMS magnitude (multiplied times N_r/N_s turns ratio) and phase angles equal to the actual a-phase slip-frequency rotor voltage phase angle plus the rotor position angle at time zero and the actual a-phase slip-frequency rotor current phase angle plus the rotor position angle at time zero.

One very important difference between this DFIM and the traditional squirrel-cage machine is the torque vs. speed characteristic. In the squirrel-cage machine the torque to the shaft is positive for all speeds less than synchronous and negative for all speeds greater than synchronous. In the DFIG case, for any rotor speed (or slip), the powers from the rotor to the stator are determined by the rotor voltage $\bar{V}_{ar}$ and are used to control the real and reactive power into the energy supply grid. To see how this works, consider Example 8.10 below.

Example 8.10 Consider a three-phase, 60 Hz (stator current frequency), six-pole DFIG spinning at 1280 RPM (assumed to be counter-clockwise). Call this case II.

The stator voltage is 1.0pu angle zero and for convenience with this example we assume the controls are such that P_s =1.0pu and Q_s = 0.0pu. The unknown rotor voltage (referred to the stator) will be V_r (dropping the *a* subscript) angle δ. From the previous example (with the same conditions) we see that slip is -0.0667. The equivalent circuit parameters are $R_r =$ 0.005pu , $X_{lr} = 0.16$pu, $X_m = 3.0$pu, $R_s = 0.007$pu, $X_{ls} = 0.17$pu.

(a) What are the values of V_r and δ for this machine?

(b) How much power is coming into the machine rotor from the converters?

(c) How much real power is coming from the shaft of the machine?

(d) How much reactive power is lost in the machine?

Solution:

(a) From the circuit above, the complex power out of the stator is equal to 1 angle zero in per unit and the stator voltage is equal to 1 angle zero in per unit, so the stator current (out of the stator) is equal to 1 angle zero in per unit. The midpoint voltage (top +) is the stator current times the stator impedance plus the stator voltage. The rotor current (from the converter source) is then equal to the stator current plus the midpoint voltage divided by the shunt magnetizing impedance (midpoint current down). The rotor voltage is then equal to the rotor current times the rotor resistance plus slip times the rotor leakage impedance plus the midpoint voltage times slip. In this case those calculations give $V_r =$ 0.0698pu and $\delta =$ -2.7862 rad (-159.64 degrees).

(b) Using the results above, the power into the rotor from the rotor source is equal to the rotor voltage times the conjugate of the rotor current which in this example gives $P_{r,in} =$ -0.061pu and $Q_{r,in} =$ -0.0476pu.

(c) To see how much power is coming from the shaft of the machine, it necessary to examine two basic equations.

From conservation of power, P_m from the shaft must be equal to $P_{lost,total} + P_{s,out} - P_{r,in}$.

From the circuit, $P_{s,out} - P_{r,in}/s = -P_{lost,total} - I_r I_r R_r (1-s)/s$.

This gives Pm from the shaft $= P_{r,in}/s + P_{r,in} - I_r I_r R_r (1-s)/s$ which is the same as

$$P_m = (P_{r,in} - I_r I_r R_r)(1-s)/s = P_{s,out} - P_{r,in} + P_{lost,total}.$$

We first compute the total real power lost in the machine as the stator resistance times the stator current magnitude squared which is 0.007pu plus the rotor resistance times the rotor current magnitude squared which is 0.0061pu. These total to 0.0131pu. Using this last equation above, the real power coming from the shaft (Pm) is equal to the stator real power out of the machine (into the grid) which is 1.0pu minus the rotor real power out of the machine (into the grid) which is 0.061pu plus the real power losses which are 0.0131. This gives 1.0741pu.

(d) The reactive power lost in the machine is the sum of the three reactive powers consumed by the three reactances. In this case that is 0.7143pu. Since the stator does not provide any vars in this case, the vars coming from the rotor voltage divided by slip is 0.7143pu which means that the vars coming into the rotor from the converters are -0.0476pu. This is bothersome at first because the stator does not supply any vars to the machine and the rotor converter supplies negative vars to the machine (sends VARs into the AC grid)! This is because a portion of the reactive power from the converter sources into the rotor is due to the negative slip. In this case that number is about 0.76pu. This number can also be computed by taking the reactive power from the converters (-0.0476) and multiplying it time $(1 - s)/s$ which in this case is about 0.76pu. An argument for why this occurs could be that it is necessary to ensure conservation of reactive power and preserve the circuit consumptions. In the circuit, this reactive power comes from the reactive part of the voltage source named V_r/s–V_r which gives $Vr \cdot (1 - s)s$ inside the dotted box labeled "Mechanical power" in the alternative equivalent circuit shown in Figure 8.21 below. The other dotted box represents the converters which are connected to the stator grid.

There are typically two control systems that drive the converter connected to the rotor – active power and reactive power. The control of reactive power can be thought of as voltage control. There are other control mechanisms to allow variable pitch control and to limit turbine speed. A gear mechanism is also common to facilitate optimal rotor speeds. As is the case for induction motors, single phase units are rarely used for large bulk power operation. The majority of wind turbines are 3-phase machines.

It is appropriate to discuss the prime mover for these generators. As the name implies, the wind is the source of energy and it is used to turn a

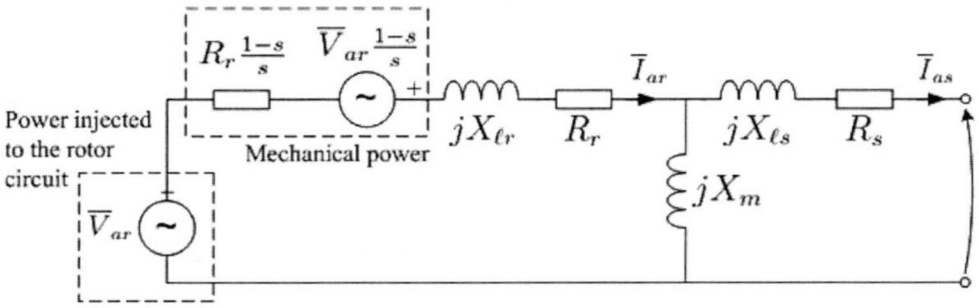

Figure 8.21: Alternative steady-state equivalent model of the doubly-fed induction machine (Hector Pulgar Dissertation, 2010 [15])

turbine which provides the torque for the generator. A fraction of the wind kinetic energy is captured by controlling the pitch on the turbine blades.

In the DFIG wind turbine case II, quantities such as maximum torque are computed in a normal way. In this case, the rotor is connected through a bi-directional back-to-back voltage source converter to the stator. The converter uses power electronics to create rotor currents of any frequency and magnitude desired. A schematic of this operation is shown in the figure below. Several examples show that the real power flow from the converter into the rotor (Pr) is positive for motors operating at speeds above synchronous speed (super-synchronous) and for generators operating at speeds below synchronous speed (sub-synchronous). Similarly, the real power flow from the converter into the rotor (P_r) is negative for motors operating below synchronous speed (sub-synchronous) and generators operating above synchronous speed (super-synchronous). It is also interesting to note that the DFIM can be a motor with either positive slip or negative slip and a generator with either a positive or negative slip.

Including these features, another basic circuit for the DFIG connected machine is given in Figure 8.22. Typically, below the wind speed rating of the turbine, the generator torque control is active and the blade pitch is held at a constant angle for maximum power and above the wind speed rating of the machine, the generator torque is typically held constant while the blade pitch is adjusted accordingly. Albert Betz of Germany showed that there is an upper limit of about 60% on how much of the wind kinetic energy can be captured and turned into electricity [16]. In reality, only about 40% or so can be captured with typical turbines. This is similar to the upper limit on efficiency of thermal power plants and gasoline engines.

In addition to wind turbines, there is an interesting application of DFIGs

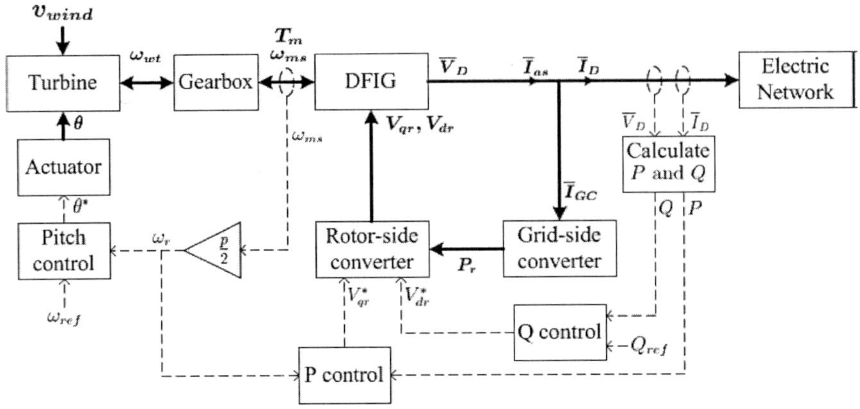

Figure 8.22: Type C wind turbine scheme with controls (Hector Pulgar Dissertation, 2010 [15])

where the wound rotor machine can be used to exchange energy between two electric grids that might not be exactly the same frequency. Such devices are traditionally called variable frequency transformers (VFTs). For example, consider the United States Eastern Interconnect (with frequency exactly 60 Hz) being connected to the stator and the Canadian province of Quebec being connected to the rotor (with frequency exactly 59.4 Hz). For a 60 Hz system with 4 poles, this corresponds to 1782 RPM. If these two systems were directly interconnected, this difference in frequency could cause a separation by protective devices.

Example 8.11 Consider a three-phase, 60 Hz (stator current frequency), four-pole induction machine spinning at a speed corresponding to the difference between the two frequencies (18 RPM) in the same direction as the magnetic field produced by the stator currents (assumed to be counter-clockwise). Call this case III.

(a) What is the speed of the magnetic field produced by the stator currents (N_{ss}) relative to the stator?

(b) What is the speed of the magnetic field produced by the rotor currents relative to the rotor (N_{rr})?

(c) In this case, is the real power P flowing from the stator to the rotor or from the rotor to the stator?

Solution:

(a) The speed of the magnetic field produced by the stator currents relative to the stator (Nss) is 120*60/4 = 1800 RPM (counter-clockwise).

(b) The speed of the magnetic field produced by the rotor currents relative to the rotor (Nrr) is 120*(59.4)/4 = 1782 RPM (counter-clockwise) and 1800 RPM relative to the stator (Nrs).

Clearly the two speeds of revolution of the magnetic fields produced by the stator currents and those produced by the rotor currents are the same (1800 RPM) relative to the stator, thus producing constant torque.

(c) Actually, it is not possible to precisely determine the answer to this question from the given information. Only one thing is known for sure – the shaft drive motor needed to increase the rotor speed from zero to 18 RPM counter-clockwise in order to have both magnetic fields turn together. This is because the magnetic field produced by the rotor currents had a counter-clockwise speed of only 1782. It would be necessary to increase the rotor speed to make the two fields turn together. The control system for the rotor converter voltage can then make the real and reactive power to be sent from the stator to the rotor or the rotor to the stator. The fact that the rotor circuit has the lower frequency is an indication that it would probably be good to send real power from the stator circuit to the rotor circuit.

Now, consider the reverse situation where the rotor network has a frequency greater than the stator network by the same amount (0.6 Hz).

Example 8.12 Consider the same three-phase, 60 Hz (stator current frequency), four-pole induction machine spinning at a speed corresponding to the difference between the two frequencies (18 RPM) in the direction opposite to the magnetic field produced by the stator currents (assumed to be counter-clockwise). Call this case IV.

(a) What is the speed of the magnetic field produced by the stator currents relative to the stator (N_{ss})?

(b) What is the speed of the magnetic field produced by the rotor currents relative to the rotor (N_{rr})?

(c) In this case, is the real power P flowing from the stator to the rotor or from the rotor to the stator?

Solution:

(a) The speed of the magnetic field produced by the stator currents (N_{ss}) relative to the stator is $120 \cdot 60/4 = 1800RPM$ counter-clockwise.

(b) The speed of the magnetic field produced by the rotor currents relative to the rotor (N_{rr}) is $120 \cdot (60.6)/4 = 1818RPM$ (in the counter-clockwise direction) and $1800RPM$ relative to the stator (N_{rs}).

Clearly the two speeds of revolution of the magnetic fields produced by the stator currents and those produced by the rotor currents are the same (1800 RPM) relative to the stator, thus producing constant torque.

(c) Again, it is not possible to precisely determine the answer to this question from the given information. Only one thing is known for sure – the shaft drive motor needed to decrease the rotor speed from zero to -18 RPM (counterclockwise) in order to have both magnetic fields turn together. This is because the magnetic field produced by the rotor currents had a counter-clockwise speed of 1818. It would be necessary to decrease the rotor speed to make the two fields turn together.

The control system for the rotor converter voltage can then make the real and reactive power to be sent from the stator to the rotor or the rotor to the stator. The fact that the rotor circuit has the higher frequency is an indication that it would probably be good to send real power from the rotor circuit to the stator circuit.

The unique feature of these VFT devices is that they can transfer power between grids of two different frequencies. One possible scenario is to determine the frequency of both systems and set the real power transfer between them to zero. From this, determine the slip between the two systems and use a VFT simulation to determine the rotor voltages needed in the VFT to obtain that slip. Then, simply use a synchronizing device between the rotor of the VFT and the network to be connected to the rotor set to the rotor side frequency. Similarly, use a synchronizing device between the stator of the VFT and the network to be connected to the stator set to the stator side frequency. Once both sides are synchronized, the controls can be set to transfer power in either direction.

8.9 Conclusion

In this chapter, we have discussed the induction machine principles, derivation of the equivalent circuit, and the analysis of the machine. Since the equivalent circuit is similar to a transformer, the parameters can also be estimated using two tests similar to the open-circuit test and short-circuit tests. These are called the no-load test and the blocked-rotor test. These are not discussed here.

8.10 Problems

8.1 A three-phase, six-pole, 60 Hz, induction motor has an input of 240 kW. It runs at a slip of 0.05. Find the

(a) rotor electrical frequency,

(b) rotor copper loss,

(c) mechanical speed in rpm, and

(d) mechanical output in kW.

8.2 The rotor copper loss of a three-phase 60 Hz, four-pole induction motor is 6 kW and the rotor frequency $f_r = 3$ Hz. Find P_{ag}, P_m, and N_{act}. (Answer: 120 kW; 114 kW; 1710 rpm).

8.3 A three-phase synchronous generator driven at 8000 rpm supplies an induction motor at 400 Hz. The induction motor has eight poles and runs at 5800 rpm. Find (i) the number of poles on the generator, (ii) slip of induction motor, and (iii) frequency f_r of rotor current in the induction motor. (Answer: number of poles = 6, $s = 0.0333$, and $f_r = 13.333$ Hz).

8.4 A three-phase, four-pole, 30 hp, 220 V (line-line), 60 Hz, wye-connected induction motor draws a current of 77 A from the line source at a PF of 0.88 lagging. The slip is $s = 0.05$. At this operating point, we have

$$P_{\text{scl}} = 1033 \text{ W and } P_c = 485 \text{ W}$$

Find:

(a) the power across air gap P_{ag}.

(b) T^e torque of electric origin. (Answer: 128.9 N-m)

(c) P_m mechanical power delivered to shaft. (Answer: 23087 W)

(d) actual horsepower developed and the overall efficiency if rotational losses are 540 W. (Answer: $\eta = 0.87$)

8.5 A three-phase, 440 V (line-line), 60 Hz, wye-connected eight-pole, 100 hp induction motor has the following parameters expressed per phase.

$$R_a = 0.06\Omega, x_{\ell s} = x'_{\ell r} = 0.26\Omega$$
$$R'_r = 0.048\Omega, R_c = 107.5\Omega$$
$$X_m = 8.47\Omega$$

The rotational losses are 1600 W. Use the approximate equivalent circuit for $s = 0.03$. Find:

(a) $\overline{I}'_r$, $\overline{I}_a$. (Answer: $\overline{I}_a = 160.5\angle - 27.53°$),

(b) input PF,

(c) efficiency. (Answer: 90.49%)

8.6 A three-phase, four-pole, 480 V (line-line) 50 hp, 60 Hz, wye-connected induction motor has the following parameters per phase.

$$R_a = 0.1\Omega \; x_{\ell s} = 0.35\Omega$$
$$x'_{\ell r} = 0.4\Omega \; R'_r = 0.12\Omega$$

At no load (i.e., $s = 0$), the motor draws a current of 19.64 at a PF of 0.089 lagging. For $s = 2.5\%$, find input line current $\overline{I}_a$ and PF, T^e, P_m, and efficiency assuming rotational losses as 950 W. (Answer: $\overline{I}_a = 63.6\angle - 26.2°$, $T^e = 239.32$ N-m, $P_m = 43982.2$ W)

8.7 For Problem 8.6, neglecting R_a, find s_{mT}, $T^e_{\max}$.

8.8 A three-phase, wye-connected, two-pole, 60 Hz, 110 V induction motor has the following parameters: $L_{\ell s} = 2.1$ mH, $L'_{\ell r} = 2.1$ mH, $M = 40$ mH, and $R'_r = 0.4\Omega$. Assume $a = 1$. It runs at 3000 rpm. Find (i) frequency in Hz of current in the rotor, (ii) torque, and (iii) slip for maximum torque $T^e_{\max}$ and value of $T^e_{\max}$. (Answer: $f_r = 10.0$ Hz; $T^e = 8.81$ N-m; $s_{mT} = 0.258$; and $T^e_{\max} = 9.665$ N-m)

8.9 A three-phase, wye-connected, four-pole, 60 Hz, 693 V induction motor has $L_{\ell s} = L'_{\ell r} = 7.96$ mH, $M = 26.52$ mH, and $R'_r = 0.275\Omega$. Assume

$a = 1$. Find (i) s_{mT}, (ii) $T^e_{\max}$, (iii) speed at maximum torque, and (iv) the new value of R'_r so that $T^e_{\max}$ occurs at 1620 rpm. (Answer: $s_{mT} = 0.5$; $T^e_{\max} = 160.75$; $N_{\text{act}} = 1710$ rpm; $R'_{r(\text{new})} = 0.55\Omega$)

8.10 Suppose in Problem 8.8, it is connected for four-poles and $R'_r = 2.4\Omega$ and the motor runs at 1000 rpm. Other parameters remain the same. Find T^e, s_{mT}, and $T^e_{\max}$. (Answer: $s_{mT} = 0.84$; $T^e_{\max} = 9.88$ N-m; $T^e = 7.94$ N-m)

8.11 A three-phase, 60 Hz, 866 V (line-line), six-pole induction motor is operating at rated voltage at a speed of 1170 rpm. The per-phase equivalent circuit is shown in Figure 8.23.

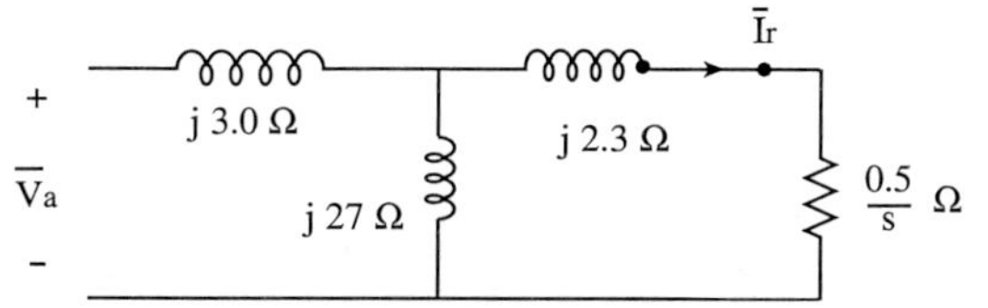

Figure 8.23: Equivalent circuit for Problem 8.11

(a) Find P_{ag}, P_m, and the torque in N-m. (Answer: $T^e = 227.5$ N-m)

(b) Find $T^e_{\max}$ and the slip at which maximum torque occurs. (Answer: $T^e_{\max} = 483.4$ N-m)

8.12 The per-phase equivalent circuit of a three-phase, four-pole 60 Hz induction motor is given in Figure 8.24.

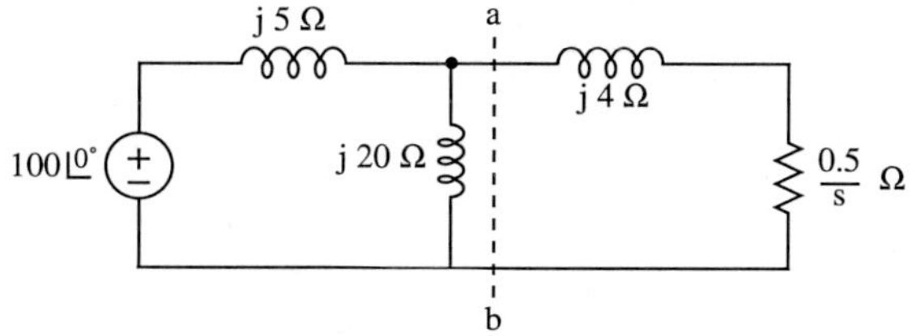

Figure 8.24: Equivalent circuit for Problem 8.12

(a) Find $T^e(s)$ using the Thevenin equivalent to left of the dotted line. (Answer: $\frac{50.93s}{64s^2+0.25}$)

(b) Find $T^e_{\max}$, s_{mT}, and T^e_{starting}. (Answer: $T^e_{\max} = 6.366$ N-m)

8.13 A three-phase, 60 Hz, 866 V, six-pole, wye-connected induction motor has the following parameters:

$$x_{\ell s} = 1.5\Omega, x'_{\ell r} = 1.25\Omega, \text{ and } R'_r = 0.6\Omega, X_m = 15\Omega$$

Both R_a and R_c are neglected. The motor operates at rated voltage and has torque of electric origin $T^e = 200$ N-m. Using the approximate equivalent circuit, find

(a) slip s, speed N_{act} in rpm, and f_r (the frequency of rotor current)

(b) $T^e_{\max}$ and T^e_{starting}.

8.14 The *per-phase* equivalent circuit of an induction motor is shown in Figure 8.25. It is a three-phase, 400 V, 60 Hz, four-pole motor running

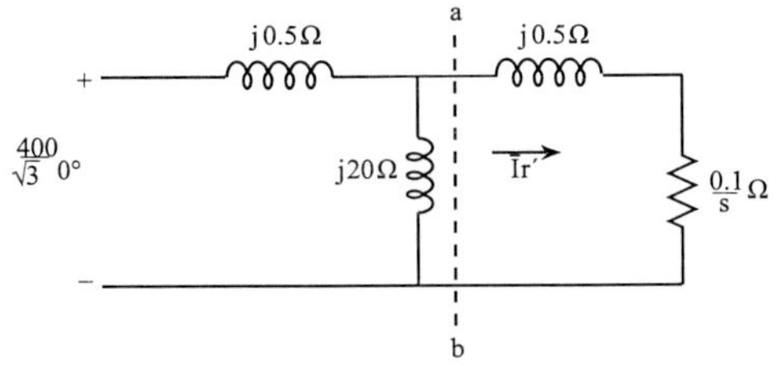

Figure 8.25: Equivalent circuit for Example 8.14

at 1755 rpm.

Find the Thevenin equivalent at ab, $\overline{I}'_r$, P_{ag}, P_m, and T^e. (Partial answer: $P_m = 35$ kW; $T^e = 190.47$ N-m)

8.15 The per-phase equivalent circuit of a three-phase, four-pole, wye-connected, 60 Hz induction motor is shown in Figure 8.26.

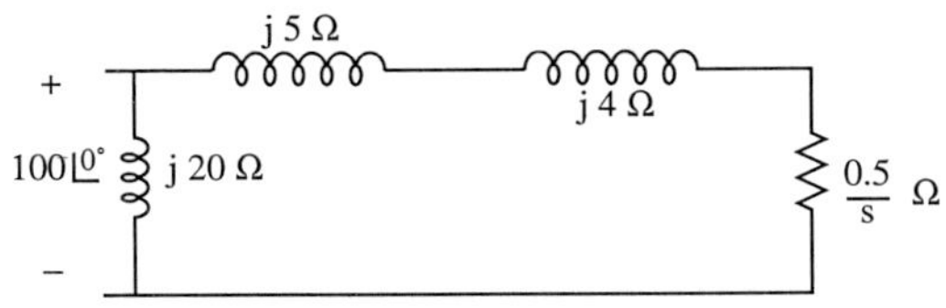

Figure 8.26: Equivalent circuit for Example 8.15

(a) Find s_{mT} and $T^e_{\max}$.

(b) Find T^e_{starting}.

(c) Compare results with Problem 8.12.

8.16 A three-phase, six-pole, 440 V (line-line), 60 Hz, wye-connected induction motor has $L_{\ell s} = L'_{\ell r} = 20$ mH. $aM = 20$ mH. Find:

(a) R' such that $s_{mT} = 0.1$.

(b) T^e_{starting}.

8.17 Parameters of a three-phase, wye-connected, four-pole, 60 Hz, 693 V (line-line) induction motor are $x_{\ell s} = 3\Omega$, $x'_{\ell r} = 3\Omega$, $X_m = 15\Omega$, and $R'_r = 0.275\Omega$. Use the exact equivalent circuit.

(a) Find s_{mT}, $T^e_{\max}$ and the speed at maximum torque.

(b) If $T^e_{\max}$ were to occur at 1620 rpm, what should the value of R'_r be? (Partial answer: $T^e_{\max} = 160.75$ N-m; $R'_{r(\text{new})} = 0.55\Omega$)

8.18 A three-phase, six-pole, 440 V (line-line), 60 Hz, wye-connected motor has $x_{\ell s} = x'_{\ell r} = 12\Omega$, $X_m = 40\Omega$. Using approximate equivalent circuit, find

(a) R'_r such that $s_{mT} = 0.1$.

(b) $T^e(s)$ and T^e_{starting}.

8.19 A three-phase, wye-connected, 380 V, 100 hp, six-pole induction motor is operating at *rated* voltage. The equivalent circuit parameters are

$$x_{\ell s} = 0.25\Omega, X_m = 1.0\Omega, x'_{\ell r} = 0.2\Omega, \text{ and } R'_r = 0.1\Omega$$

(a) Find the slip when the motor delivers its rated mechanical power assuming no losses. (Hint: Substitute $z = R'_r/s$, so that $(1 - s)R'_r/s = z - R'_r$). (Partial answer: $s = 0.106$)

(b) If now the mechanical losses are 1600 W, calculate the torque of the load. (Answer: 433.2 N-m)

(c) What is the efficiency of the motor? (Answer: 87.5%)

8.20 A three-phase, 440 V, 60 Hz, wye-connected, four-pole induction motor operating at *rated* voltage has the following parameters:

$$x_{\ell s} = 0.6\Omega, X_m = 3.4\Omega, x'_{\ell r} = 0.49\Omega, \text{ and } R'_r = 0.2\Omega$$

(a) Calculate s_{mT}, maximum torque, and starting torque. (Partial answer: $s_{mT} = 0.2$)

(b) Draw the torque-slip characteristic of the motor.

(c) The motor drives a mechanical load with the following torque characteristic:

$$T_m = k\omega_m^2$$

where ω_m is the speed of the motor in mechanical radians power per second and $k = 3.4 \times 10^{-3}$ N-m/(r/s)^2. Find the value of the slip at the equilibrium point where $T^e = T_m$. (Partial answer: $s \simeq 0.031$)

8.21 A three-phase, two-pole, 866 V, wye-connected, 60 Hz induction motor has per-phase stator leakage reactance $x_{\ell s} = 0.5\Omega$, rotor leakage reactance $x'_{\ell r} = 0.5\Omega$, and magnetizing reactance $X_m = 5\Omega$. $R'_r = 0.2\Omega$. For a slip of $s = 0.08$ and rated voltage, find rotor current $\overline{I}_r$ and T^e. (Answer: 170 A; 575 N-m)

8.22 A three-phase, 60 Hz, 866 V, wye-connected, six-pole induction motor is operating at *rated* voltage at a speed of 1170 rpm. The motor parameters are

$$x_{\ell s} = 3\Omega, x'_{\ell r} = 2.3\Omega, X_m = 27\Omega, \text{ and } R'_r = 0.5\Omega$$

(a) Find the slip and the frequency of rotor currents.

(b) Calculate the power P_g consumed by the motor, the mechanical power P_m, the rotor copper losses P_r, and the torque of electrical origin T^e.

(c) Find the starting torque, maximum torque, and the slip at maximum torque.

8.23 A three-phase, 60 Hz, 1 kW, 200 V, six-pole induction motor is absorbing its rated power running at a 3% slip.

(a) Calculate rotor copper losses and the mechanical power,

(b) Calculate speed and torque of electrical origin,

(c) Under constant terminal voltage and constant load torque, the rotor resistance is doubled. Find slip, speed, mechanical power, rotor losses, and absorbed active power.

(d) Find the frequency of rotor currents for cases (b) and (c).

(Partial answer: $N_{\text{act}} = 1164$ rpm; $T^e = 7.97$ N-m; $P_{ag} = 1000$ W)

8.24 A three-phase, 60 Hz, 25 kW, 866 V, four-pole, wye-connected, wound rotor induction motor has the following parameters

$$x_{\ell s} = 2\Omega, x'_{\ell r} = 2.2\Omega, X_m = 18\Omega, \text{ and } R'_r = 0.5\Omega$$

(a) The motor is absorbing its rated power at rated voltage. Find slip, speed, and mechanical power developed. ($P_m = 24,475$ W)

(b) Find the maximum torque developed by the motor. Determine the resistance that has to be added to each phase of the rotor winding so that the starting torque becomes equal to the maximum torque. (Additional resist $= 3.5\Omega$, $T^e_{\max} = 402.85$ N-m)

(c) If the mechanical losses are 500W, find the useful mechanical load power and torque and the overall efficiency η of the motor. ($\eta = 0.959$)

8.25 Figure 8.27 gives the per-phase equivalent of a two-pole, 60 Hz, 110 V

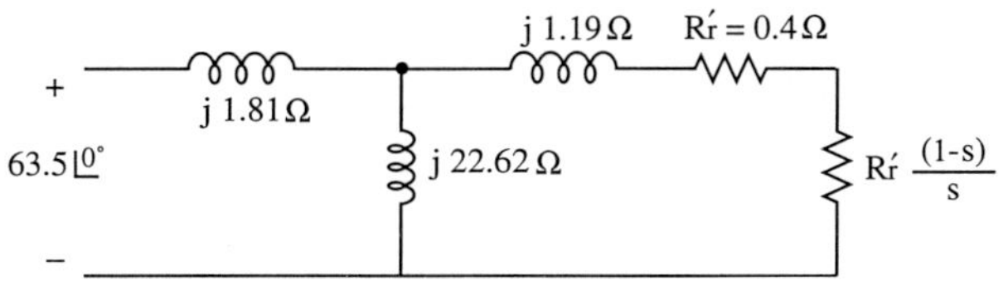

Figure 8.27: Equivalent circuit for Example 8.25

(line-line) induction motor. The motor runs at 1600 rpm. Find (i) f_r in Hz, (ii) T^e, (iii) s_{mT}, and (iv) $T^e_{\max}$.

Chapter 9

DC MACHINES

9.1 Introduction

The DC machine is a versatile device having superior torque-speed characteristics. The speed control is one of its positive features, although these days the induction machine can be made to have similar characteristics with suitable front-end power electronics. Both the stator (field winding) and the rotor (armature winding) have DC currents at the terminals. Compared to AC machines, DC machines are costly. For lower ratings, the field windings on the stator can be a permanent magnet.

9.2 Features of a DC Machine

A DC machine has a single concentrated winding on the stator which is usually excited by a DC current. Alternatively, the stator field can have a DC field using permanent magnets. The rotor currents are supplied through brushes and a commutator. The commutator is designed so that the rotor and stator fields are perpendicular to each other. This serves two functions: the torque is maximized for a given current level and no complicated control electronics is needed to operate the machine. An important feature of DC machines is that they can run well at any speed up to some maximum rated value. Because of the commutator, they are larger and more expensive than equivalently sized induction or synchronous machines. For variable speed operation, this is offset by a much simpler excitation structure. While we use the term DC machines, there is a class of small commutator machines, called universal motors, which are operated on AC voltages at power system frequencies (50 or 60 Hz).

9.3 Principles of Operation

In order to build up an understanding of how DC machines operate, we will start with the simplest possible commutator machine. Next, we will discuss a more realistic machine and develop its equivalent circuit. Finally, we will discuss how the machine operates with various winding connections.

The simplest DC machine is shown in Figure 9.1(a).

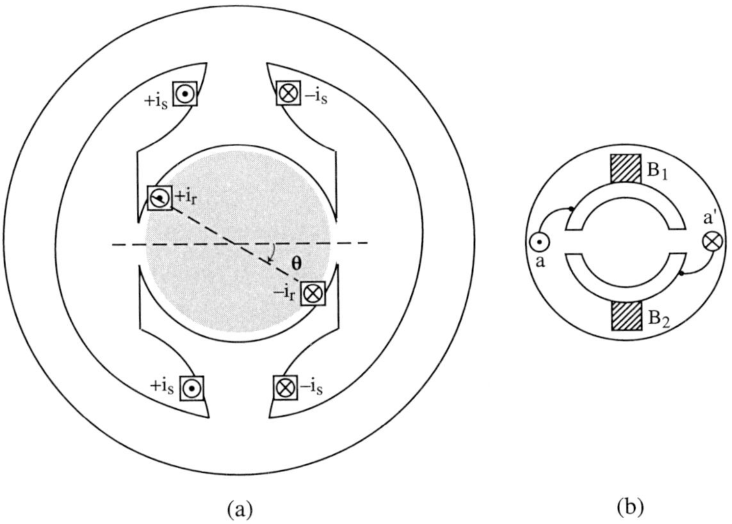

Figure 9.1: DC machine with only one rotor winding

The flux linkage equations for the two windings are

$$\lambda_r(i_r, i_s, \theta) = L_{sr}(\theta)i_s + L_r i_r \tag{9.1}$$

$$\lambda_s(i_r, i_s, \theta) = L_s i_s + L_{sr}(\theta)i_r \tag{9.2}$$

Because of the symmetry of the rotor, the self inductance of the stator coil is not a function of angle. The rotor self inductance is a function of θ. However, it is approximated as being independent of the position. This approximation is good if the reluctance of the iron is negligible compared to that of the air gap. The mutual inductance between the rotor and the stator coils is shown in Figure 9.2. The rotor coil is assumed to be confined to a single slot in the rotor to give this result.

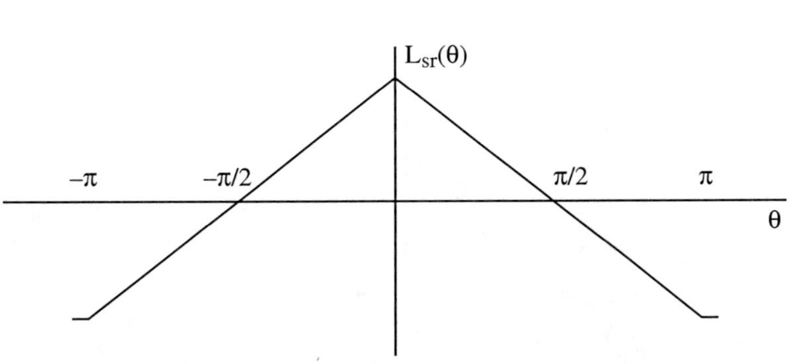

Figure 9.2: Mutual inductance as a function of angle

The calculation of the torque of electric origin is done using the co-energy. The co-energy is given by

$$W'_m(i_r, i_s, \theta) = \frac{1}{2}L_s i_s^2 + \frac{1}{2}L_r i_r^2 + L_{sr}(\theta) i_r i_s \tag{9.3}$$

The torque of electric origin is

$$T^e(i_r, i_s, \theta) = \frac{\partial W'_m(i_r, i_s, \theta)}{\partial \theta} = i_r i_s \frac{dL_{sr}(\theta)}{d\theta} \tag{9.4}$$

The DC machine is excited by DC currents on both the stator and rotor. However, the rotor winding is connected through a commutator segment as shown in Figure 9.1(b), so that

$$i_r = \begin{cases} I_r; & -\pi < \theta \leq 0 \\ -I_r; & 0 < \theta \leq \pi \end{cases} \tag{9.5}$$

I_r is the external DC current, whereas i_r is the current in the winding that is alternating. If the stator current is set equal to I_s, then the torque is only a function of angle and is

$$T(\theta) = \begin{cases} I_r I_s \frac{dL_{sr}(\theta)}{d\theta}; & -\pi < \theta \leq 0 \\ -I_r I_s \frac{dL_{sr}(\theta)}{d\theta}; & 0 < \theta \leq \pi \end{cases} \tag{9.6}$$

The torque as a function of angle is as shown in Figure 9.3. The torque is almost a constant except when the rotor and stator coils are almost lined up. The "dead band" around 0 and π is a function of the angle subtended by the stator salient pole. In this case, we have assumed a small dead band.

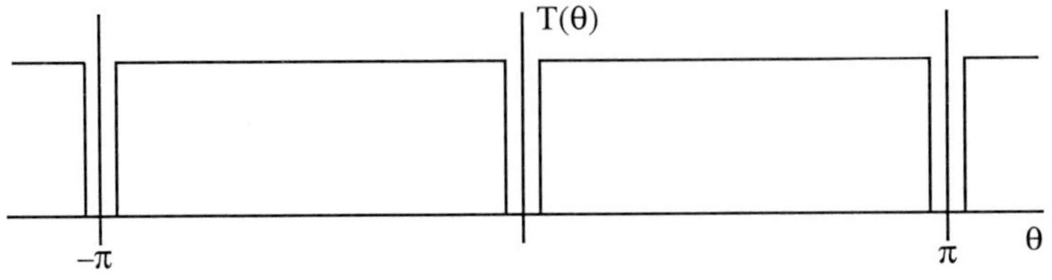

Figure 9.3: The torque as a function of angle for a single rotor coil DC machine

Clearly this machine will operate at any speed since the average torque will always be non-zero whenever the currents are not zero. To operate the machine in reverse, one only needs to reverse the current in one of the coils. In many smaller DC machines, the stator coil is replaced by permanent magnets.

The voltage induced in the rotor winding with DC excitation is

$$v_r(t) = I_r R_r + I_s \left| \frac{dL_{sr}(\theta)}{d\theta} \right| \frac{d\theta}{dt} \tag{9.7}$$

The second term is the time rate of change of flux in the rotor coil. The absolute value is used in the equation, since the terminals are switched every time the motor rotates through 180 degrees.

9.4 A More-realistic DC Machine

An actual DC machine operates with many commutator segments and rotor coils, as shown in Figure 9.4. The currents in the coils are indicated by the dots and crosses in the figure. The rotor coils (also called the armature coils) are supplied with current through two brushes and a commutator. The commutator acts to produce an armature magnetic axis that is always (approximately) perpendicular to the stator (also called field) magnetic axis. The way this is done is shown in Figure 9.5. In Figure 9.5, the machine is unrolled into a linear geometry for the purposes of illustration.

In the figure, the peak of the armature flux density is displaced 90 degrees from the peak of the field flux density. The bottom of the figure shows how the armature currents are supplied by the commutator. All the currents on one side of the brush are in one direction, and all those on the other side are in the opposite direction. The brushes are stationary and the commutator

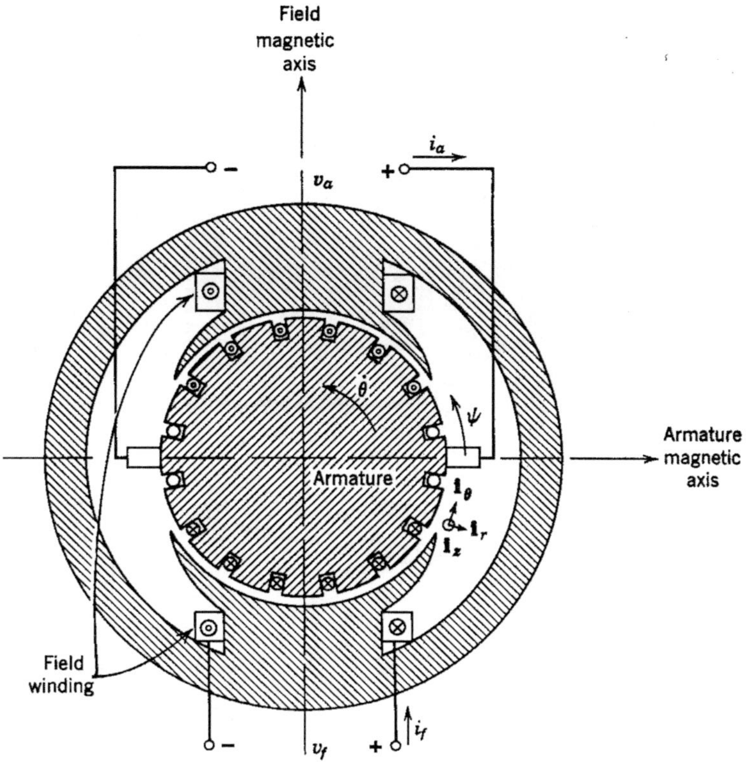

Figure 9.4: Two-pole commutator machine [14]

segments rotate with the armature. Both the brushes and the commutator segments are conducting. As the rotor rotates, the current in a winding segment goes to zero and then reverses sign. The reversal occurs twice every revolution so that the current in each winding segment is at the frequency of the mechanical rotation. The net effect of this is that the currents in the windings adjacent to the upper field pole are out of the paper, and those adjacent to the lower field pole are into the paper. Therefore, the torque on the machine is not a function of armature position but only proportional to the product of field and armature currents. This makes the DC machine very useful if variable speed operation is desired. This gives

$$T^e = G i_a i_f \tag{9.8}$$

where G is a constant, i_a is the armature (rotor) current, and i_f is the field (stator) current. It is easy to verify from (9.8) that the units of G

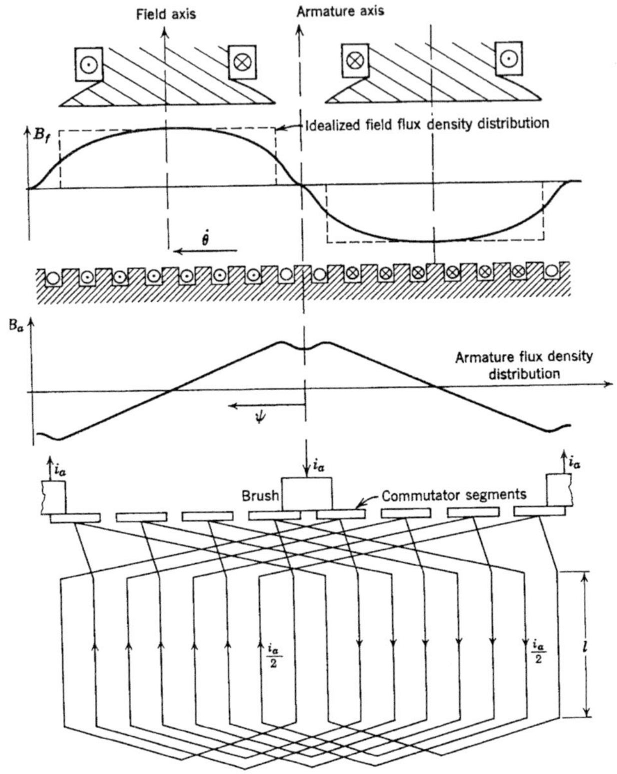

Figure 9.5: Developed views of a two-pole commutator machine showing flux distributions and armature connection [14]

are Henrys.

Equivalent circuit

The equivalent circuit for the DC motor is shown in Figure 9.6. In the figure, ω_m is the rotational speed of the motor in mechanical rad/sec.

From the equivalent circuits, we can see that the dynamic equations governing the machine are

$$v_f = i_f R_f + L_f \frac{di_f}{dt} \tag{9.9}$$

$$v_a = i_a R_a + L_a \frac{di_a}{dt} + G\omega_m i_f \tag{9.10}$$

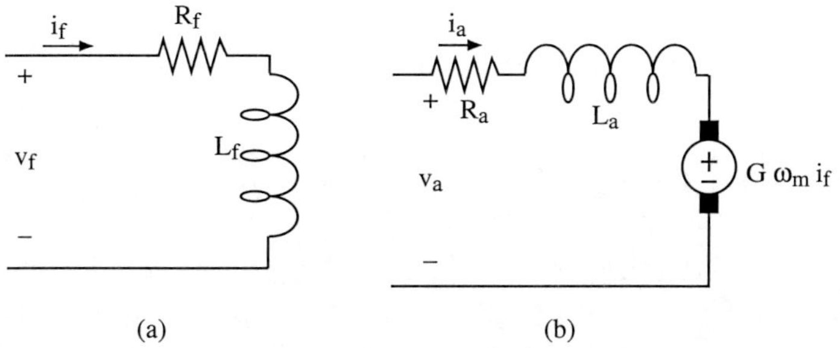

Figure 9.6: (a) Field-coil equivalent circuit, (b) armature-equivalent circuit

The G in (9.10) is the same as the G in the torque expression (9.8). Note that the elementary machine considered in Section 9.3 had the same basic behavior as the realistic machine considered in this section. One question one might ask is, "Why is there not a term in the field equations that depends on the armature current?" The answer is that the armature flux is kept approximately constant by the commutator. The field coil sees only small perturbations due to the changes in the armature field, and these perturbations are at the rotation frequency times the number of commutator segments. These perturbations have a negligible effect on the field current.

If we consider constant voltage-constant speed operation and steady state conditions, we get

$$\text{Armature power } P_a = V_a I_a = I_a^2 R_a + G\omega_m I_a I_f \tag{9.11}$$

$$\text{Field power } P_f = V_f I_f = I_f^2 R_f \tag{9.12}$$

$$\text{Mechanical power } P_m = T^e \omega_m = G I_a I_f \omega_m \tag{9.13}$$

It can also be easily verified from (9.13) that G has units of Henrys. These equations show that the mechanical power comes from the armature terminals and not the field terminals. This allows the replacement of the field coils with a permanent magnet, as is often done on small machines. Contrast this to the synchronous machine where the power flows in through the stator windings, and the rotor winding can be replaced by a permanent magnet.

Equations (9.8), (9.9), and (9.10) govern the dynamics of the machine. Compared to those for other machines, the equations are quite simple, and the DC machine lends itself to a variety of applications where the speed needs to be varied. With two separate windings, the machine can be connected in a variety of ways for various applications. The three main ways are called the *separate excitation*, *series excitation*, and the *shunt excitation*. We will now consider each of these connections and the resulting machine performance.

9.5 Separately Excited Machines

In separately excited machines, the field and armature windings are connected to separate power sources. The field winding is almost always connected to a constant voltage source, resulting in a constant field current (and constant magnetic field). The speed and torque of the machine are then controlled by the armature current. In this section, we therefore assume that the field current is equal to I_f, a constant. This is the same situation that exists if the field windings are replaced by permanent magnets, as is often done with small motors. The dynamical equations with constant field current are then given by

$$J\frac{d\omega_m}{dt} + B\omega_m = GI_f i_a - T_{load} \tag{9.14}$$

where J is the moment of inertia, B is the friction coefficient, and T_{load} is the load torque. The armature electrical equation is given by

$$v_a = i_a R_a + L_a\frac{di_a}{dt} + G\omega_m I_f \tag{9.15}$$

One important property of these equations are that they are linear, so that a linear control system can be used for speed or torque control. For example, a proportional integral derivative (PID) controller to control the speed might have an armature voltage given by

$$v_a(t) = \propto (\omega_o - \omega_m) + \beta\frac{d\omega_m}{dt} + \gamma\int(\omega_o - \omega_m)dt \tag{9.16}$$

where ω_o is the desired speed.

If we look for steady-state operation with constant armature voltage V_a, we got from (9.15)

$$I_a = \frac{V_a - G\omega_m I_f}{R_a} \tag{9.17}$$

$$P_a = V_a I_a = V_a\frac{(V_a - G\omega_m I_f)}{R_a} \tag{9.18}$$

The mechanical output is

$$T^e = GI_f \frac{(V_a - G\omega_m I_f)}{R_a} \tag{9.19}$$

$$P_m = T^e \omega_m = GI_f \omega_m \frac{(V_a - G\omega_m I_f)}{R_a} \tag{9.20}$$

The expressions for I_a, P_a, T^e, and P_m given by (9.17–9.20) are sketched as functions of ω_m for constant V_a and I_f in Figure 9.7. There are three regions of operation as defined by energy flow.

Brake operation occurs when power is put into the machine both from the armature terminals ($P_a > 0$) and from the mechanical terminals ($P_m < 0$), and all of this power is dissipated in the armature resistance R_a. Generator operation has $P_m < 0$ and $P_a < 0$ and occurs when $(V_a - G\omega_m I_f) < 0$, i.e., $G\omega_m I_f > V_a$ or equivalently $\omega_m > \frac{V_a}{GI_f}$. Motor operation occurs when $P_a > 0$ and $P_m > 0$, i.e., when $V_a > G\omega_m I_f$ or $\omega_m < \frac{V_a}{GI_f}$. In the motor operation regime, the maximum power is obtained by taking the derivative of P_m and setting it equal to zero. Thus,

$$\frac{dP_m}{d\omega_m} = 0 = GI_f \frac{(V_a - 2GI_f \omega_m)}{R_a} \tag{9.21}$$

i.e.,

$$\omega_m = \frac{V_a}{2GI_f}. \tag{9.22}$$

P_m^{max} is obtained by substituting this value of ω_m in (9.20). Thus,

$$P_m^{max} = \frac{V_a}{2}\left(V_a - \frac{V_a}{2}\right) \Big/ R_a = \frac{V_a^2}{4R_a} \tag{9.23}$$

The torque-speed characteristics of the separately excited DC motor are given by (9.19). It is plotted in Figure 9.8 for various values of V_a. For practical machines, the small value of R_a leads to a nice steep slope of the curve. This is undoubtedly the most desired feature of a DC machine. The speed can therefore be controlled very smoothly by varying the armature voltage V_a.

A separately operated DC machine can also operate as a generator with the power being delivered at the rotor terminals. Figure 9.6b gives the generator parameters. The open-circuit voltage is $G\omega_m i_f$, and the generator has internal impedance of the armature resistance in series with the armature inductance.

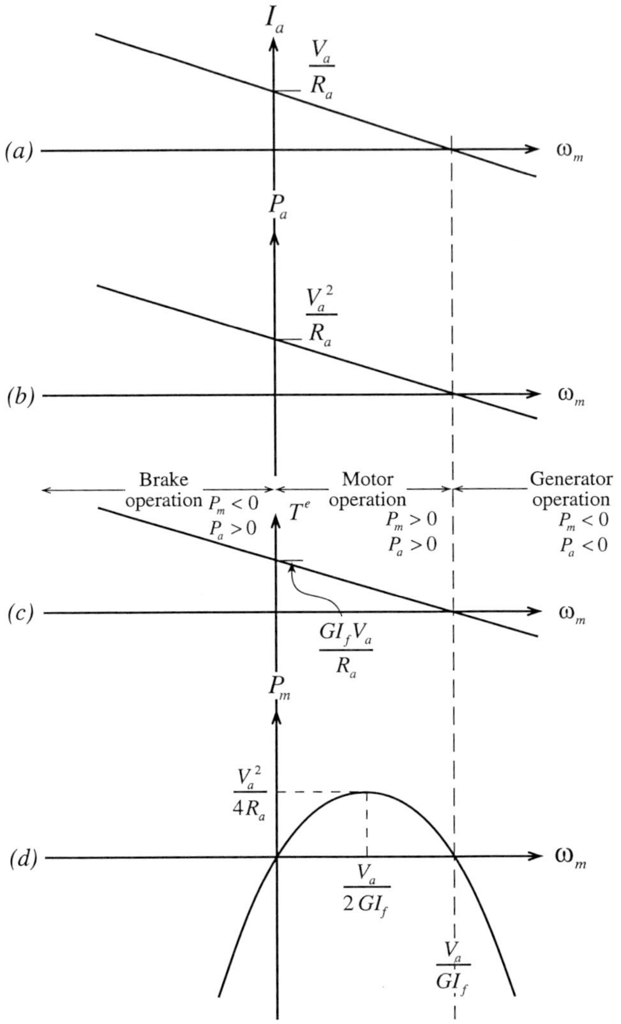

Figure 9.7: Torque-speed curves for a separately excited DC machine for various armature voltages [14]

Example 9.1 A certain separately excited DC motor has $V_a = 300$ V and rated armature current of 60 A. $R_a = 0.2\Omega$. The field current is $I_f = 2$ A and $G = 1.5$ H. Find the speed and the horsepower developed.

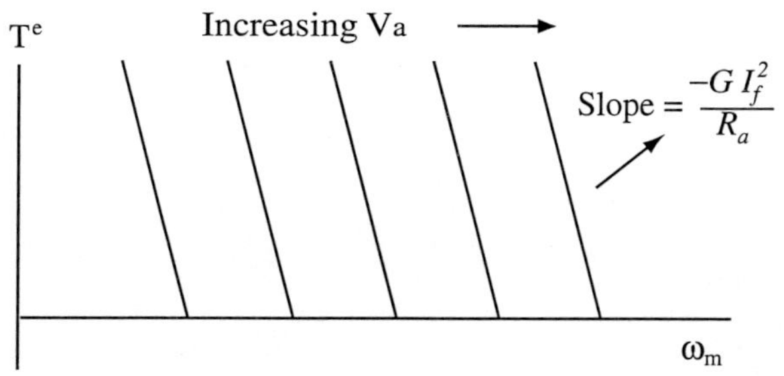

Figure 9.8: Maximum torque as a function of speed for a separately excited motor

Solution: From (9.17),

$$\begin{aligned} 300 - 60(0.2) &= G\omega_m I_f \\ 288 &= (1.5)\omega_m(2) \end{aligned}$$

Hence,

$$\omega_m = 96 = 2\pi\frac{N}{60}\text{mechanical rads/sec}$$

Therefore,

$$\begin{aligned} N &= 916.8 \text{ rpm} \\ T^e = GI_fI_a = (1.5)(2)(60) &= 180 \text{ N-m} \\ P_m = T^e\omega_m &= (180)(96) \text{ W} = \frac{(180)(96)}{746} = 23.16 \text{ hp} \end{aligned}$$

since

$$1 \text{ hp} = 746 \text{ W}$$

9.6 Series-excited Machines

Series excitation means that the stator coil and the rotor coil are in series. This results in the equivalent circuit shown in Figure 9.9 where $i = i_a = i_f$ and $v = v_a + v_f$. From the equivalent circuits, it is obvious that the electrical equation is

$$v = iR_t + L_t\frac{di}{dt} + G\omega_m i \tag{9.24}$$

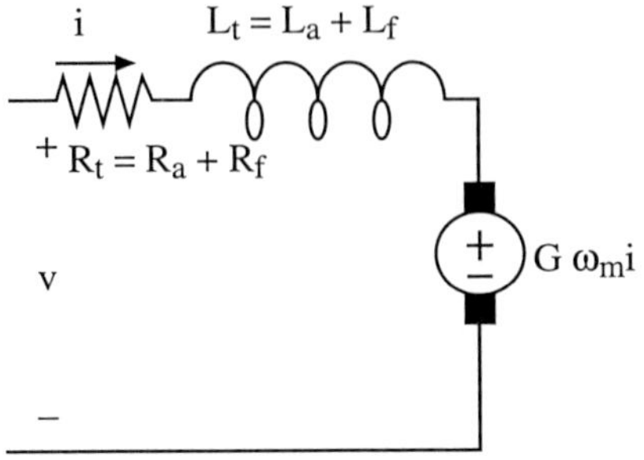

Figure 9.9: Equivalent circuit for series excitation of a DC machine

and the torque equation is now

$$T^e = Gi^2 \tag{9.25}$$

The field coil usually has far more inductance and resistance than the armature winding, so it tends to dominate the parameters for series excitation.

9.7 Constant-voltage Excitation (Series Motor)

The steady-state solution for the electrical and torque equations for a constant voltage $v = V$ in the series configuration is obtained from (9.24) as

$$I = \frac{V}{R_t + G\omega_m} \tag{9.26}$$

$$T^e = GI^2 = G\frac{V^2}{(R_t + G\omega_m)^2} \tag{9.27}$$

The power P_m is obtained as

$$P_m = T^e\omega_m \tag{9.28}$$

The torque-speed curves are as shown in Figure 9.10(a). The current-speed curves are shown in Figure 9.10(b). If the machine is lightly loaded, it is possible for the speed to exceed the maximum speed ratings. The speed at which a series-excited motor runs is a strong function of the load torque. The characteristics of the series motor are suitable for railway traction application.

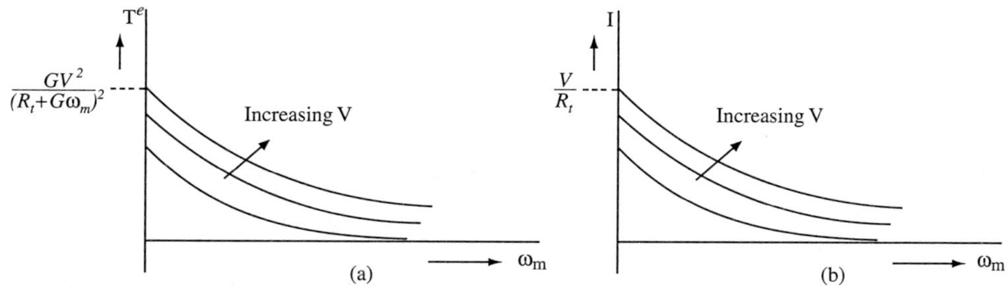

Figure 9.10: Torque-speed curves (a) and current as a function of speed (b) for series-excited motors

Example 9.2 A 220 V DC series motor draws $I = 25$ A and runs at 300 rpm. $R_a = 0.6\Omega$ and $R_f = 0.4\Omega$. Find power delivered to shaft that drives a fan and also compute the torque.

Solution: From (9.26),

$$\begin{aligned} G\omega_m I &= V_a - I(R_a + R_f) \\ &= 220 - 25(0.6 + 0.4) \\ &= 195 \text{ V} \end{aligned}$$

From (9.27) and (9.28),

$$\begin{aligned} P_m &= (G\omega_m I)(I) \\ &= (195)(25) = 4880W = 6.54 \text{ hp} \\ T^e &= \frac{P_m}{\omega_m} = \frac{4880}{2\pi \times \frac{300}{60}} = 155.2 \text{ N-m} \end{aligned}$$

9.8 Universal Motors

A series-excited commutator motor can be run using either AC or DC and is often called a universal motor, especially if it is run on AC. If it is excited with an AC voltage, the steady-state voltage and current are

$$v(t) = \sqrt{2}V\cos(\omega_e t) \tag{9.29}$$

where ω_e is the electrical frequency

$$\begin{aligned} i(t) &= \sqrt{2}I\cos(\omega_e t + \theta_i) \\ I &= \frac{V}{\sqrt{(R_t + G\omega_m)^2 + L_t^2\omega_e^2}} \end{aligned} \tag{9.30}$$

The torque is now

$$\begin{aligned} T^e &= Gi^2(t) = 2GI^2\cos^2(\omega_e t + \theta_i) \\ &= GI^2[1 + \cos(2\omega_e t + 2\theta_i)] \end{aligned} \tag{9.31}$$

The machine cannot respond to the term corresponding to twice the electrical frequency in (9.31), and hence only the time-average torque is considered. The expression for that is

$$T^e_{av} = GI^2 = \frac{GV^2}{(R_t + G\omega_m)^2 + L_t^2\omega_e^2} \tag{9.32}$$

The shape of the torque-speed curves is similar to but not exactly the same as that of Figure 9.10. If variable-speed operation of an AC-driven universal motor is desired, the input voltage is chopped using SCR or Triac circuits to effectively reduce the rms input voltage.

9.9 Shunt-excited Machines

Shunt excitation means that the field coil and the armature windings are in parallel. This results in the equivalent circuit shown in Figure 9.11 where

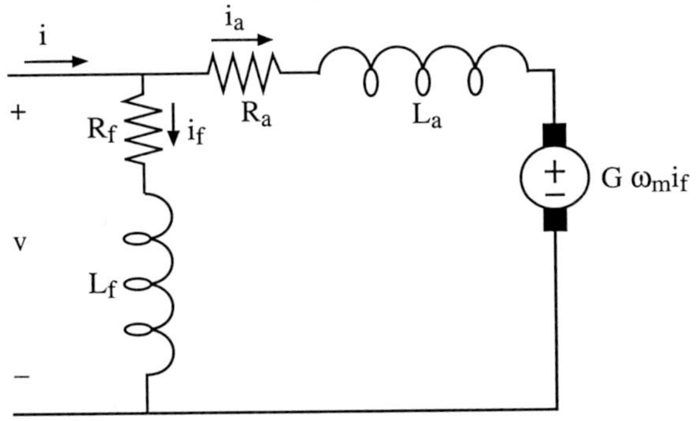

Figure 9.11: Equivalent circuit for shunt excitation of a DC machine

the current orientations are shown for motor operation $i = i_a + i_f$ and $v = v_a = v_f$. From the equivalent circuit, it is obvious that the electrical equation is

$$v = i_f R_f + L_f \frac{di_f}{dt} = i_a R_a + L_a \frac{di_a}{dt} + G\omega_m i_f \tag{9.33}$$

and the torque equation is now

$$T^e = G i_a i_f \tag{9.34}$$

9.9.1 Shunt motor constant-voltage excitation

The steady-state solution for the electrical and torque equations for a constant voltage $v = V_t$ in the shunt configuration is

$$I_f = \frac{V_t}{R_f}$$

$$I_a = \frac{V_t - G\omega_m I_f}{R_a} = V_t \frac{(R_f - G\omega_m)}{R_a R_f} \tag{9.35}$$

$$T^e = G I_a I_f = G V_t^2 \frac{R_f - G\omega_m}{R_f^2 R_a} \tag{9.36}$$

The torque-speed curve is shown in Figure 9.12. Speed control is best effected by controlling the field current. This is done by adding an external resistance to the field winding.

Example 9.3 A shunt motor connection is made for the motor of Example 9.1 and shown in Figure 9.13. $R_f = 100\Omega$ and $I = 30$ A. Find the rpm and the horsepower developed by the motor.

Solution: From Figure 9.13,

$$\begin{aligned}
I_f &= \frac{300}{100} = 3 \text{ A} \\
I_a &= I - I_f = 30 - 3 = 27 \text{ A} \\
G\omega_m I_f &= 300 - (27)(0.2) \\
&= 294.6 \text{ V} \\
\omega_m &= \frac{294.6}{(1.5)(3)} = 65.5 \text{ rad/sec } = \frac{2\pi N}{60}
\end{aligned}$$

Hence $N = 625$ rpm.

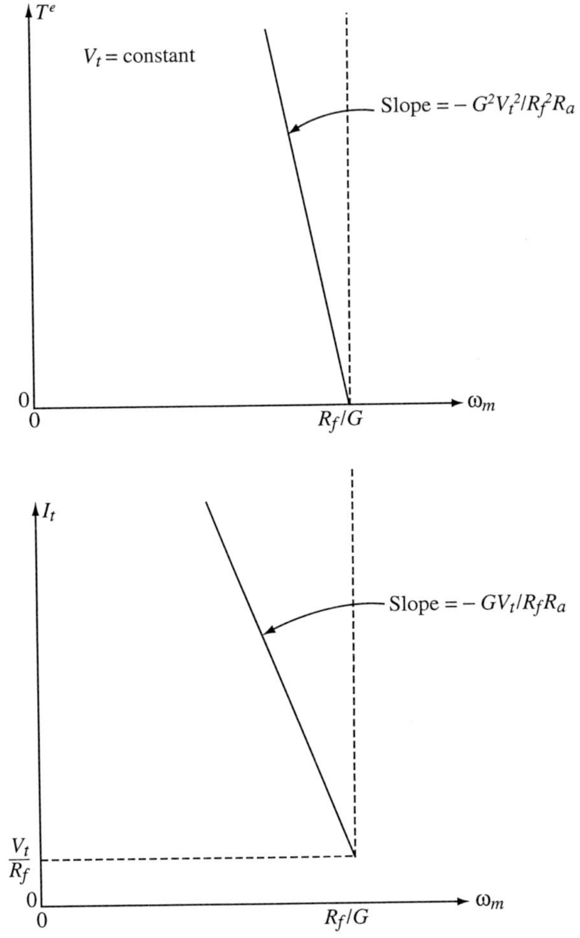

Figure 9.12: Torque and terminal current as functions of speed in a shunt-excited DC motor with constant terminal voltage [14]

9.9.2 Shunt generator

A shunt-excited machine can be operated as a generator but is rarely used as such. The direction of the currents i_a and i are reversed, and a load R_L is connected across the terminals as shown in Figure 9.14. The dynamic

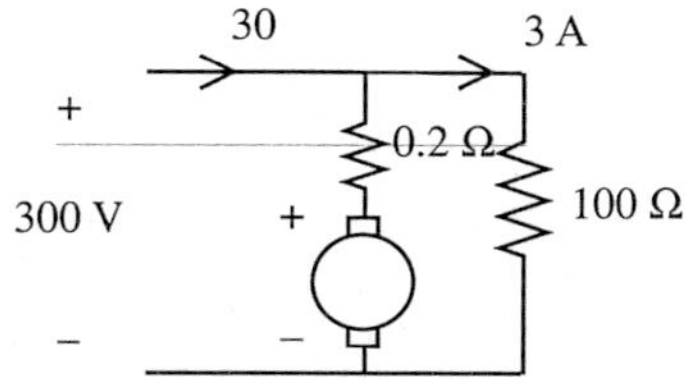

Figure 9.13: Shunt motor example

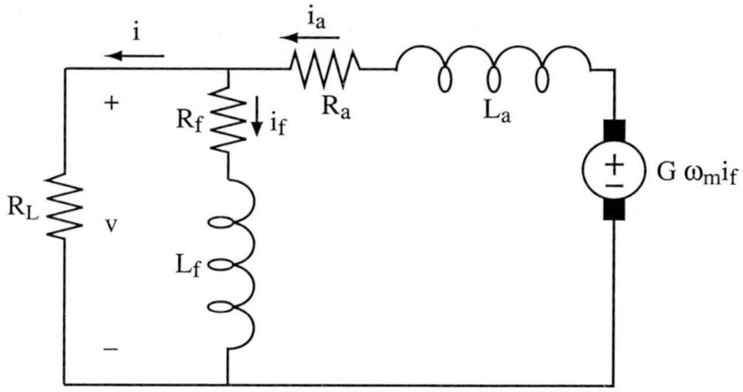

Figure 9.14: Shunt generator

equations are

$$
\begin{aligned}
v &= i_f R_f + L_f \frac{di_f}{dt} = G\omega_m i_f - L_a \frac{di_a}{dt} - i_a R_a \\
&= (i_a - i_f) R_L
\end{aligned}
$$

In steady state, we have $v = V_t$, $i_f = I_f$, $i = I$, and $i_a = I_a$, so that

$$
\begin{aligned}
V_t &= I_f R_f \\
V_t &= G\omega_m I_f - I_a R_a \\
V_t &= (I_a - I_f) R_L = I R_L
\end{aligned}
$$

The expression for P_m and T^e can be derived easily.

9.9.3 Compound-excited machines and starting

In a machine with compound excitation, part of the field coil is in series with the armature, and the rest is in parallel. The main use of this configuration is to limit the current flowing in the armature circuit. At low speeds, the back EMF is quite small, and very large currents can flow in the armature circuit. One way around this is to start the machine with series excitation and switch to compound excitation at an intermediate speed. Finally, at higher speeds, the machine can be switched to shunt excitation. Another way of preventing excessive armature current is by connecting a resistor in series with the winding. The resistance can be reduced and eventually eliminated as the speed increases. The disadvantage of this method is the waste of energy in the resistor.

The best method of starting and speed control of a DC machine is to use a power electronic circuit to chop the DC input to either the field or armature winding or both. This technique is called pulse-width modulation (PWM). The voltage is switched between some value, say V_o, and zero at a high frequency. The machine currents cannot respond to the high frequency and is only approximately DC. The effective terminal voltage is then V_o times the duty ratio. The torque-speed curves given above then apply for this situation using the effective voltages.

9.10 Discussion

DC machines are mainly useful because they provide an easy way of varying the speed of a motor. Historically, they have been the machine of choice when a speed variation was required. In recent years, however, power electronics has advanced to the stage where variable-frequency sources are readily available. A variable-frequency source allows an induction machine or a synchronous machine to be run at varying speeds. The advantage of DC machines is that the control circuitry is far simpler than that of a variable-speed induction or synchronous motor drive. The disadvantage is that the DC machine itself is more complicated, heavier, and more expensive. In addition, the commutator and brushes are high-maintenance items due to mechanical wear and tear. While DC machines will continue to be used in large numbers, they are being replaced by induction or synchronous machines in variable-speed applications. For example, most new diesel-electric locomotives now use induction motors to drive the wheels; whereas ten years ago, they all used DC motors.

In generator applications, DC machines are obsolete. Synchronous machines with rectifiers are so much simpler that they have totally replaced DC generators. A synchronous machine run as a generator is often called an alternator. In automobiles, the DC generators were replaced by alternators around 1960.

In the literature, there is a class of motors called DC brushless motors. The terminology is unfortunate since a brushless DC machine is really a form of synchronous machine with a permanent magnet on the rotor. The rotor position is measured, and the stator currents are switched electronically on the basis of the rotor position. The control of this motor is somewhat similar to a standard DC motor, and hence the name.

9.11 Conclusion

In this chapter, the working of a DC machine is explained in the co-energy framework. Practical connections in the form of separately excited, series-excited, and shunt-excited are discussed. Their torque-speed characteristics are explained when operating as a motor.

9.12 Problems

9.1 A 100 kW, 200 V shunt generator has $R_a = 0.05\Omega$ and $R_f = 57.5\Omega$. If the generator operates at rated voltage, calculate the induced voltage at (a) full load and (b) half load. (Answer: (a) 252 V, (b) 241 V).

9.2 A separately excited motor runs at 1045 rpm with a constant field current while the armature current is 50 A. The voltage at the terminals of armature is 120 V. $R_a = 0.1\Omega$. If the load on the motor changes such that it now takes 95 A at 120 V, determine the speed at this load. (Answer: 1004 rpm).

9.3 A series motor, having an armature resistance of 0.1Ω and a field resistance of 0.15Ω, takes 48 A at 230 V and runs at 720 rpm. The total core and friction losses are 650 W. Neglecting stray-load and brush-contact losses, calculate (a) the developed torque, (b) output power at the shaft, and (c) the motor efficiency. (Answer: (a) 138.8 N-m; (b) 13.16 hp; (c) 88.9%).

9.4 A DC machine is a nice system for speed control since it is naturally a linear system if the field current is held constant. This problem is designed to illustrate some principles of speed control. The DC machine in this problem is held at constant field current so that the back EFM in the armature circuit is $0.6\omega_m$ V. (That is, $GI_f = 0.6$.) The armature resistance is $R_a = 1.2\Omega$ and the armature inductance is $L_a = 100$ mH. The moment of inertia is 1 N-m^2. The friction and all other mechanical losses are negligible, and the load torque is linearly proportional to the speed.

(a) If the load torque $T_L = 0.5\omega$ and the armature voltage $V_a = 200\ V$, what is the equilibrium speed of the motor?

(b) The machine is operating in steady state under the conditions of part (a) when the load torque is suddenly doubled to $T_L = 1.0\omega$. Assuming this change takes place at $t = 0$, find the speed as a function of time for $t \geq 0$. What is the new steady-state speed?

(c) In order to maintain constant speed, a feedback system is added to the armature circuit. This feedback system changes the armature voltage according to the relation $\frac{dv_a}{dt} = \alpha(\omega_{\text{ref}} - \omega)$ where ω_{ref} is the desired speed. If $\alpha = 1$(V/sec)/(rad/sec) and the desired speed is the solution to part (a), repeat part (b).

(d) An engineer suggests that the response time to changes in torque can be improved by increasing α. Can this produce an unstable system? If yes, find the maximum value of α for which the system is stable. If no, show that the system is stable for all α.

9.5 In Example 9.2, find the value of G. The speed is to be reduced to 200 rpm by adding an external resistance R_{ext} in series with R_f and R_a. Torque required by the fan is proportional to the square of the speed. Find the value of this resistance and the new power. (Answer: $G = 0.248$ H; $T^e = 68.98$ rpm; $R_{\text{ext}} = 7\Omega$; $P_m = 1445$ W)

9.6 A DC machine has $R_f = 10\Omega$ and $R_a = 2\Omega$. When run as a separately excited generator with $I_f = 3$ A and $N = 1000$ rpm, the open-circuit voltage is 250 V.

(a) If the machine is connected as a series motor and applied voltage is 350 V DC, find the steady-state torque at a speed of 2000 rpm. (Answer: $T^e = 3.054$ N-m)

(b) If the load torque is 10 N-m, what is the value of the applied voltage necessary to produce a speed of 800 rpm? (Answer: 278.8 V)

Chapter 10

SINGLE-PHASE INDUCTION MACHINES

10.1 Introduction

While three-phase induction machines provide optimum designs, most residential and some commercial locations are served by only single-phase power. In these locations, running three-phase machines would require using a three-phase inverter, an expensive solution for low-power applications. Even in locations wired for three phase, it may be economically advantageous to use single-phase motors for small applications. Single-phase motors are widely used up to about 5 horsepower. In 1998, it is estimated that more than 120,000,000 single-phase induction motors were manufactured.

10.2 Principles of Operation

The basic principle of operation of a three-phase induction motor is that the stator windings produce a rotating magnetic field. This rotating magnetic field induces rotor currents that lag the stator field. The interaction between the stator field and the rotor currents causes a torque, which drives the mechanical load. In a single-phase machine, the idea is to duplicate this rotating magnetic field to produce torque.

The magnetic field in the air gap of a three-phase two-pole motor has the form

$$B_{3\phi} = B_{3m} \cos(\theta - \omega t) \tag{10.1}$$

This field is sinusoidal in space at any time. The peak of the field rotates with angular velocity ω (377 rads/sec at 60 Hz). At time $t = 0$, the peak is at $\theta = 0$; a quarter of a cycle later, the peak is at $\theta = 90°$, and so on. This is the ideal field for an induction motor. If we have a pure single-phase two-pole induction motor with one winding, the field is

$$B_{1\phi} = B_{1m} \cos\theta \cos\omega t \tag{10.2}$$

This field is stationary in space and sinusoidal in time. Using trigonometric identities, this becomes

$$B_{1\phi} = \frac{1}{2} B_{1m}[\cos(\theta - \omega t) + \cos(\theta + \omega t)] \tag{10.3}$$

Equation (10.3) represents two rotating magnetic fields, one rotating in the $+\theta$ direction and one rotating in the $-\theta$ direction. The net torque is the sum of the two torques produced by the two rotating fields. This is shown in Figure 10.1. It can be seen from Figure 10.1 that the pure single-phase

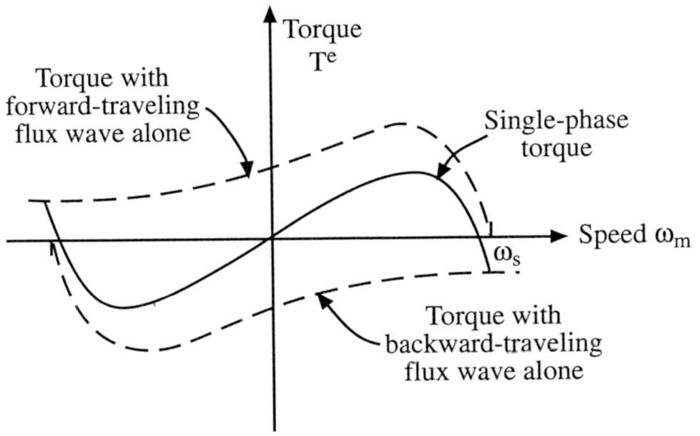

Figure 10.1: Torque-speed curve for a single-phase motor [14]

motor can run in either direction at speeds less than 3600 rpm (for two-poles and 60 Hz). However, it has no starting torque. (The torque is zero when the speed is zero.) Thus, the machine cannot start by itself. In most applications, a specific direction of rotation is required, so the design of practical single-phase motors is directed toward producing a motor that will start and run in the desired direction. This is done by adding a second (auxiliary) winding displaced in angle from the first (main) winding. The current in the auxiliary winding is out of phase from the main winding.

10.3 General Two-Phase Machines

In this analysis, the main winding is directed so as to produce a sinusoidal field that peaks at $\theta = 0$. The current in this winding will be the zero-phase reference. Thus, the field is

$$B_{main} = B_m \cos\theta \cos\omega t \tag{10.4}$$

$$B_{main} = \frac{1}{2} B_m \left[\cos(\theta - \omega t) + \cos(\theta + \omega t)\right] \tag{10.5}$$

The auxiliary winding is displaced at angle θ_a with respect to the main winding, and the current has the phase ϕ_a. This produces a magnetic field:

$$B_{aux} = B_a \cos(\theta + \theta_a)\cos(\omega t + \phi_a) \tag{10.6}$$

$$B_{aux} = \frac{1}{2} B_a \left[\cos(\theta - \omega t + \theta_a - \phi_a) + \cos(\theta + \omega t + \theta_a + \phi_a)\right] \tag{10.7}$$

The effect of the field from the auxiliary winding is to enhance one of the rotating fields and decrease the other. For example, let us assume that $\theta_a = 90^\circ$, which is the most common case, and let $\phi_a = 45^\circ$ and $B_a = 0.5 B_m$. This gives

$$\begin{aligned} B &= B_{main} + B_{aux} \\ &= 0.5 B_m \left[\cos\theta - \omega t) + 0.5\cos(\theta - \omega t + 45^\circ) + \cos(\theta + \omega t) \right. \\ &\quad \left. + 0.5\cos(\theta - \omega t + 135^\circ)\right] \\ &= B_m \left[0.70\cos(\theta - \omega t + 14.6^\circ) + 0.37\cos(\theta + \omega t + 28.7^\circ)\right] \end{aligned} \tag{10.8}$$

Thus, the forward-rotating field has a magnitude approximately twice that of the backward field. The torque is roughly proportional to the square of the magnitude of the field, so that when the motor is stopped, the torque produced by the forward field is approximately four times that produced by the backward field. Therefore, the motor will start and run in the forward direction.

10.4 Practical Methods of Exciting an Auxiliary Winding

There are six different types of single-phase induction motor. The differences are in the way the auxiliary winding is constructed and excited. These six motor types are

1. Split-phase
2. Capacitor-start
3. Permanent-split capacitor
4. Two-valued capacitor or capacitor-start/capacitor-run
5. Split-phase start/capacitor-run
6. Shaded-pole

Each of these motor types will be described in turn.

10.4.1 Split-phase motors

The split-phase motor is the most widely used single-phase motor. They are relatively inexpensive and are used to drive easily started loads. They are typically used in the 1/20 to 1/2 horsepower range. In a split-phase motor (as in all the single-phase motors), the main winding is connected directly across the line voltage. The equivalent circuit of the main winding is the same as for one of the phases of a three-phase motor except for the values of the components. Thus, the main winding current lags the voltage since the motor is an inductive load. The auxiliary winding has fewer turns of smaller and therefore higher-resistance wire. It is normally displaced half a pole pitch ($\theta_a = 90°$ for a two-pole motor) from the main winding. The ratio of resistance to inductance of the auxiliary winding is higher than that of the main winding. The starting auxiliary winding is connected across the line. Because the angle of the impedance in the auxiliary winding is less than that of the main winding, the auxiliary current lags the voltage less than the main current. The phasor diagram using the voltage as zero-phase reference is shown in Figure 10.2. The angle between the currents, ϕ_a is typically about 30°. Thus, the motor has starting torque and will run.

Once the motor has started and is reaching rated speed, the auxiliary winding is switched out of the circuit, usually by a centrifugal switch. This is done because at operating speed the torque with the auxiliary winding connected is actually less than without it. Also, the auxiliary winding is small and is not rated for continuous duty. The motor would overheat if it stayed connected.

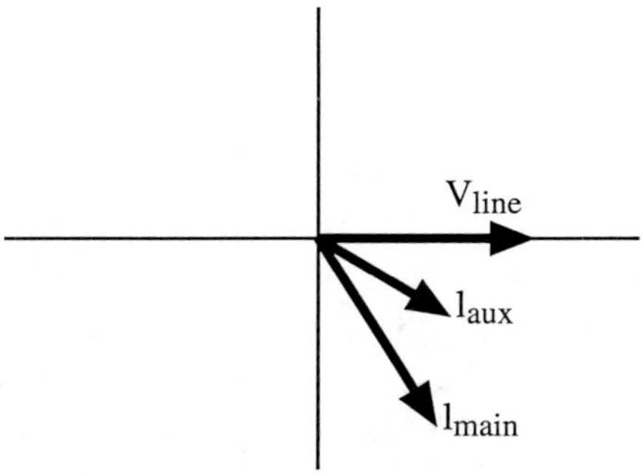

Figure 10.2: Phasor diagram for a split-phase motor

10.4.2 Capacitor-start motor

Capacitor-start motors are used for high starting loads and are by far the most popular type of single-phase motor for heavy-duty general-purpose applications. They are available in ratings from 1/8 to 3 horsepower. They are constructed with the auxiliary winding displaced 90° from the main winding, as in the split-phase motor. While starting, the auxiliary winding is connected in series with a capacitor across the line voltage. The capacitor is sized to produce a phase shift of 90° between the main and auxiliary windings. From (10.7), it can be seen that this is the optimum value. The phasor diagram is shown in Figure 10.3. Because of the capacitor, the current in

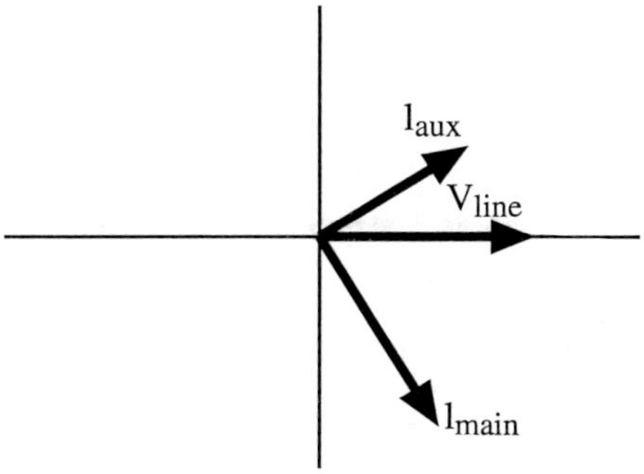

Figure 10.3: Phasor diagram for a capacitor-start motor at standstill

the auxiliary winding leads the voltage. As in the split-phase motor, the auxiliary winding is disconnected when the motor nears its running speed. Thus, the capacitor and auxiliary winding need to have a short time rating.

10.4.3 Permanent-split capacitor motor

Permanent-split capacitor motors are used extensively in fan and blower applications and in refrigeration compressors for refrigerators, freezers, and air conditioners. These motors are designed for optimum efficiency. The construction is the same as for a capacitor-start motor except that the auxiliary winding and capacitor are always in the circuit. For optimum efficiency, the capacitor should be sized so that the current in the auxiliary winding is 90° out of phase with the current in the main winding at normal running speed. The angles of the currents are functions of speed since the resistance in the equivalent circuit is proportional to $1/s$, where s is the slip. Having the capacitor optimized at running speed means that the starting torque is reduced. In some applications, it is necessary to compromise and trade some efficiency for starting torque.

A permanent-split capacitor motor is more efficient than a capacitor-start motor because the backward field is reduced. The backward field causes core losses and rotor losses while actually slightly reducing the torque at running speed. All types of single-phase motors except the shaded-pole motor can be reversed by reversing the connections on the auxiliary winding.

10.4.4 Two-valued capacitor or capacitor-start/capacitor-run motor

The capacitor-start/capacitor-run motor is used in applications requiring both high starting torque and high running efficiency. As can be surmised from the name, they use two capacitors, one for starting and one for running. Therefore, each capacitor can be sized for its purpose. Usually, the running capacitor is permanently connected in series with the auxiliary winding. At starting, the starting capacitor is switched in parallel with the run capacitor. A centrifugal switch can be used to switch out the starting capacitor, typically at about 80% of the synchronous speed. Obviously, this motor would be more expensive than either of the other two-capacitor motors.

10.4.5 Split-phase start/capacitor-run motor

This motor is similar in concept to the two-capacitor motor except that only a single-run capacitor is used. For starting, the capacitor is shorted, which causes the machine to start as a split-phase machine. This produces a higher starting torque than would be provided by the run capacitor.

10.4.6 Shaded-pole motor

A shaded-pole motor is the simplest and cheapest of the single-phase motors. However, it also is the poorest performer, having both a low starting torque and low efficiency ($< 40\%$). They are typically used in applications of less than 1/4 horsepower. Instead of having a complete auxiliary winding, they are constructed using salient poles on the stator with a shorted turn on one side of each salient pole. The field produced by the main winding induces current in these shorted turns. The current in the shorted turns is out of phase with the main winding current. This results in the forward rotating field being slightly larger than the backward one. Thus, the motor will have a little starting torque. Once the motor starts, the torque provided by the main field produces the running torque.

10.5 Conclusion

Brief discussions of the principles of operation of single-phase motors, how rotating fields are produced, methods of starting, and techniques to obtain good starting as well as running torques have been presented in this chapter.

Chapter 11

DISTRIBUTED ENERGY RESOURCES

11.1 Distributed and Renewable Energy Applications

The system that is the power grid has become more expansive than the system it was decades ago. Now, a power system means not just the bulk high-voltage transmission system backbone that power engineers have traditionally analyzed and still remains the focus in learning and applying the foundational concepts. In addition, the system now also includes interactive, customer-focused loads that act as a resource to the system. There are different names for this concept, and ways that it can occur, where *demand response* is a major one. Associated with demand response, distributed energy resources (DERs) are a way of achieving the same desired capabilities.

11.1.1 DERs and the intelligent electric grid

The upsurge of DERs, demand-side resources, and other non-traditional energy transactive sources that can be integrated either with each other on smaller scales as in *microgrids*, and/or within the bulk power system is part of the *Smart Grid* vision that became popular starting in the early 2000s. What is the smart grid? The "smart grid" is hard to rigorously define; it is more of a concept or a vision of what the grid should look like, and it means different things to different people. On the smart grid, it has been said that it is "sort of like the weather – everyone talks about it, but few understand what's really going on" (unknown source). Many power engineers have disliked the term, because it implies that the grid was not already 'smart,' an assertion which many will take offense to, and disagree with. The phrase 'a more intelligent electric grid' may be more accurate to state the goal, though now we need to describe what this vision entails.

As described by DOE [17], the smart grid is the modernization of the electrical power industry infrastructure which uses digital technology to improve the reliability, security, and efficiency of the electric system. The smart grid includes large generation, to delivery systems, to electricity consumers, and a growing number of distributed generation and storage resources.

The smart grid characteristics are described in The Energy Independence and Security Act of 2007, Title XIII, Section 1301, which characterizes a Smart Grid as follows: (1) Increased use of digital information and controls technology to improve reliability, security, and efficiency of the electric grid. (2) Dynamic optimization of grid operations and resources, with full cybersecurity. (3) Deployment and integration of distributed resources and generation, including renewable resources. (4) Development and incorporation of demand response, demand-side resources and energy-efficiency resources.

Further, it is about the deployment of 'smart' technologies (real-time, automated, interactive technologies that optimize the physical operation of appliances and consumer devices) for metering, communications concerning grid operations and status, and distribution automation. Integration of 'smart' appliances and consumer devices. Deployment and integration of advanced electricity storage and peak-shaving technologies, including plug-in electric and hybrid electric vehicles, and thermal storage air conditioning. Provision to consumers of timely information and control options. The smart grid vision also requires the development of standards for communication and interoperability of appliances and equipment connected to the electric grid, including the infrastructure serving the grid. Identification and lowering of unreasonable or unnecessary barriers to adoption of smart grid technologies, practices, and services.

The focal needs for moving the Smart Grid idea forward, that have been significantly addressed and improved within the last decade, has been to address how to better allow for the integration of intermittent, uncertain electric generation sources like wind and solar, and how to improved the customer-side interaction with the utility such as through smart meters and feedback to the customer about demand usage.

There was a major impact of the 2009 Stimulus Bill on Renewable Energy and Smart Grid, though the American Recovery and Reinvestment Act (ARRA). The 2009 stimulus bill contained several provisions related to renewable energy $32 billion to enhance the electric power grid. It also contained a three year extension to renewable production tax credits that helped propel renewable energy developmetns and applications forward. About $40 billion went toward energy efficiency development in various forms, and $2 billion went toward advanced battery manufacturing. The ARRA also funded a lot of synchrophasor-related projects including through work with the North American Synchrophasor Initiative (NASPI) (https://www.naspi.org/). Synchrophasors and renewable energy integration are related, as improved coverage of accurate phase angle difference information is key toward better understanding and designing these systems.

11.1.2 Demand response

Demand response is tightly linked with the demand side management and controllable load. Most generation sources are "supply-side resources," large-scale bulk plants owned by utility stakeholders. In addition to supplying the energy to meet real time demand, these plants also provide other necessary services such as reactive power support and inertia for system stability enhancement. Conceptually, lowering the load is a source of energy – one can either raise generation or lower load. Demand response is a broad term that encompasses the utilization of a number of different strategies that can be designed to achieve response to grid fluctuations using what is available on the customer's side, "demand-side" resources. "Demand side" can refer to any program that attempts to modify customer's energy use. These programs have been around in some form for decades. One way this has occurred historically was that a utility would offer incentives to customers to reduce demand, such as to pay customers to get rid of their old (inefficient) spare refrigerators. In addition to efficiency improvements, other demand-side management techniques target conservation of energy, management of the loads, and changing the fuel types in the generation resource mix which also changes the system response capabilities.

Demand side management can be considered in an integrated resource planning process meaning that supply and demand side resources are given equal consideration. The need for new and improved market designs to handle this is still an emerging area with a lot of open questions. It is needed to overcome the challenge related to utility sales and profits; a disincentive to manage the load can occur since more energy sold is typically higher profit. The cost-effectiveness of demand response, and DERs can be considered as part of that, is difficult to measure, and hence one must be creative in how to formulate the DSM program from a cost recovery and incentive perspective so that utilities and other stakeholders would prefer DSM. Part of this is customer involvement and it may simply be for some consumers to just be able to see and better understand how they use electricity, and trying to translate this data into strategies to save money.

11.1.3 Distribution system enhancements

Distribution system automation has been making steady advances for many years, a trend that should accelerate with smart grid funding. The term self-healing is often used to refer to automatic distribution system reconfiguration, and how to accomplish this under a wide variety of circumstances and at all different levels in the has been, and continues to be, an exciting emerging area of work. Along with this comes the need for increased visibility and monitoring, and this has been steadily improving through concerted efforts as well, such as improved control center side monitoring and modeling of the distribution system and bring together transmission and distribution system side information (expertise, data, models). Since the vast majority of electricity outages that do occur arise in the distribution system, it is an important and understudied area upon which to focus, especially to be able to achieve the realization of benefits from more ubiquitous residential level DERs.

11.1.4 Electric vehicle integration

In addition to renewable energy, another significant driver in controllable electric energy resources has been the use of pluggable hybrid electric vehicles. The US has seen a steadily increasing adoption of pluggable hybrid electric vehicles (PHEVs) in the last decade based on the volatility of gas prices and incentives from government organizations. PHEVs are a motivating driver for load control. Recharging all PHEVs when drivers return home at 5pm would be problematic, so some type of load control is a must. The infrastructure and control support for when, where, and how to best recharge the vehicles is essential and tightly related to many of the other energy system considerations that we have discussed in other chapters of this book.

11.1.5 Distributed generation

Distributed generation (DG) typically is small scale, up to about 50 MW. DG includes renewable and non-renewable sources that may be isolated from the grid or may be grid-connected. Distributed generation is usually located near the end user. In addition to solar and wind (discussed next), DG technologies can also include microturbines, reciprocating Internal combustion engines, biomass, micro-hydro, and fuel cells. DGs can be especially practical and useful in remote locations. There are environmental benefits as well, with renewable resources and reduced emissions. Some DGs can also make use of the waste heat, making the overall system more efficient. Finally, DGs can sell power back to the grid. In summary, the vision of the next-generation grid is a grid that has more comprehensively integrated generation, distribution, transmission, buildings, and vehicles.

11.2 Solar Energy Resources

An important application of power electronics and also an important emerging field in power and energy systems is solar power. Solar photovoltaic (PV) systems are a prevalent way to harness solar energy and convert it into electrical energy, initially as DC. The US Department of Energy's Solar Technology Office defines Photovoltaics (PV) as a *"material or device that is capable of converting energy contained in photons of light into an electrical voltage or current"* [18]. The power can be used as DC such as directly for loads within a home or with inverters for AC grid-connection. As engineers, we want to be able to effectively design and use solar systems as we would the other types of energy sources. Challenges with PV integration include its operational characteristics: intermittency, overgeneration during the day, fast decrease that causes other generators to need to quickly ramp up at sunset, and the decreased frequency response have been a cause of concern. Energy storage is a solution that can help, and technology to do this more cost effectively and on a larger scale has been improving; however, it has its own set of challenges. Hence this section provides some context about solar energy as both an emerging topic and as an exemplar use case of distributed energy resources (DERs) driving the importance of power electronics. More details on energy from the sun can be found in [12].

11.2.1 Solar Resource Basics

Before we talk about PV systems, some basics about the sun are important. First, we need to know how much sunlight is available. It is also important to point out that we can predict where the sun is at any given time. We also define the term "insolation," which is "incident solar radiation." The term "irradiation" is also commonly used. For any candidate solar site, we want to be able to estimate the average daily insolation. We also want to be able to choose effective locations and panel tilts for the PV panels. The sun's rays must pass through the atmosphere to reach the earth, and hence both the atmosphere and the angle of the sun with respect to the earth are important in determining how much energy actually arrives as input to the PV system. The earth's orbit and the day of the year are also factors. To understand more about going from the sun to the earth, [Ref: Reference: Gilbert Masters book, Renewable Energy Systems, Chapter 4.4 and 4.5, Chapter 7]. The solar declination angle δ is based on the day of the year, and it gives the angle between the equator plane and the line from the sun to the center of the earth. The altitude angle β_n at solar noon is the angble between the sun and the horizon. Azimuth angle ϕ_s measures how far off the sun is from solar noon, and hours are referenced from solar noon, where ϕ_s is positive in the morning when sun is in the east, negative in the evening when sun is in the west, using the reference in the Northern Hemisphere (for us) to be true south. "Solar noon" refers to when the sun is directly over the local line of longitude. During solar noon, the sun's rays are perpendicular to the collector face.

$$\delta = 23.45 \sin\left(\frac{360}{365}(n - 81)\right) \tag{11.1}$$

$$\beta_N = 90^\circ - L + \delta \tag{11.2}$$

The key point is that detailed information about the sun's position in the sky at any given location and time is something that is known and can be looked up. This information is used in PV system design, including to predict the sun's position and to pick the best locations and tilt angles for PV panels. Some PV systems track the sun, which can be automated. Tracking systems can be single-axis (usually with a rotating polar mount, parallel to earth's axis of rotation), or two-axis (horizontal [altitude, up-down] and vertical [azimuth, east-west]). PV panel tilts can also be adjusted manually, perhaps seasonably. The rule of thumb for the Northern Hemisphere is to use a south facing collector, tilted at an angle equal to the local latitude.

This can also help determine what sites will be in the shade at any time. In shading analysis, a sketch of the azimuth and altitude angles of trees, buildings, and other obstructions along with the sun path diagram is used to indicate times when the site will be in the shade. Shading analysis is important for site design since shading of a portion of a panel can greatly reduce the output for the full panel, depending upon its design.

11.2.2 Non-PV Solar Technologies

Before we discuss PV systems, it is important to note that not all solar technologies are PV. In particular, Concentrating Solar Power Technologies (CSP) operate on the basic idea of converting sunlight into thermal energy, then using that thermal energy to produce electricity. Concentration is needed to get a hot enough temperature. Three successfully demonstrated technologies for CSP include (1) parabolic trough, (2) solar central receiver, (3) solar dish/ sterling For example, in the solar dish/Stirling design, multiple mirrors approximate a parabolic dish. A receiver absorbs solar energy and converts it to heat, which is delivered to Stirling engine. For parabolic troughs, receivers are tubes called heat collection elements (HCE). Heat transfer fluid circulates in the tubes which delivers collected energy to a steam turbine/generator. The parabolic mirrors rotate east to west to track the sun.

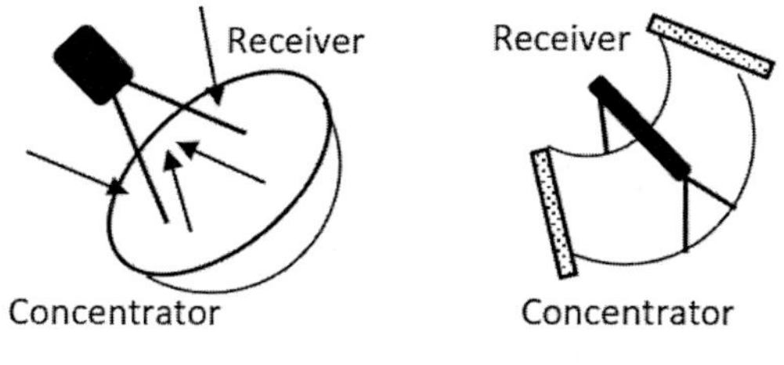

Figure 11.1: CSP technology examples [18]

CSP technologies use mirrored surfaces to concentrate sunlight onto a receiver to run a heat engine. All can be hybridized with auxiliary fuel sources, and higher temperatures corresponds to higher efficiency. These technologies are important to note, because some of the largest-scale solar plants use them. However, most of the capacity additions in recent years have been solar PV.

11.2.3 Photovoltaic systems

In photovoltaic (PV) systems, a solar cell essentially behaves like diode. Photon energy is converted into DC. Shadows and defects on the panel will reduce energy conversion. An inverter converts DC to AC, if desired. Storage systems such as batteries can be integrated to act as a buffer to smooth out changes in both generation and load.

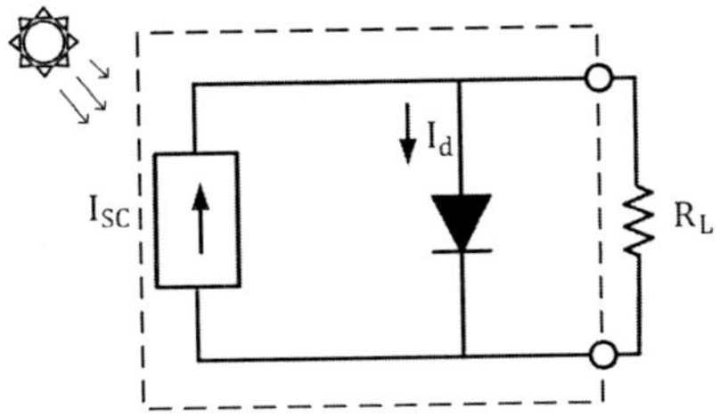

Figure 11.2: PV Cell Circuit Model

The simplest PV cell model is an ideal current source in parallel with a diode as shown in Fig. 11.2. The ideal source current I_{SC} is proportional to insolation, so if insolation drops by 50%, I_{SC} drops by 50%. The diode current is I_d:

$$I_d = I_0(e^{qV_d/kT} - 1)$$

For any I_{SC} value, we can thus calculate the relationship between cell terminal voltage and current (the I-V characteristics). The PV cell I-V characteristic is the upside-down diode I-V characteristic shifted up by I_{SC} because $I = I_{SC} - I_d$. That is, the load current I through resistor R_L, from KCL is the following, as shown in Fig. 11.3:

$$I = I_{SC} - I_d = I_{SC} - I_0(e^{qV_d/kT} - 1)$$

where the I-V curve intersects the x-axis at V_{OC} and the y-axis at I_{SC}. A junction temperature of 25°C is often used as a standard so $qV_d/kT = 38.9V_d$ at 25°C. Setting $I = 0$A and solving for V_d, the open circuit voltage is

$$V_{OC} = (kT/q)ln(I_{SC}/I_0 + 1) = 0.0257ln(I_{SC}/I_0 + 1) \text{ at } 25°\text{C}$$

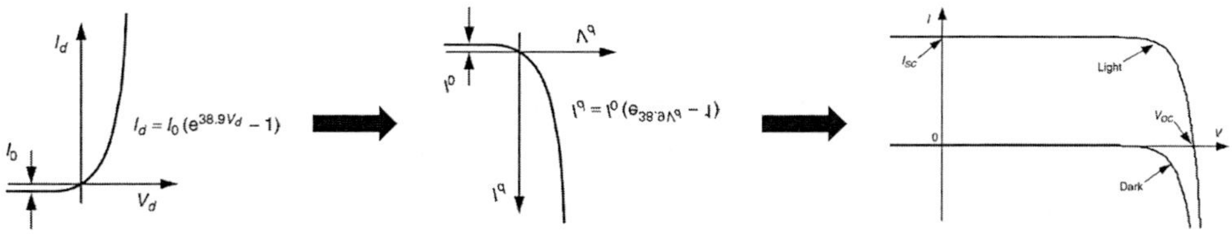

Figure 11.3: PV Cell I-V Characteristics

An I-V curve for an illuminated cell is given by the "dark" curve + I_{SC}. More light effectively shifts the curve up in I, but V_OC does not change much. By varying the insolation, we hence obtain not a single I-V curve, but a collection of them.

As discussed in Chapter 4, power electronics come into play to maintain operation at the maximum power point. Otherwise, as conditions change, the operating point deviates from MPP, and efficiency drops. Maximum power point tracking (MPPT) methods hence keep the operating point at or close to MPP at all times, and this is a standard part of many PV systems.

11.3 Wind Energy as a Resource

Wind turbines are a major use case for induction machines, discussed earlier in the text. Here, we briefly cover wind as a resource. That is, what is the power available in the wind, that can then be harnessed by machines? Then, this can be extrapolated to harness the power in the wind at different geographical locations and heights, as a function of available windspeed. More detailed coverage and derivations of the power in the wind is found in [12]. Most wind turbines are horizontal axis wind turbines, and are grouped into a wind farm. Most modern wind turbines have three blades.

The power in the wind can be derived by combining the mass flow rate through area A with the power through area A, considering a small packet of air:

$$P_W = \frac{1}{2}\rho A v^3$$

where P_W W is the power in the wind, ρ (kg/m^3) is the air density, A (m^2) is the cross-sectional area that wind passes through, and v (m/s) is the windspeed normal to A.

Some observations: (1) Power in the wind increases with the cube of windspeed. The relationship is nonlinear, so we cannot just use average windspeed. For example, the energy in 1 hour of 20 mph winds is the same as the energy in 8 hours of 10 mph winds. (2) Power in the wind is proportional to cross sectional area of the blades A. For a horizontal axis wind turbine, $A = (\pi/4)D^2$, so P_W is proportional to the blade diameter squared. This explains why larger, higher wind turbines are more cost effective. Since power increases with cube of wind speed, we can expect a significant economic impact from even a moderate increase in wind speed.

Windspeeds increase as the height above ground level increases. How much wind speed increases with height depends on the terrain. Forests and buildings slow the wind significantly when compared to smooth surfaces like water. The equation below is an approximation of the variation in wind speed due to elevation and roughness (where a better characterization would be based on actual measurements),

$$\frac{v}{v_0} = \left[\frac{H}{H_0}\right]^{\alpha}$$

where α is the friction coefficient, v is the windspeed at height H, v_0 is the windspeed at height H_0 (typically 10 m). A typical value of α in open terrain is 1/7. For a large city, α=0.4; for a small town α=0.3; for high crops α=0.2; for smooth water or ground α=0.1.

Not all of the power in the wind can be extracted by a wind turbine. Even theoretically, this is not possible. To see this, consider that there are two extreme cases, and neither makes sense practically: (1) The downwind velocity is zero: the turbine extracted all of the power; (2) The downwind velocity is unchanged from the upwind velocity: the turbine extracted no power.

In 1919, Albert Betz theorized that there must be some ideal slowing of the wind such that the turbine can extract the maximum power. As wind passes through the wind turbine, it will slow down, and the power extracted by the blades P_b is related to how much the wind slows down:

$$P_b = \frac{1}{2}\dot{m}\left(v^2 - v_d^2\right)$$

where $\dot{m}$ is the mass flow rate of air within the stream tube, v is the upwind wind speed, and v_d is the downwind wind speed.

The mass flow rate, measured at the plane of the rotor with cross section A, is the following

$$\dot{m} = \rho A v_b$$

where we assume that the velocity at the blades v_b is the average of the upwind and downwind velocities, v and v_d, respectively.

Then,

$$P_b = \frac{1}{2}\rho A \left(\frac{v + v_d}{2}\right)\left(v^2 - v_d^2\right)$$

The windspeed ratio is the ratio of the downwind to the upwind velocities, commonly denoted $\lambda = \frac{v_d}{v}$ (be careful with notation, as λ has other meanings in the rest of this book).

Considering the windspeed ratio as a variable that we want to control using the rotor, we rewrite the power extracted by the blades as two multiplicative terms, P_W (power in the wind, defined earlier) and C_P (rotor efficiency):

$$P_b = \left(\frac{1}{2}\rho A v^3\right)\left(\frac{1}{2}(1+\lambda)((1-\lambda^2))\right)$$
$$P_b = (P_W)(C_P)$$

We can find the wind speed ratio λ that maximizes rotor efficiency C_P as $\lambda = 1/3$, which corresponds to $C_P = 59.3\%$. We invite the interested reader to show the proof of this as a homework assignment. That is, maximum efficiency occurs when the wind is slowed to 1/3 of its upstream speed. This ideal slowing of the wind to achieve maximum efficiency is called "Betz efficiency" or "Betz' law."

To understand how much energy we can expect from a wind turbine, we need to estimate the probability that the wind is between any given windspeeds. A Rayleigh probability distribution is often a typical starting point when little is known about the wind distribution at a particular site. Detailed discussions of wind farm siting considerations are beyond the current scope of this text, but it can be found in the literature.

11.4 Microgrid Applications

Microgrids are a considered a key building block of the smart grid. The DOE definition of a microgrid is the following [19]: "A group of interconnected loads and distributed energy resources within clearly defined electrical boundaries that acts as a single controllable entity with respect to the grid. A microgrid can connect and disconnect from the grid to enable it to operate in both grid-connected or island-mode." In 2011 and 2012, related to the Smart Grid Demonstration Program and the American Recovery and Reinvestment Act of 2009 (ARRA), two workshops were held that identified a significant number of research and development needs and challenges for microgrids, based on a wide range of input. The US Department of Defense (DoD) has been a major supporter of microgrid research, assessment and demonstration, as well as DOE, and other organizations. These seem to coincide well with the current trajectory. An overview of microgrid benefits, research challenges, and applications is laid out clearly in [19, 20, 21] in detail, where we briefly highlight some of the key points here. A major benefit of microgrids can be summarized as 'flexibility' or adaptability, to enable integration of different types of technologies to support many different types of end-user needs. Most microgrid applications are focused on demonstration to meet specific application needs. Example application objectives include peak load reduction, renewable energy mandates, and energy reliability at critical facilities. Microgrids comprise energy resources, switchgear protection and transformers, communications and controls, and also require site engineering and operations planning.

The R&D areas for microgrids include component-level and systems-level: Component-level research areas for microgrids includes switch technologies, control and protection, advanced power electronics technologies, topologies and control algorithms for multiple inverters. Systems-level research areas for microgrids include safety, communications and control, protection and coordination, system design, multi-objective optimization frameworks for system design and economic analysis, and optimal operations under uncertainty. Microgrid planning and design research includes development of system architectures with associated interfaces, applications, and services. Other important areas are the modeling and analysis of microgrids, steady state control and coordination, transient state control and protection, interaction with utility and other microgrids, and operational optimization. Ensuring the trustworthiness of the data and control information flow is essential, especially under the stressed situations where microgrids' performance are most needed, and achieving situational awareness of security as well as physical state in a highly flexible environment like a microgrid presents its own unique research challenges. at both component and system levels.

Today, detailed data for microgrids in the US are maintained in a database, accessible online [22]. Microgrids in this database meet the definition of "a network of distributed energy resources and loads that can disconnect and re-connect to the larger utility grid as a single entity, allowing the connected loads to be served during utility outages." It is noted that backup generators are not included in the database unless they are part of a microgrid that serves other functions, such as daily power requirements or participation in utility markets. The database is updated on a semi-annual basis from a variety of sources.

Currently, the database includes data for 170 microgrids in Texas, 3 in Arizona, and 5 in Illinois. Most of the microgrids in Texas can be seen to be put in place by medical facilities, Buc-ee's (gas station/convenient stores), and H-E-B (grocery stores); this makes sense given the essential nature of these types of services in conjunction with Texas's proclivity to hurricanes and other severe weather.

11.5 Conclusion

Brief discussions of distributed energy resources, particularly with respect to their integration for smarter electric grids, renewable energy from sun and wind, and microgrids are presented in this chapter to serve as a starting point on where one can look for the emerging technologies and possibilities arising from the first principles of power and energy systems that have been converted in this textbook. These give a small sampling of a few areas that have shown tremendous opportunities for power engineers to make energy systems more efficient and power the next generation of humanity.

Index

List of Figures

Bibliography

[1] U.S. Energy Information Administration (EIA), "Independent Statistics and Analysis." https://www.eia.gov/, 2021.

[2] J. A. Cohn, *The grid: biography of an American technology.* MIT Press, 2017.

[3] National Renewable Energy Laboratory (NREL), "Community Choice Aggregation: Challenges, Opportunities, and Impacts on Renewable Energy Markets." https://www.nrel.gov/docs/fy19osti/72195.pdf, 2019.

[4] American Public Power Association, "2021 Statistical Report." https://www.publicpower.org/system/files/documents/2021-Public-power-Statistical-Report.pdf, 2021.

[5] M. B. Cain, R. P. O'neill, A. Castillo, *et al.*, "History of optimal power flow and formulations," *Federal Energy Regulatory Commission*, vol. 1, pp. 1–36, 2012.

[6] F. Capitanescu, J. M. Ramos, P. Panciatici, D. Kirschen, A. M. Marcolini, L. Platbrood, and L. Wehenkel, "State-of-the-art, challenges, and future trends in security constrained optimal power flow," *Electric Power Systems Research*, vol. 81, no. 8, pp. 1731–1741, 2011.

[7] I. Ivanova, "Texas winter storm costs could top $200 billion - more than hurricanes Harvey and Ike," *CBS News*, 2021.

[8] National Academies of Sciences, Engineering, and Medicine, *Enhancing the Resilience of the Nation's Electricity System.* Washington, DC: The National Academies Press, 2017.

[9] K. Bhattacharya, M. H. Bollen, and J. E. Daalder, *Operation of Restructured Power Systems.* Power Electronics and Power Systems, New York, NY: Springer, 2001.

[10] P. T. Krein, *Elements of Power Electronics.* The Oxford Series in Electrical and Computer Engineering, New York, NY: Oxford University Press, Oct. 1997.

[11] D. Hart, *Power Electronics.* New York, NY: McGraw Hill, Jan. 2010.

[12] G. M. Masters, *Renewable and Efficient Electric Power Systems.* Wiley-IEEE Press, 2 ed., June 2013.

[13] W. M. Grady, "Course Notes, Electrical and Computer Engineering, University of Texas at Austin: Understanding Power System Harmonics," 2012.

[14] H. H. Woodson and J. R. Melcher, *Electromechanical Dynamics Part I: Discrete Systems.* Nashville, TN: John Wiley & Sons, Jan. 1968.

[15] H. A. P. Painemal, *Wind farm model for power system stability analysis.* PhD thesis, University of Illinois at Urbana-Champaign, 12 2010.

[16] T. Ackermann, *Wind power in power systems*, vol. 200. Wiley New York, 2005.

[17] US Department of Energy (DOE) Office of Electricity, "Grid Modernization and the Smart Grid." https://www.energy.gov/oe/activities/technology-development/grid-modernization-and-smart-grid.

[18] US Department of Energy (DOE) Solar Energy Technologies Office, "Solar Energy Technologies Office." https://www.energy.gov/eere/solar/solar-energy-technologies-office.

[19] D. T. Ton and M. A. Smith, "The US Department of Energy's microgrid initiative," *The Electricity Journal*, vol. 25, no. 8, pp. 84–94, 2012.

[20] US Department of Energy (DOE) Office of Electricity, "Microgrid Workshop Report August 2011," 2011.

[21] US Department of Energy (DOE) Office of Electricity, "2012 DOE Microgrid Workshop Summary Report," 2012.

[22] US Department of Energy (DOE), "U.S. Department of Energy Combined Heat and Power and Microgrid Installation Databases," 2022.

[23] M. Ilic, F. Galiana, and L. Fink, eds., *Power Systems Restructuring: Engineering and Economics.* Boston: Kluwer Academic Publishers, Jan. 1998.

[24] L. L. Lai (ed.), *Power System Restructuring and Deregulation.* London: John Wiley & Sons, 2001.

[25] J. W. Nilsson and S. Riedel, *Electric Circuits.* Upper Saddle River, NJ: Pearson, 11 ed., 2018.

[26] Z. A. Yamayee and J. L. Bala, Jr., *Electromechanical Energy Devices and Power Systems.* New York: John Wiley & Sons, 1994.

[27] S. S. Venkata and F. R. Bergseth, *Introduction to Electric Energy Devices.* Englewood Cliffs, NJ: Prentice Hall, 1987.

[28] S. J. Chapman, *Electric Machinery Fundamentals.* Power & Energy, New York: McGraw-Hill, 2 ed., 1991.

[29] V. Del Toro, *Basic Electric Machines.* Englewood Cliffs, NJ: Prentice Hall, 1989.

[30] A. E. Fitzgerald, etc., C. Kingsley, and S. D. Umans, *Electric Machinery.* New York, NY: McGraw-Hill, 5 ed., 1990.

[31] L. W. Matsch and J. D. Morgan, *Electromagnetic and Electromechanical Machines.* Nashville, TN: John Wiley & Sons, 3 ed., 1986.

[32] S. A. Nasar, *Electric Energy Conversion and Transmission.* New York: Macmillan, 1985.

[33] P. C. Sen, *Principles of Electric Machines and Power Electronics.* Nashville, TN: John Wiley & Sons, 3 ed., 2013.

[34] P. C. Krause and O. Wasynczuk, *Electromechanical Motion Devices.* New York, NY: McGraw-Hill, 1989.

[35] R. D. Shultz and R. A. Smith, *Introduction to Electric Power Engineering.* New York: John Wiley & Sons, 1988.

[36] P. Krause, O. Wasynczuk, S. D. Sudhoff, and S. Pekarek, *Analysis of Electric Machinery.* IEEE Press Series on Power Engineering, Nashville, TN: John Wiley & Sons, 3 ed., Aug. 2013.

[37] J. R. Cogdell, *Foundations of Electric Power.* Upper Saddle River, NJ: Pearson, 1999.

[38] D. E. Johnson, J. L. Hilburn, and J. R. Johnson, *Basic Electric Circuit Analysis.* London, England: Prentice-Hall, 1978.

[39] J. J. Cathey, *Electric Machines: Analysis and Design Applying MATLAB.* New York, NY: McGraw-Hill, 2001.

[40] N. N. Rao, *Elements of Engineering Electromagnetics.* Upper Saddle River, NJ: Pearson, 5 ed., 1999.

[41] American Public Power Association, "2021 Statistical Report." https://www.publicpower.org.